BOTTOM SOILS, SEDIMENT, AND POND AQUACULTURE

BOTTOM SOILS,

SEDIMENT,

AND POND

AQUACULTURE

CLAUDE E. BOYD

Department of Fisheries and Allied Aquacultures
at Auburn University, Alabama

CHAPMAN & HALL

 New York • Albany • Bonn • Boston • Cincinnati • Detroit • London • Madrid • Melbourne
Mexico City • Pacific Grove • Paris • San Francisco • Singapore • Tokyo • Toronto • Washington

Library of Congress Cataloging-in-Publication Data

Boyd, Clause E.
    Bottom soils, sediment and pond aquaculture / Claude E. Boyd.
        p.   cm.
    Includes bibliographical references and index.
    ISBN 0-412-06941-5
    1. Pond aquaculture. 2. Pond soils 3. Pond sediments.
    I. Title.
    SH137.4.B69 1995   94-44496
    639.3'11--dc20   CIP

British Library Cataloguing in Publication Data available

Table of Contents

Preface

Aquaculture pond managers measure water-quality variables and attempt to maintain them within optimal ranges for shrimp and fish, but surprisingly little attention is paid to pond soil condition. Soil–water interactions can strongly impact water quality, and soil factors should be considered in aquaculture pond management.

The importance of soils in pond management will be illustrated with an example from pond fertilization and another from aeration. Pond fertilization may not produce phytoplankton blooms in acidic ponds. Total alkalinity is too low to provide adequate carbon dioxide for photosynthesis, and acidic soils adsorb phosphate added in fertilizer before phytoplankton can use it. Agricultural limestone application can raise total alkalinity and neutralize soil acidity. The amount of limestone necessary to cause these changes in a pond depends on the base unsaturation and exchange acidity of the bottom soil. Two ponds with the same total alkalinity and soil pH may require vastly different quantities of limestone because they differ in exchange acidity.

Aeration enhances dissolved oxygen concentrations in pond water and permits greater feed inputs to enhance fish or shrimp production. As feeding rates are raised, organic matter accumulates in pond soils. In ponds with very high feeding rates, aeration may supply enough dissolved oxygen in the water column for fish or shrimp, but it may be impossible to maintain aerobic conditions in the surface layers of pond soil. Toxic metabolites produced by microorganisms in anaerobic soils may enter the pond water and harm fish or shrimp.

Aquacultural education has traditionally been slanted toward basic animal biology. Teachers normally emphasize the biology and husbandry of individual species, and environmental aspects of pond management are somewhat neglected. The situation is improving, because in recent years more attention has been given to water quality and water-supply-related topics. Soil science still is largely ignored in the training of aquaculturists. In my opinion, aquacultural education is incomplete without a basic course in soil science, and a course in soil fertility is highly desirable. Even after students have taken basic soils courses, they need to learn how soil science is related to aquaculture.

I hope that this book will help aquaculturists obtain a better understanding of those principles of soil science that are important in pond management. Readers with no prior knowledge of soil science will need to supplement information in this book by studying selected topics in a general soil science book. I am certain that a little effort devoted to studying pond soils can be worthwhile to students, pond managers, extension workers, and researchers interested in pond aquaculture.

I am indebted to Dr. Ben Hajek and Dr. Fred Adams of the Department of Agronomy and Soils, Auburn University who have given me advice for a great many years. I am grateful for the opportunities that I have had to collaborate with several researchers—Drs. Yoram Avnimelech, Craig S. Tucker, Harry Daniels, H. S. Schmittou, Yont Musig, Bartholomew Green, and David Teichert-Coddington—and several shrimp farm operators—Gilberto Escobar, Jeffrey Peterson, Kenneth Corpron, and Dan Fegan—on various pond soil projects. I learned much from these associations. Most of all, I am indebted to the many graduate students who worked with me over the years. They have been a source of never-ending inspiration and help.

This book could not have been completed without the assistance of June Burns, who typed the entire manuscript, Margaret Tanner, who made the final copies of all of the artwork, and Prasert Munsiri, who provided considerable editorial assistance. I greatly appreciate their efforts.

Symbols

Chapter 2

A	= area (length^2)
F	= force
g	= gravitational acceleration (9.8 m sec^{-2})
h	= height
i	= hydraulic gradient (m m^{-1})
k	= hydraulic conductivity (cm sec^{-1})
K_w	= equilibrium constant for water
m	= mass
pH	= negative logarithm of the hydrogen ion activity
Q	= discharge (volume time^{-1})
r	= radius (length)
V	= volume (length^3)
v	= seepage velocity (m sec^{-1})
w	= weight
γ	= viscosity of water (0.073 N m^{-1} at 20°C)
η	= porosity
θ	= angular measurement
π	= pi (3.1416)
ρ	= specific weight (kg m^{-3})

Chapter 3

a	= ion activity (molar)
I	= ionic strength (molar)
K	= equilibrium constant
ln	= natural logarithm

log	= common logarithm
m	= measured molar concentration
mole	= 1 gram molecular weight
pH	= negative logarithm of hydrogen ion activity
Q	= *reaction quotient*
R	= *universal gas law constant (0.001987 kcal °A^{-1})*
S_i	= effective size of ion i
Z_i	= charge of ion i
γ	= activity coefficient
ΔF_f°	= standard free energy of formation (kcal)
ΔF_r°	= standard free energy of reaction (kcal)

Chapter 4

b	= empirical coefficient for Langmuir adsorption isotherm
C	= concentration of substance (weight volume^{-1})
D	= diffusion coefficient
dQ/dt	= rate of loss of substance from water
K_f	= Freundlich adsorption coefficient
K_{SD}	= soil distribution coefficient
$1/n$	= slope of line
log	= common logarithm
m	= mass of adsorbent
pH	= negative logarithm of hydrogen ion activity
r^2	= coefficient of determination
V_e	= equilibrium concentration
X	= amount of substance adsorbed
x	= thickness of oxidized layer
X_m	= maximum amount of substance adsorbed

Chapter 5

B	= organic matter influx to soil (g m^{-2} yr^{-1})
C	= final organic matter concentration (%)
C_i	= initial organic matter concentration (%)
e	= base of natural logarithm (2.303)
g	= generation time (hours)
K	= rate constant for organic matter decomposition (time^{-1})
ln	= natural logarithm
n	= number of bacterial generations
N_0	= initial number of bacterial cells
N_t	= number of bacterial cells at time t
t	= time

Chapter 6

E	= electrode potential at nonequilibrium conditions (V)
E_7	= redox potential adjusted to pH 7 (V)
E_h	= redox potential (V)
$E°$	= *standard electrode potential (V)*
e^-	= electron
F	= Faraday constant (23.06 kcal V-g equivalent^{-1})
K	= equilibrium constant
n	= number of electrons transferred in reaction
pH	= negative logarithm of hydrogen ion activity
Q	= equilibrium quotient
R	= universal gas law constant (0.001987 kcal °A^{-1})
T	= temperature (degrees absolute; 273.15 + °C)
ΔF_r	= free energy of reaction at nonequilibrium condition (kcal)
$\Delta F_r°$	= standard free energy of reaction (kcal)
$\Delta F_f°$	= standard free energy of formation (kcal)

Chapter 7

A	= surface area (length^2)
B	= buoyant force
C	= crop management factor
C_D	= drag coefficient
D	= drag force
DE_w	= dielectric constant of water
D_p	= pond depth (m)
d_p	= particle diameter
E	= amount of erosion
G	= gravitation force
g	= gravitational acceleration (9.8 m sec^{-2})
K	= soil-erodibility factor
m_p	= mass of settling particle
N	= Newton (kg·m sec^{-2})
P	= conservation practice factor
Q	= inflow rate (volume time^{-1})
R	= erosion index
R_e	= Reynolds number
S_L	= degree and length of slope (topographic factor)
t	= time
V	= volume of pond

V_p	= particle volume
v_{cs}	= critical settling velocity of particle (m sec^{-1})
v_s	= velocity of settling particle (m sec^{-1})
ζ	= zeta potential
μ	= dynamic viscosity of water (1.022×10^3 N· sec m^{-2} at 20°C)
σ	= thickness of zone of influence of charge on particle
ρ_p	= particle density
ρ_w	= density of water (998.2 kg m^{-3} at 20°C)

Chapter 8

pH	= negative logarithm of hydrogen ion activity

Chapter 9

°	= degree of angle
pH	= negative logarithm of hydrogen ion activity

Chapter 10

B	= titration volume for control vials (mL)
D	= diameter
H	= height
L	= length
pH	= negative logarithm of hydrogen ion activity
r	= *radius*
V	= volume (length^3)
W	= weight
π	= pi (3.1416)
ρ_w	= density of water (0.9982323 g cm^{-3} at 20°C)

Abbreviations

A	= cross-sectional area
°A	= degrees absolute temperature
ADP	= adenosine diphosphate
(aq) as subscript	= aqueous
ATP	= adenosine triphosphate
°C	= degrees centigrade
(c) as subscript	= crystalline
cal	= calorie
CEC	= cation-exchange capacity
cm	= centimeter
C:N	= carbon:nitrogen ratio
CoA	= coenzyme A
COLE	= coefficient of linear expansion
DO	= dissolved oxygen
EPA	= Environmental Protection Agency
EPC	= equilibrium phosphorus concentration
FAD	= flavin adenine dinucleotide
g	= gram
(g) as subscript	= gas
h	= hour
ha	= hectare
HTH	= high-test hypochlorite
kcal	= kilocalorie
kg	= kilogram
L	= liter
(l) as subscript	= liquid
M	= molar (1 gram molecular weight per liter)
m	= meter

meq	= milliequivalent
mg	= milligram
min	= minute
mL	= milliliter
mm	= millimeter
mole	= gram molecular weight
mμ	= millimicron
N	= normal (1 gram equivalent weight per liter)
NAD	= nicotinamide adenine dinucleotide
NAD$^+$	= nicotinamide adenine dinucleotide phosphate
NPS	= net phosphorus sorption capacity
NTU	= nephelometer turbidity unit
OM	= organic matter
P	= probability level
PAC	= phosphorus adsorption capacity
PBC	= phosphorus buffering capacity
ppm	= parts per million
R	= soil respiration
RQ	= respiratory quotient
(s) as subscript	= solid
SAR	= sodium adsorption ratio
SCS	= Soil Conservation Service
sec	= second
SS	= suspended solids
T	= time
US	= United States
USA	= United States of America
USDA	= United States Department of Agriculture
V	= volts
yr	= year
μg	= microgram
μm	= micrometer

Atomic Weights

Atomic Weights of Selected Elements

Element	Symbol	Atomic Weight	Element	Symbol	Atomic Weight
Aluminum	Al	26.9815	Manganese	Mn	54.9380
Arsenic	As	74.9216	Mercury	Hg	200.59
Barium	Ba	137.34	Molybdenum	Mo	95.94
Boron	B	10.811	Nickel	Ni	58.7
Bromine	Br	79.904	Nitrogen	N	14.0067
Cadmium	Cd	112.40	Oxygen	O	15.9994
Calcium	Ca	40.08	Phosphorus	P	30.9738
Carbon	C	12.01115	Platinum	Pt	195.09
Chlorine	Cl	35.453	Potassium	K	39.1
Chromium	Cr	51.996	Selenium	Se	78.96
Cobalt	Co	58.9332	Silicon	Si	28.086
Copper	Cu	63.546	Silver	Ag	107.868
Fluorine	F	18.9984	Sodium	Na	22.9898
Gold	Au	196.967	Strontium	Sr	87.62
Helium	He	4.0026	Sulfur	S	32.064
Hydrogen	H	1.00797	Thallium	Tl	204.37
Iodine	I	126.9044	Tin	Sn	118.69
Iron	Fe	55.847	Tungsten	W	183.85
Lead	Pb	207.19	Uranium	U	238.03
Lithium	Li	6.939	Vanadium	V	50.942
Magnesium	Mg	24.31	Zinc	Zn	65.37

Customary Metric Conversion Factors

Length

	in.	ft	yd	mile	mm	m	km
1 in.	1	0.083	0.0278	—	25.4	0.0254	—
1 ft	12	1	0.33	—	305	0.305	—
1 yd	36	3	1	—	910	0.91	—
1 mile	—	5,280	1,760	1	—	1,610	1.61
1 mm	0.0394	—	—	—	1	0.001	—
1 m	39.37	3.28	1.1	—	1,000	1	0.001
1 km	—	3,280	1,093	0.620	—	1,000	1

Area

	in.^{2}	ft^2	yd^2	acre	cm^2	m^2	ha
1 in.^{2}	1	0.007	—	—	6.45	0.000645	—
1 ft^2	144	1	0.11	—	—	0.093	—
1 yd^2	1,296	9	1	—	—	0.836	—
1 acre	—	43,560	4,840	1	—	4,050	0.405
1 cm^2	0.155	—	—	—	1	0.0001	—
1 m^2	1,555	10.8	1.2	—	10,000	1	.0001
1 ha	—	107,600	11,955	2.47	—	10,000	1

Volume

	in.^{3}	ft^3	gal	liter	m^3	ac-ft	ha-m
1 in.^{3}	1	0.00058	0.00433	0.0164	—	—	—
1 ft^3	1,728	1	7.48	28.3	0.0283	—	—
1 gal	231	0.134	1	3.8	0.0038	—	—
1 liter	60.5	0.035	0.264	1	0.001	—	—
1 m^3	—	35.3	264.0	1,000	1	—	—
1 ac-ft	—	43,560	325,850	—	1,233	1	0.1233
1 ha-m	—	353,300	—	—	10,000	8.11	1

BOTTOM SOILS,

SEDIMENT,

AND POND

AQUACULTURE

I

Soils in Pond Aquaculture

1.1 Introduction

Water is the medium in which aquaculture is conducted, and much has been written about water supply, water quality, and water quality management for aquaculture ponds. Soil also is a key factor in aquaculture, but much less attention is given to soil condition than to water supply and water quality. Most ponds are built from and in soil. Many dissolved and suspended substances in water are derived from contact with soil. Pond bottom soils are the storehouse for many substances that accumulate in pond ecosystems, and chemical and biological processes occurring in surface layers of pond soils influence water quality and aquacultural production. An understanding of soil properties and the reactions and processes in soils can be useful in pond aquaculture.

1.2 Pond Soils

Materials composing the bottoms of streams, lakes, and ponds are known as sediment, mud, or soil. These terms, often used interchangeably, describe the bottom material in ponds, and they probably are equally acceptable to most readers. The original pond bottom usually is made of terrestrial soil, and when the pond is filled with water the bottom becomes wet. According to the dictionary, a mixture of solid material and water can be called mud. Solids settle from the water and cover the pond bottom with sediment, and in older ponds, the sediment layer may be several to many centimeters thick. Thus, the term *sediment* often is used in referring to the solids in the bottom of a pond. Soil scientists and other agriculturists refer to the part of the earth's surface in which plants grow as soil. The soil at a given site may have formed in place. At another site the soil may have formed somewhere else and deposited as sediment. Nevertheless, the surface layer of the earth's surface at both sites is called *soil* by agriculturists. Aquaculture is a type of agriculture, and I will follow the convention of agriculturists and use the term *pond soil* for general references to the pond bottom. However, at some places in the text, it is necessary to identify the upper layer of pond soil as sediment to distinguish it from the original pond soil.

Basic functions of pond soils are illustrated in Figure 1.1. Embankments that impound water and form ponds are made of soil; soil also forms a barrier to seepage so that ponds will hold water. The pond in Figure 1 receives water from runoff and a well. Some ponds may be filled entirely by runoff, water pumped from streams or lakes, or water from estuaries. Surface water and ground water used to fill and maintain water levels in ponds acquire dissolved and suspended substances through contact with soil and other geologic materials. Further contact and exchange between water and soil occurs while water stands in ponds.

Substances continually settle from pond water onto the pond bottom, for example, suspended solids in surface waters that enter ponds, particles of soil and organic matter that are eroded from the pond bottom and insides of levees by water currents and wave action, manure and uneaten feed from management inputs, and remains of plants and animals produced within the pond. Substances also may enter the solid phase of the soil from the aqueous phase through ion exchange, adsorption, and precipitation. For example, potassium in the water can be exchanged for other cations on the soil; phosphorus can be adsorbed by soil; calcium carbonate may precipitate from solution and become a part of the bottom soil matrix.

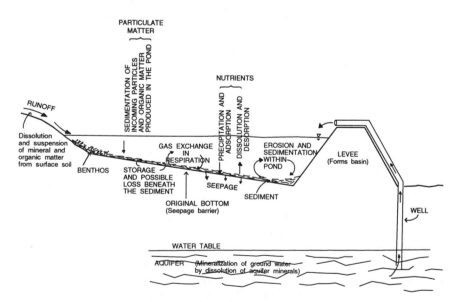

FIGURE 1.1. Basic functions of pond soils.

Substances that enter the soil may be stored permanently, or they may be transformed to other substances by physical, chemical, or biological means and lost from the pond ecosystem. To illustrate, phosphorus adsorbed by a pond soil can become buried in the sediment and lost from circulation within the pool of available phosphorus. Organic matter deposited on the pond bottom usually is decomposed to inorganic carbon and released to the water as carbon dioxide. Nitrogen compounds may be denitrified by pond soil microorganisms and lost to the atmosphere as nitrogen gas.

Bacteria, fungi, algae, higher aquatic plants, small invertebrates, and other organisms known as benthos live in and on the bottom soil. Crustaceans and even some species of fish spend much of their time on the bottom, and many fish species lay eggs in nests built in the bottom. Benthos serve as food for some aquaculture species; it also is involved in gas exchange, primary and secondary productivity, decomposition, and nutrient cycling.

Substances stored in the pond soil can be released to the water through ion exchange, dissolution, and decomposition. Release of inorganic ions or compounds from the soil into the water through exchange or dissolution only occurs until equilibrium is obtained between the solid and solution phases of the substances in question. The equilibrium concentration of a dissolved substance is an important consideration in pond management. The equilibrium concentration of a nutrient may be too low for optimal phytoplankton growth, or the equilibrium concentration of a heavy metal may be high enough to cause toxicity to aquatic animals. Microbial decomposition is extremely important because organic matter is oxidized to carbon dioxide and ammonia and other mineral nutrients are released. Thus, in decomposition, carbon, nitrogen, and other elements are mineralized or recycled. Carbon dioxide and ammonia are highly soluble and quickly enter the water. They may be used as nutrients, but if their concentrations are too high they can be toxic to aquatic animals.

Metabolic activities of microorganisms in pond soils are critical factors in pond dynamics. Microorganisms use molecular oxygen in oxidizing organic matter to carbon dioxide. During decomposition, soluble organic compounds are produced that may enter the pond water before being completely oxidized and contribute to the dissolved-organic-matter fraction in the water. Soils with large inputs of organic matter have high levels of microbial activity, and the microbial community may use oxygen faster than it can penetrate the soil from the water above. This leads to anaerobic conditions and the development of a microbial flora that can use organic compounds or oxidized inorganic compounds instead of molecular oxygen as electron and hydrogen acceptors in metabolism. This process is often called *anaerobic respiration*. Denitrification, organic acid and alcohol production, and formation of nitrite, hydrogen sulfide, ferrous iron, manganous manganese, and methane are the results of anaerobic respiration. Products of anaerobic respiration can enter the water, and some are toxic to aquatic animals.

In summary, the pond soil is beneficial to the pond ecosystem as a basin to hold water, a storehouse of various chemical substances, a habitat for plants and animals, and a nutrient recycling center. It also can exert a large oxygen demand, become anaerobic, and be a source of toxic dissolved substances.

1.3 Critical Soil Properties

Soils consist of weathered minerals and organic matter. They are a product of interactions among parent material, climate, and biological activity. It is well known that soils differ from place to place on the earth's surface, and beneath a given site the soil profile consists of horizontal layers that change in characteristics with depth below the land surface. Soils have been studied in great detail, and desirable soil characteristics for agricultural and engineering purposes have been identified and classified. Much less is known about desirable characteristics of soils for aquaculture ponds, but there is enough information available to identify several critical factors (Table 1.1).

The most active fractions of the soil are clay particles, because of their electrical charge and large surface area, and the organic matter, because of its biological availability and high chemical reactivity. The pore spaces among the mineral fragments and organic matter in soil are filled with air and water. In flooded soils the air is displaced completely by water. This greatly impedes the movement of oxygen into a pond soil, because the oxygen must move by diffusion or be carried along with water that seeps through the soil. Coarse-textured submerged soils normally are better oxygenated than fine-textured ones. Pond soils often are fine-textured because they usually have at least 20–30% clay content to limit seepage, and they usually have a higher percentage of organic matter than

Table 1.1. Selected Soil Properties that Influence Aquaculture Pond Management

Property	Processes Affected in Pond
Particle size and texture	Erosion and sedimentation, embankment stability, seepage, suitability of bottom habitat
pH and acidity	Nutrient availability, microbial activity, benthic productivity, hydrogen ion toxicity
Organic matter	Embankment stability, oxygen demand, nutrient supply, suitability of bottom habitat
Nitrogen concentration and C:N ratio	Decomposition of organic matter, nutrient availability
Redox potential	Toxin production, mineral solubility
Sediment depth	Reduction in pond volume, suitability of bottom habitat
Nutrient concentrations	Nutrient availability and productivity

terrestrial soils in the surrounding area. Pond soils normally are highly reactive, have a high oxygen demand, and tend to become anaerobic.

In addition to texture and organic matter content, specific chemical compounds in soil may have a pronounced effect on aquaculture. Soils that have been highly weathered and contain appreciable quantities of aluminum oxides and hydroxides are acidic, and waters in contact with acidic soils have low total alkalinity concentrations and are poorly buffered against pH change. The presence of iron pyrite in aerobic pond soil can result in extreme acidity because sulfuric acid is produced by pyrite oxidation. Free calcium carbonate in soil leads to high concentrations of total alkalinity and total hardness in overlaying water. The concentrations and proportions of major ions in surface water are governed by the kind of rocks and soil with which water has been in contact and the amount of rainfall relative to evaporation. In areas with easily weathered rocks or fertile soils, surface waters usually are more highly mineralized than in areas with resistant rocks or infertile soils. Surface waters of arid climates are more highly mineralized than those in humid climates. Within a given climatic region, ponds built on different kinds of soils will have different degrees of mineralization and different proportions of major ions, as illustrated in Figure 1.2 for Alabama ponds.

The four most important soil features for aquacultural production are texture, organic matter content, pH, and presence or absence of particular soluble compounds that may be beneficial or harmful to water quality. When soils in ponds are flooded, the most marked change in their composition is loss of air from the pore spaces and the gradual accumulation of organic matter.

1.4 Reactions and Processes

Some important reactions and processes controlling pond soil–water interactions are summarized in Table 1.2. Although these phenomena occur in ponds, all have not been thoroughly studied. We often must rely upon findings from soil science and other disciplines when describing many of them. The following are great obstacles to an understanding of pond soil–water interactions: (1) most of the reactions and processes occur continuously and simultaneously; (2) two or more reactions may be competing for the same reactants; (3) different processes and reactions that are interdependent may occur at different rates; and (4) it is difficult to study more than one or two reactions or processes at a time, and frequently the necessary components can only be isolated and studied in laboratory systems. In this text, most of the reactions and processes are discussed individually, but interrelationships are stressed. Readers will want to understand the individual reactions and processes, but they also should try to develop an overview of the pond soil–water system as a whole. Such a view is essential in assessing the influence of pond soils in practical pond management situations.

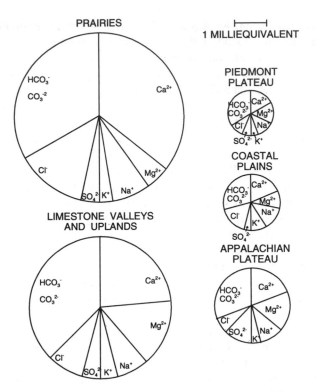

FIGURE 1.2. Pie diagrams for the proportions of major ions in pond waters from five soil areas in Alabama. The diameters of the circles are proportional to the degree of mineralization of water in milliequivalents per liter. (*Source: Arce and Boyd.*[6])

1.5 Soils and Aquacultural Production

Research and practical observations clearly reveal that pond soils influence water quality and production. In extensive aquacultural production, yields are better in areas where soils are fertile than in places where soils are infertile.[1] Fertilized ponds with acidic soils typically have lower fish production than do fertilized ponds with near-neutral or slightly alkaline soils. Ponds with high concentrations of soil organic matter develop anaerobic zones in their bottoms, and toxic metabolites from microbial activity retard growth or even cause mortality of aquacultural crops.[2] Banerjea made an extensive study of soil condition in ponds in India and found that pH and concentrations of carbon, nitrogen, and phosphorus in soils affected the potential of ponds for fish production.[3]

1.6 Pond Soil Management

Techniques available for overcoming limitations of site soil properties in pond construction and seepage control are discussed by Yoo and Boyd.[4] After a pond

Table 1.2. Selected Physical, Chemical, and Biological Reactions that Occur in Pond Soil or at Soil–Water Interface

Phenomena	Example
Reactions	
Dissolution	$CaCO_3 + CO_2 + H_2O = Ca^{2+} + 2HCO_3^-$
Precipitation	$Al^{3+} + H_2PO_4^- + 2H_2O = Al(OH_2)H_2PO_4 + 2H^+$
Hydrolysis	$Al^{3+} + 3H_2O = Al(OH)_3 + 3H^+$
Neutralization	$HCO_3^- + H^+ = H_2O + CO_2$
Oxidation	$NH_4^+ + 2O_2 = NO_3^- + 2H^+ + H_2O$
Reduction	$SO_4^{2-} + 4H_2 = S^{2-} + 4H_2O$
Complex formation	$Cu^{2+} + CO_3^{2-} = CuCO_3^0$
Adsorption	Adsorption of phosphorus on soil colloids
Cation exchange	K (soil) $= K^+$ (water)
Hydration	$Al_2O_3 + 3H_2O = Al_2O_3 \cdot 3H_2O$
Processes	
Sedimentation	Soil particles in runoff settle to pond bottom.
Decomposition	Microorganisms break down soil organic matter: $CH_2O + O_2 \rightarrow CO_2 + H_2O$.
Photosynthesis	Benthic algae produce organic matter and release oxygen: $6CO_2 + 6H_2O \rightarrow C_6H_{12}O_6 + 6O_2$.
Diffusion	Oxygen diffuses into bottom soil from water above.
Seepage	Water carrying dissolved substances seeps downward into pond soil.
Erosion	Water currents in pond erode the bottom soil.
Suspension	Particulate matter eroded from the bottom is suspended in pond water.

has been constructed and aquacultural production initiated, a few techniques are widely used for mitigating deficiencies in pond soil conditions. Fertilizer may be periodically applied to replace phosphorus removed by bottom soil. Ponds are limed to increase the base content and pH of bottom soils so that pond water will have adequate total alkalinity and total hardness for aquatic animal production. Aeration and water circulation devices can be used to improve dissolved oxygen concentrations in the water and at the pond bottom. Sedimentation basins can be used to remove suspended soil particles from water before it is added to ponds. Between crops, pond bottoms are often dried and tilled to improve aeration and accelerate the decomposition of organic matter. These and other procedures for managing pond soils will be discussed.

1.7 Pond Soil Analysis

Many procedures are available for analyzing soil properties for agricultural and engineering purposes. Soil tests often are made in pond site selection.[5] There has been some use of soil analyses in pond management. Bottom samples are sometimes analyzed for type and abundance of benthic organisms, and pond soil respiration measurements have been made. The most popular chemical analyses are pH, organic matter, and lime requirement. Soil nutrient analyses have not provided reliable guidelines for pond fertilization. Analyses of pond soils for heavy metals have not been useful for identifying toxicity levels. However, there is an increasing interest in soil analyses in pond aquaculture, so selected methods of soil analysis are provided in the final chapter.

1.8 Prospectus

When I became involved in aquacultural research in the 1960s, there was almost no interest in water-quality management. Most researchers were concerned with selection of suitable species, general culture techniques, disease and parasites, and nutrition. Production was relatively low, and water quality was not a serious limitation in most culture systems. As production levels increased, water quality became a more serious consideration. Much research has been conducted on water quality in aquaculture ponds over the past 10–15 years, and some highly effective management procedures have been developed. We now have a situation where water quality can be maintained within a desirable range, but above a certain level of production, fish and shrimp grow poorly and have low survival. I think that one of the main reasons for these newly imposed limitations lies in the condition of the pond soil. Also, as restrictions are placed on discharge of pond effluents, water is retained in ponds longer and waste products from aquaculture must be degraded within ponds rather than discharged to natural waters. Maintenance of good soil condition will be more difficult in ponds that cannot be drained after each crop. I suspect that there will be much concern over pond soils during the next 10 to 20 years. Although material in this book should be useful to the aquaculture community, knowledge of pond soils is inadequate to provide answers to many of the practical questions that pond managers have about soils in aquaculture. Hopefully, this book will stimulate interest and research on pond soil management and thus serve as a true stimulus to commercial aquaculture.

References

1. Toth, S. J., and R. F. Smith. 1960. Soil over which water flow affects ability to grow fish. *New Jersey Agriculture* **42**:5–11.

2. Boyd, C. E. *Water Quality in Ponds for Aquaculture*. 1990. Alabama Agricultural Experiment Station, Auburn University, Ala.

3. Banerjea, S. M. 1967. Water quality and soil condition of fish ponds in some states of India in relation to fish production. *Indian J. Fish.* **14**:113–144.

4. Yoo, K. H., and C. E. Boyd. *Hydrology and Water Supply for Aquaculture.* 1994. Chapman & Hall, New York.

5. Hajek, B. F., and C. E. Boyd. 1994. Rating soil and water information for aquaculture. *Aquacultural Eng.* **13**:115–128.

6. Arce, R. G., and C. E. Boyd. *Water Chemistry of Alabama Ponds.* 1980. Bulletin 522, Alabama Agricultural Experiment Station, Auburn University, Ala.

2

Physical, Chemical, and Mineralogical Properties of Soils

2.1 Introduction

Soils used for constructing ponds originally occurred at or near the land surface. Some were well-drained soils, and others were situated in wetlands and covered with water all or part of the time. There are three basic types of aquaculture ponds: watershed ponds, levee ponds, and excavated ponds.[1] Watershed ponds are formed by making embarkments across natural watercourses to impound runoff. Soils for building embankments may come from both inside and outside the areas that will become pond bottoms. Surfaces of pond bottoms may be the original land surfaces with vegetation removed, or the original land surfaces may be cut to obtain fill material for embankments and subsurface soil layers exposed. Sometimes, where topography is irregular, depressions in bottoms may be filled to prevent excessively deep areas. Final pond bottom surfaces will be compacted. Levee ponds are constructed in fairly level areas, and fill material for levees usually is cut almost entirely from pond bottom areas. Bottoms of new levee ponds are normally subsurface layers of soil that have been compacted. Excavated ponds are made by digging pits, and pond bottoms are formed by compacting subsurface layers of soil. Where site soils are highly permeable, clay may be hauled in from other areas, spread over bottoms, and compacted to form barriers against seepage in watershed and levee ponds. Bottom soils of such ponds may have physical and chemical features much different than those of native site soils. Once ponds are filled with water, soils change in characteristics as a result of continuous flooding and inputs of substances from pond waters.

Several major factors cause changes in soil characteristics in new ponds. Flooding greatly reduces soil aeration by cutting off direct contact with air. Aquaculture ponds receive applications of lime, manures, fertilizers, and feeds, but a given pond seldom receives all of these substances. Residues of materials applied to ponds accumulate in soils. Organic matter originating in photosynthesis by the pond flora is deposited on bottoms. Suspended particles of inorganic and organic matter enter ponds in inflow and settle to bottoms. Wave action, activity of aquatic animals, and pond management procedures (e.g., inflowing water, mechanical aeration, induced water circulation, and seining) erode pond bottoms and suspend soil particles. Suspended particles settle again, causing changes in soil-particle-size distribution across the pond bottom. When ponds are drained, soil is eroded from bottoms and removed in effluents. Rain falling on empty pond bottoms erodes soil particles from higher elevations and deposits them in lower areas. Sometimes, empty ponds are dried and tilled between crops. Features of pond soils result from initial characteristics of surface soils, movement of soil during construction, and alterations induced by the aquatic environment and aquaculture operations.

This chapter discusses physical, chemical, and mineralogical properties of soils and soil classification. The discourse is limited to those aspects of soil science needed as background for topics covered in other chapters. More complete treatments of physical, chemical, and mineralogical soil properties and soil classification may be found in introductory texts on soil science.

2.2 Soil

Soil constitutes the solid, surface layer of the earth upon which humans conduct most of their activities, and everyone has some familiarity with features of the land surface and the soil. The dictionary reveals that soil is "the weathered, unconsolidated, upper layer of the earth's crust which may be dug or plowed and in which plants grow." The term *soil* is a general reference to the surface layer of the land surface. At a particular location on the land surface, there usually is *a soil* of definite characteristics.

2.2.1 Development of Soils

Soils developed over geologic time through weathering of the original rocks that formed the earth's mantle. Weathering consists of disintegration of rocks and minerals by mechanical means and decomposition of minerals through chemical reactions. In soil formation, biological activity also is involved because it affects weathering and adds organic matter to soils.

Heating and cooling and freezing and thawing of rocks cause them to break into small pieces. These small pieces can be transported by water. Fragments of rock are abraded during transport and sorted during deposition. Glaciation disintegrates rocks and minerals, and glaciers also mix, transport, and sort fragments. Wind transports small fragments and causes abrasion when the airborne fragments strike rocks and minerals on the land surface. Plants trap fine fragments, and roots of large plants fracture rocks by exerting pressure as they grow into small cracks in rocks. Rocks are mechanically "broken down" and left in place or transported by gravity, water, or winds and deposited at other places.

Chemical weathering includes effects of dissolution, acidity, hydrolysis, oxidation, hydration, and other reactions. Original minerals in rocks are altered chemically to become secondary minerals. Some important primary and secondary minerals in soils are given in Table 2.1. More complete lists can be found in texts on soil science and geochemistry. Particle size of minerals continues to decrease during chemical weathering, and soluble components are lost through leaching in high-rainfall environments. The smallest mineral particles formed by weathering are clays. The formation of clay minerals through weathering is illustrated in Figure 2.1. Rocks are broken down to small pieces called *primary*

Table 2.1. *Some Common Primary and Secondary Minerals in Soils*[a]

Name	Compound
Primary Minerals	
Quartz	$(SiO_2)n$
Muscovite	$KAl_3Si_3O_{10}(OH)_2$
Microcline	$KAlSi_3O_8$
Orthoclase	$KAlSi_3O_8$
Biotite	$KAl(Mg,Fe)_3Si_3O_{10}(OH)_2$
Albite (Na-feldspar)	$NaAlSi_3O_8$
Hornblende	Contains Ca, Al, Mg, Fe, Si, O, and H, but the mineral has several forms.
Augite	Same elements as hornblende except there is no H. It is also variable in composition.
Anorthite (Ca-feldspar)	$CaAl_2Si_2O_8$
Olivine	$(Mg,Fe)_2SiO_4$
Secondary Minerals	
Iron and aluminum oxides (smallest particles are clays)	
Goethite	$FeOOH$
Hematite	Fe_2O_3
Gibbsite	$Al(OH)_3$, sometimes given as $Al_2O_3 \cdot 3H_2O$
Layered-silicate clays [b]	
Vermicullite	$(Ca,Mg)(Mg_{3-x}Fe_x)_2[(Si_6Al_2)_8O_{20}](OH)_4 \cdot 8H_2O$
Kaolinite	$Al_2Si_2O_5(OH)_4$
Ca-Montmorillonite	$Ca_{0.33}Al_{2.33}Si_{3.67}O_{10}(OH)_2$
Na-Montmorillonite	$Na_{0.33}Al_{2.33}Si_{3.67}O_{10}(OH)_2$
Sedimentary minerals	
Calcite	$CaCO_3$
Dolomite	$CaMg(CO_3)_2$
Gypsum	$CaSO_4 \cdot 2H_2O$
Siderite	$FeCO_3$
Pyrite	FeS_2

[a]Resistance to weathering generally decreases in descending order within the list.

[b]Also may contain Zn.

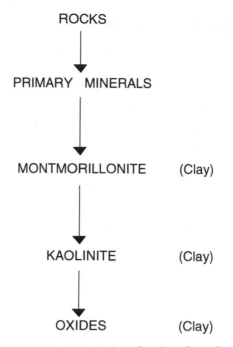

FIGURE 2.1. Weathering of rock to form clay.

minerals, which are physically and chemically weathered into silicate clays, such as montmorillonite and kaolinite, and finally to free oxides such as $Al(OH)_3$ and FeOOH.

Dead plants and animals are the major inputs of organic matter in native soils. As remains of plants and animals decompose, they become unrecognizable in form and finally are reduced to small particles of organic matter called *humus.* Most soils contain no more than a few percent of organic matter and are called *mineral soils.* However, soils formed in swamps, marshes, or other wetlands may contain high percentages of organic matter and are known as *organic soils.* Aquaculture ponds usually are constructed of mineral soils, but there are places where ponds are built on organic soils.

Major factors affecting development of soils are composition of original rocks, climate, topography, biological activity, and time. The composition of the original rocks governs the kinds of minerals available to form soil at a given place. High temperature and rainfall accelerate both mechanical and chemical weathering, leaching, erosion, and transport. Biological activity in soil also is favored by high temperature and rainfall. Topography affects both erosion and transport, because water flowing over steep surfaces has more energy to suspend soil particles than does water flowing over gentler slopes. Low, poorly drained areas are ideal environments for wetland development and formation of organic soils.

Some soils are formed in place, and others form of sedimentary materials transported from another area.

Climate is a dominant factor in soil formation. In cold climates, soil formation may be quite slow compared to warm climates. Soil is much more highly weathered and leached in a warm, humid climate than in a cold climate or warm, arid climate. In wetlands, flooding of soil reduces aeration, and microbial decomposition of organic matter occurs slowly. This makes it possible for large amounts of partially decomposed organic matter to accumulate.

Time is an important factor, because soil formation processes are very slow. Thousands of years have passed during the development of soils seen at a particular place on the land surface. Unfortunately, poor soil management can result in great damage to surface soils within a period of a few years. For example, destruction of vegetation on soil surfaces in humid regions can accelerate erosion and the fertile topsoil can be lost quickly. An analogous situation may occur in ponds. If ponds are improperly managed, the deeper areas may quickly fill with sediment derived from external or internal sources. This sediment greatly reduces the value of the pond for aquaculture.

2.2.2 Soil Profiles

Weathering is most intense near the land surface and decreases in intensity with increasing soil depth. A soil usually is layered or stratified. These layers develop because of organic matter input at the surface, a decrease of weathering with depth, downward movement of fine or soluble particles by leaching, and accumulation of leached components in particular layers. The individual layers are known as *horizons,* and vertical sections through the horizons are called *soil profiles.* The upper horizons are called *surface soil,* and the deeper, but still highly weathered, horizons are known as *subsoil.* Surface soils normally contain more organic matter than subsoils, because most biological activity occurs in the surface horizons. The upper part of a subsoil is called the *transition zone,* and the lower part is the *accumulation zone.* Clay, iron and aluminum oxides, and, in some places, calcium carbonate and other minerals are leached from the surface soil and transition zone and accumulate in the lower zone of subsoil.

Major horizons that may be encountered in a mineral soil profile are described in Table 2.2. Some soils will not exhibit all horizons. If there is a surface layer of organic matter, it is termed the *O horizon.* Organic matter decreases rapidly with depth in a mineral soil. The *A horizon* is made up of mineral soil mixed with humus. The zone of maximum leaching of clay and iron and aluminum oxides occurs in the *E horizon.* Materials leached from the upper layers tend to accumulate in the *B horizon.* In humid regions, silicate clays and oxides of iron and aluminum tend to accumulate in the B horizon; in arid regions, calcium carbonate and other salts may accumulate in the B horizon. The E horizon of

Table 2.2.   Major Soil Horizons found in Soil Profiles (all horizons are not present in all soils)

Horizons	Description
O	Organic layers of plant and animal residues above the mineral soil
Oi	Fresh to slightly decomposed residues
Oe	Partially decomposed residues
Oa	Highly decomposed residues
A	Upper layer of mineral soil mixed with organic matter (humus). Darker in color than deeper layers of mineral soil. Usually brown or black.
E	Layer subjected to leaching of silicate clays and iron and aluminum oxides. Lighter in color than the A horizon and has a greater percentage of silt and sand than the B horizon below.
B	Clays and oxides leached from the E horizon tend to accumulate. In semiarid and arid regions, salts accumulate.
C	Consists of unconsolidated material. Little weathering occurs and biological activity is not important.
R	Unweathered, consolidated bedrock
Transition	Occurs between major horizons. For example, an AE horizon has properties of A and E horizons, but it would be more like A than E. If the transition horizon between A and E was more like E than A, it would be called an EA horizon.

highly leached soil often comprises resistant fragments of quartz of silt-sized particles and larger, and the B horizon tends to contain the most clay. The beginning of the subsoil does not correspond to a specific horizon, but it usually begins in the A, E, or B horizons. The materials contained in the O through B horizons are called the *solum*. The *C horizon* rests upon the consolidated bedrock, and it is much less weathered than the horizons above.

One often hears references to topsoil and subsoil. The topsoil is the plow layer. Topsoil also is commonly used to identify fertile soil placed on roadbanks, lawns, or flower beds to provide a better soil for cultivating grass or highly valued plants. The subsoil is the soil below the plow layer. Topsoil and subsoil are used more often in practical discussions of construction or agriculture than in the description of soil profiles.

Soil profiles are illustrated in Figure 2.2 with vertical sections of soil that have been glued to mounting boards for exhibit. A good place to easily view soil profiles is at sites where roads have been cut through hills. The resulting banks often provide excellent examples of soil profiles. Soils differ from place to place with respect to properties such as soil depth, particle-size distribution, amount and type of clay, organic matter concentration, and development of individual horizons within the profile. These differences have been used to develop procedures for classifying soils. A popular procedure is called *soil taxonomy*,[2] and it will be discussed later in this chapter.

Ponds are usually built in places where soils have a discernible profile. How-

FIGURE 2.2. Soil profile prepared for exhibition to students.

ever, the upper horizons often are removed or covered with fill material during construction.[3] A new profile tends to form in ponds. Soft organic and mineral sediments accumulates over the harder original pond bottom. There is often a flocculent layer of particulate and colloidal matter, which is somewhat gel-like in appearance and 1–10 cm or more thick above the soft sediment. This flocculent layer can be disrupted by management procedures that cause strong water currents, and it is carried away in outflowing water when ponds are drained. As in native soils, the concentration of organic matter in pond bottom soil decreases with soil depth. Layers of sediment and underlaying soil in a pond bottom are illustrated in Figure 2.3.

Recent research by Munsiri and Boyd (unpublished) on ponds at Auburn

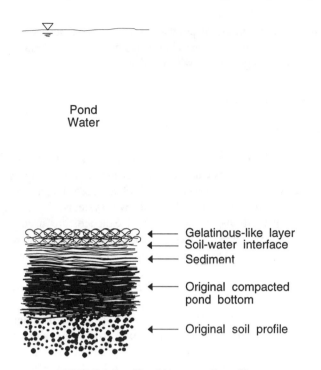

Pond
Water

Gelatinous-like layer
Soil-water interface
Sediment

Original compacted
pond bottom

Original soil profile

FIGURE 2.3. Pond bottom soil profile.

University have resulted in a proposed system for naming horizons in pond soil profiles. In this system, the flocculent layer is called the F horizon. The upper layer of sediment (5–10 cm thick) is highly fluid and stirred by biological and physical processes. It is called the S horizon; the S indicates stirring. The upper, oxidized part of the S horizon is termed the S_o horizon and the lower, reduced part of the S horizon is called the S_r horizon. Below the S horizon, the sediment is less fluid. This part of the sediment is called the M horizon with the M standing for mature, bulk sediment. Between the sediment and original pond bottom is a transitional stratum that is called the T horizon. The original pond soil is the P horizon. The upper part of the T horizon is called the MT horizon indicating that it is similar in characteristics to the M horizon. The lower part of the T horizon is more similar to the P horizon and is called the PT horizon.

The bulk density of the sediment was <0.3 g/cm^3 in the S horizon, 0.3–0.7 g/cm^3 in the M horizon, 0.7–1.4 g/cm^3 in the T horizon, and >1.4 g/cm^3 in the P horizon. Organic carbon concentrations were 3–4% in the S horizon, 2–3% in the M horizon, and less than 0.5% in the P horizon. Concentrations of organic carbon declined rapidly across the T horizon.

The depths of the S and M horizons were much less in shallow water edges of ponds than in the central, deeper water areas. The thickness of the M horizon increased with pond age, but the chemical and physical characteristics of the horizons did not change greatly with pond age.

2.3 Particle Size in Soils

2.3.1 Soil Texture Classes

The *texture* of a soil refers to the distribution of the different sizes of soil particles. Soil particles are separated mechanically and grouped into categories with specific particle diameter limits called *separates*, which are identified as coarse fragments and gravel, sand, silt, and clay. One of two systems, U.S. Department of Agriculture or International Society of Soil Science, are used in identifying separates (Table 2.3). The U.S. Department of Agriculture system will be used here. An idea about the size of soil particles in the different separates can be obtained from Figure 2.4.

Coarse fragments 2–75 mm in diameter are called *gravel;* larger fragments usually are referred to as *stone.* Sand grains range in size from coarse to very fine, and the term *sand* refers to particles 0.05–2 mm in diameter. Sand grains usually are not sticky, and they retain very little water. Air and water can move freely among them. *Silt* particles are 0.002–0.05 mm in diameter. They stick together slightly and are more resistant to air and water movement than is clay. *Clay* particles are 0.002 mm and less in diameter, and the smallest particles are colloidal in nature. Because of the small size of clay particles, the clay fraction in soil has a tremendous surface area. Depending upon the kind, clays have surface areas of 10–800 m^2 g^{-1}. Clay particles adsorb water, inorganic ions, organic matter, and gases on their surfaces. Clays can retain comparatively large amounts of water and have marked resistance to water and gas movement. When wet, clays are sticky and plastic, but when dry, they often form hard clods.

A method for determining the particle-size distribution in soil samples is provided in Chapter 10. Once the distribution is known, a soil may be given a

Table 2.3. United States Department of Agriculture (USDA) and International Society of Soil Science (ISSS) Classifications of Soil Separates

Particle Fraction Name	USDA	ISSS
Gravel	>2 mm	>2 mm
Very coarse sand	1–2 mm	
Coarse sand	0.5–1 mm	0.2–2 mm
Medium sand	0.25–0.5 mm	
Fine sand	0.1–0.25 mm	0.02–0.2 mm
Very find sand	0.05–0.1 mm	
Silt	0.002–0.05 mm	0.002–0.02 mm
Clay	<0.002 mm	<0.002 mm

FIGURE 2.4. Soil separates.

textural class name. The simplest classification of textural class has four groups: sands (≥70% sand, ≤15% clay), silts (≥80% silt, ≤12% clay), clays (>50% clay), and loams (all other particle-size distributions). The U.S. Department of Agriculture recommends a system for naming textural classes that includes more divisions (Table 2.4). A *soil triangle* (Fig. 2.5) may be used to name a soil textural class from any particle-size distribution. For most aquaculture purposes it is adequate to classify soils as sands, silts, clays, or loams and to specify the percentage clay.

Example 2.1: Determination of Textural Class
    A soil contains 30% clay, 30% silt, and 40% sand. Assign a textural class.

Table 2.4.   United States Department of Agriculture System of Describing Soil Texture

Common Name	Texture	Textural Class Name
Sandy soils	Coarse	Sands
		Loamy sand
Loamy soils	Moderately coarse	Sandy loam
		Fine sandy loam
	Medium	Very fine sandy loam
		Loam
		Silt loam
		Silt
	Moderately fine	Sandy clay loam
		Silty clay loam
		Clay loam
Clayey soils	Fine	Sandy clay
		Silty clay
		Clay

Solution: On the soil triangle (Fig. 2.5) locate 30% clay on the clay axis and 40% sand on the sand axis. Project a line inward from 30% clay and parallel to the sand axis. Project a line inward from 40% sand and parallel to the silt axis. The intersection of the two lines occurs in the clay loam section. Soil texture is clay loam.

The soil triangle identified the soil texture as clay loam. The simple classification system would identify this soil as a loam containing 30% clay.

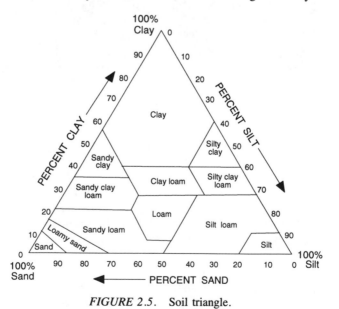

FIGURE 2.5.   Soil triangle.

2.3.2 Soil Texture in Pond Soils

Soil texture is clearly important in pond construction, for the soil must contain the proper distribution of particles to permit construction of stable embankments and watertight pond bottoms.[1,3] A soil material composed of a mixture of different-sized particles and containing at least 30% clay is ideal for pond construction. Such a soil can be compacted to form a watertight bottom, and it can be readily dried and tilled during the fallow period between aquaculture crops. Heavy clay soils are sticky and difficult to dry and till.

There has not been much interest in the texture of pond bottom soils in pond management, because the importance of soil texture in pond management has not been elucidated. Bottoms of properly constructed ponds will contain at least 20% and usually 30–40% clay to reduce the possibility for excessive seepage, and bottoms of new ponds usually will be clays or loams. Under natural conditions, soil textural class is considered to be a conservative and unchanging soil property. However, erosion and sedimentation within a pond may cause marked changes in soil texture from one place to the other in the bottom. These changes in texture normally do not increase seepage losses from ponds because they affect only surface layers.

2.4 Organic Matter

2.4.1 Organic Fraction of Mineral Soils

Most soils are mineral soils, but they contain some organic matter. The surface layer of woodland and grassland soils comprises almost entirely organic matter, but the lower horizons are much lower in organic matter concentration. The furrow slice (upper 15–20 cm) of mineral soils used for agriculture seldom contains more than 5–6% organic matter, and in tropical and subtropical areas organic matter concentrations usually are much lower. Soils with poor drainage and soils developed in cold climates contain more organic matter than well-drained soils or soils in warmer climates. Of course, soils in arid regions are low in organic matter because of lack of vegetation.

The organic matter in soils includes material in all stages of decay from fresh additions of new organic matter to highly resistant remains of old organic matter. There is a relatively constant input of fresh organic matter to an undisturbed soil. Simple organic compounds such as starches, sugars, cellulose, and proteins break down very quickly. Lignins, waxes, resins, fats, and oils in plant residues are more resistant to microbial degradation. As organic matter decomposes, its original form becomes unrecognizable; as decomposition continues, compounds released from decomposing residues or synthesized by microbial activity are converted into dark-colored organic complexes known as *humus*. Both degradation and synthesis are involved. Original material is broken down, but organic matter that is not respired away as carbon dioxide is synthesized into new compounds. Usually, 60–80% of the organic matter in mineral soils is humus.

Lignin is thought to be the major source of humus. Lignin decomposes into phenols and quinones, and through bacterial activity polyphenols and polyquinones are formed. These aromatic ring structures have a high molecular weight and react with amino compounds in the soil. The complex molecules, which are colloidal in nature, closely associated with clay surfaces, and contain many hydroxy, carboxylic, and phenolic groups, are considered the basic component in humus. According to Alexander, many classes or organic compounds can be extracted from humus as follows: amino acids, purines, pyrimidines, aromatic molecules, uronic acids, amino sugars, pentose sugars, hexose sugars, sugar alcohols, methyl sugars, and aliphatic acids.[4]

An undisturbed soil has a fairly stable organic matter concentration. This suggests that an equilibrium level of organic matter is reached in a soil so that inputs of new organic matter are roughly balanced by decomposition of older soil organic matter, and the percentage organic matter remains essentially unchanged. This also implies that organic matter, including humus, in mineral soils is highly dynamic.

In a mineral soil the organic residue and associated microbial community are associated closely with mineral particles and bind them into aggregates. This gives the soil structure, improved water movement, and aeration. Organic matter increases the water-holding capacity of the soil because it can absorb large amounts of water. Organic matter is highly reactive. As it decomposes, plant nutrients are released. It can bind cations through ion exchange and complex formation. In spite of its low concentration in mineral soils, organic matter is an important soil property.

Pond soils differ from normal mineral soils because they are flooded most of the time. Inundation prevents direct contact between soil and air. This diminishes the availability of oxygen and suppresses the rate at which microorganisms aerobically decompose organic matter. Pond soils tend to acquire a greater organic matter concentration than surface soils from which they were formed. Nevertheless, organic matter in pond soils is quite reactive, and the decomposition of organic matter in pond soils has a major role in pond ecology.

2.4.2 Organic Soils

Organic soils usually contain mineral matter, but there is a much greater proportion of organic matter in organic soils than in mineral soils. Soil not saturated with water for long periods each year is considered to be organic soil if the organic carbon concentration is above 20%.[5] Soil saturated with water most of the time is classified as organic soil at organic carbon concentrations of 12–18% (Fig. 2.6). Organic matter in soil contains 48–58% carbon.[6] A factor of 2 times organic carbon concentration often is used to estimate organic matter

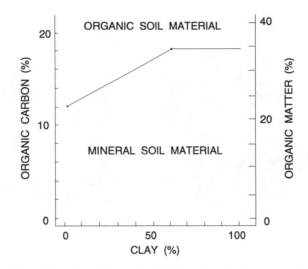

FIGURE 2.6. Separation of waterlogged soils into mineral soils or organic soils based on clay and organic carbon concentrations.

concentration from organic carbon concentration in surface soils of terrestrial ecosystems. A mineral soil is contrasted with an organic soil in Figure 2.7.

There are three basic types of organic soil material.[5] *Fibric materials* consist largely of plant fibers. A fiber is a fragment of plant tissue, excluding live roots, large enough to be retained on a 0.15-mm sieve. Fibric soil materials have a fiber content, after hand rubbing, of 75% or more of the soil volume. Coarse fragments of wood and mineral layers are excluded when determining the volume of fiber. Peat is a fibric soil material. *Sapric materials* contain highly decomposed organic matter. After rubbing, the fiber content is less than 18–20% of the soil volume. Sapric soil materials are dark gray to black and relatively stable when drained and exposed to the air. Colloquially, these materials are called muck soils. Soil materials with fiber contents between sapric and fibric are called *hemic soil materials*.

Lack of drainage is the usual reason for formation of organic soils. Microorganisms cannot decompose organic matter efficiently where soils are saturated and anaerobic conditions exist. Additionally, decomposition products that accumulate in poorly drained organic soils are acidic, and low pH also retards microbial activity. If an organic soil is drained, its decomposition rate increases and there may be a noticeable decrease in the soil's bulk volume.

Construction of ponds on organic soils should be avoided when possible. Organic soils will not form stable embankments, because, when exposed to the air and dried, a large part of the organic matter decomposes. A noticeable decrease in the height and width of embankments made from organic soils usually can be detected within one to two years. Organic soils do not provide good bottom

FIGURE 2.7.   A mineral soil (left) contrasted with an organic soil (right).

habitats for plants and animals, are highly acidic, have a high oxygen demand, and do not form firm substrates.

As indicated, organic matter tends to accumulate in mineral soils of pond bottoms. However, bottom soil seldom contains more than 5–6% organic carbon, and it should not be concluded that mineral soils in aquaculture ponds tend to be transformed into organic soils.

2.5 Soil Density and Pore Space

Soil density may be expressed as bulk density or particle density. *Bulk density* is the dry weight of a given volume of soil. A known volume of soil is dried and weighed, and the bulk density is reported as grams per cubic centimeter, tons per cubic meter, or megagrams per cubic meter, all of which are numerically equal. The range in bulk density of mineral soils is 1–2 g cm^{-3}. The lowest values are for fine-grained surface soils, and the highest are for tightly compacted subsoils. Most surface, mineral soils have bulk densities of 1.25–1.50 g cm^{-3}. Organic soils are much lighter than mineral soils; bulk densities of peat soils may be 0.2–0.4 g cm^{-3}. The bulk density of pond soils exhibits the same range as that of terrestrial soils, but in a single pond the bulk density may vary greatly from one sampling point to another (Table 2.5). Bulk density is usually lowest in deeper parts of a pond where a sediment layer has formed.

Particle density is expressed in terms of the volume of the soil particles only.

Table 2.5. *Bulk Density (g cm^{-3}) of Pond Soil from Different Soil Depths in a Fish Pond at Auburn, Alabama*

Soil Layer (cm)	Water Depth above Sampling Site (cm)					
	10	20	40	60	80	100
0–5	1.170	1.101	0.563	0.288	0.313	0.302
5–10	1.279	0.964	0.647	0.298	0.387	0.403
10–15	1.350	1.002	0.944	0.471	0.560	0.597
15–20		1.218	1.179	0.775	0.555	0.777
20–25		1.393	1.154	0.848	0.775	0.887
25–30			1.326	1.252	1.168	1.235
30–35			1.422	1.424	1.222	1.312
35–40				1.309	1.299	1.338
40–45					1.466	

Source: Masuda and Boyd (unpublished data).

Think of the solid particles in a soil sample being compressed so completely that there are no open spaces among them; the actual space occupied by particles in the sample is the volume of this hypothetical, compressed mass. Particle density is determined from the mass and volume of a soil sample. Sample mass is obtained by weighing, and the volume of particles is calculated from the mass and density of water displaced by the sample. Results are reported in the same dimensions as bulk density. The particle density of mineral soils usually falls within a fairly narrow range (2.60–2.75 g cm^{-3}). Particle density of organic matter averages about 1.25 g cm^{-3}, and particle density of soils decreases with increasing organic matter concentrations. Particle density for pond soils exhibits a range similar to that of terrestrial soils.

Soil is made up of solid particles and open spaces called *pores* (Fig. 2.8).

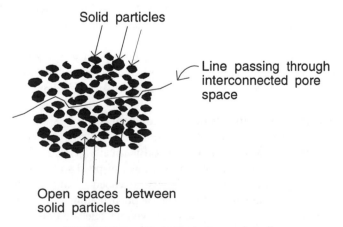

FIGURE 2.8. Illustration of pores in soil.

The pores are interconnected and form conduits for movement of air and water in a soil. Particles in a soil normally vary greatly in size and shape, and the interconnected pores provide a somewhat random pathway for air and water movement. Pores also vary greatly in size, and water and air move more freely in large macropores than in micropores.

It is sometimes desirable to know the amount of pore space in a soil. This can be calculated from particle and bulk densities. If the volume of a soil cube is expressed as 100%, then the percentage of the total volume occupied by pore space is

$$\% \text{ pore space} = 100 - \left[\frac{\text{bulk density}}{\text{particle density}} \times 100 \right] \qquad (2.1)$$

Example 2.2: Calculation of Percentage Pore Space

A fish pond soil has a bulk density of 1.07 g cm^{-3} and a particle density of 2.61 g cm^{-3}. Compute the pore space.

Solution:

$$\% \text{ pore space} = 100 - \left(\frac{1.07}{2.61} \times 100 \right) = 41.0\%$$

Percentage pore space is a function of particle fineness; fine-grained soils are more porous than coarse-grained ones. A sandy soil might have 45% pore space, whereas a clay soil might contain more than 60% pore space. Organic soils have more pore space than mineral soils. Pore size in soil tends to decrease as particle size decreases. Pores in a sandy soil are larger than those in clay or organic soils. As a result, gas and water can move more readily in a coarse-textured soil than in a fine-textured one even though the fine-textured soil has a greater percentage of pore space.

The pore space in the surface 0–15-cm layer of pond soils is often as high as 90%. This is apparently related to the high moisture content, which tends to separate the soil particles. Deeper layers of pond soil often have 30–40% pore space.

When making soil treatments or calculating amounts of certain substances in agricultural soil, one must know the weight of soil that is plowed. The weight of a 15-cm-deep furrow slice of surface soil is usually given as 2,200 ton ha^{-1}. Boyd and Cuenco showed that the surface layer of pond soil is lighter than the furrow slice in agricultural soils.[7] They recommended a value of 1,500 ton ha^{-1} for the weight of the upper 15-cm layer of pond bottom soil. However, their study involved ponds under 10 years old without thick sediment layers. Bulk density of the 0–15-cm layer in a 22-year-old pond (Table 2.5) averaged 0.767 g

cm^{-3}, and the weight of a layer 15 cm deep $\times$ 1 ha is 1,150 ton ha^{-1}. The bulk density of pond soil changes with depth, as illustrated by data in Table 2.5. Surface layers have a much lower bulk density than deeper layers, and the bulk density of the surface layer is lower in deeper parts of the pond than in shallower parts. Ponds with mineral soils often have a low bulk density in the surface soil layer. For example, data in Table 2.5 are for a pond with soil carbon concentrations of 2–3%. The low bulk density was caused by a high moisture content (200–400%) in the 0–15-cm soil layer. The bulk density of the 0–15-cm soil layer increases when the pond is drained and its bottom dried.

2.6 Soil Water

Pond soils are saturated with water most of the time. Water can exit ponds by seepage beneath the levees and directly through bottom soils; in some instances, water can seep into ponds. Water that seeps through pond soils contains dissolved substances. However, because pond soil is saturated and seepage is a slow process, diffusion of dissolved substances within the soil water also may be important in moving dissolved substances downward through the soil. When ponds are drained for harvest, many pond managers like to dry pond soils. During this drying period, they hope to aerate the soil, oxidize reduced substances, and stimulate microbial decomposition of organic matter. Thus, the amount and movement of water in pond soils is important, and a brief discussion of soil water will be provided.

2.6.1 Classification and Energy Relationships

The water in soil is held around the soil particles and in the pore space. Soil water may be divided into several categories.[8] A soil completely saturated with water is said to be at its *maximum retentive capacity*. If such a soil is permitted to drain until all of the drainable water has left the larger pores, it is said to be at *field capacity*. Soil water can be removed by plants and lost to the air as water vapor through transpiration or direct evaporation from soil surfaces. When the water content of a soil drops so low that plants remain wilted at night, the soil is at its *permanent wilting percentage*. The water remaining at the permanent wilting percentage is held in very small pores and around the particles. If soil at the permanent wilting percentage is held in air that is not quite saturated with water vapor, further water is lost by evaporation. The moisture status at this point is known as the *hydroscopic coefficient*. Water remaining at the hydroscopic coefficient is tightly attached to colloidal soil particles.

If saturated soil is placed on a porous plate in a sealed chamber and pressure is applied to force water from the soil, field capacity occurs at about 0.1–0.5 atm pressure, the permanent wilting percentage at about 12–18 atm, and the hydroscopic coefficient at around 30 atm. A tremendous amount of pressure is required to remove all of the water remaining at the hydroscopic coefficient. As the soil moisture content declines, the remaining water is held tighter and more energy is required to remove it.

The maximum retentive capacity of a soil depends on the volume of pore space alone. Fine-textured soils have higher maximum retentive capacities than do coarse-textured soils. The relationship of texture to the other categories of soil water is shown in Figure 2.9. The amount of unavailable water (water remaining at the permanent wilting percentage) increases in fine-textured soils, but they still contain more available water than do coarse-textured ones.

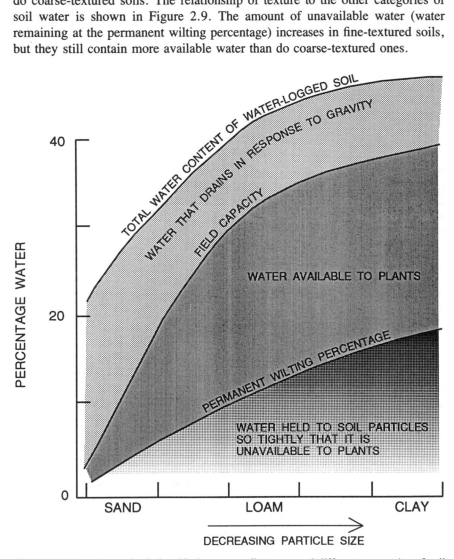

FIGURE 2.9.   General relationship between soil texture and different categories of soil water.

2.6.2 Water Movement

Water can move through soil pores as vapor, and liquid water may move by both saturated and unsaturated flow. *Saturated flow* involves seepage or infiltration of water through soil in which all of the pore spaces are filled with water. This is obviously a common means of water movement in pond bottom soils. *Unsaturated flow* involves many factors, but most unsaturated flow probably occurs through micropores of soil, because in an unsaturated soil macropores are filled with air. Unsaturated flow only occurs in the bottoms of empty ponds during the fallow period or immediately after dry ponds are refilled. Capillary movement of water will be discussed here, because it is a common process in dry pond bottoms.

Seepage

Soils have interconnected pore spaces among their constituent particles, and water flows through the pores in response to gravity. The pores do not provide a straight channel for water movement. A molecule of water moving through soil pores exhibits a random pattern of movement (Fig. 2.10). In going from one point to another in the soil, the length of the flow path is unknown. The area of flow is not the cross-sectional area of the flow path, for flow only occurs through the pores. The porosity of soils (η) is defined as

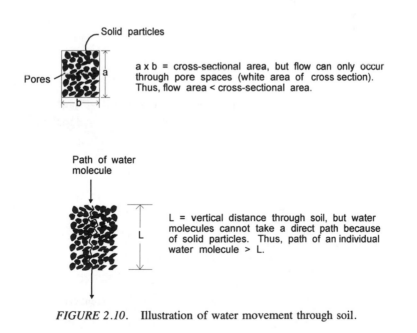

Solid particles

Pores

a x b = cross-sectional area, but flow can only occur through pore spaces (white area of cross section). Thus, flow area < cross-sectional area.

Path of water molecule

L = vertical distance through soil, but water molecules cannot take a direct path because of solid particles. Thus, path of an individual water molecule > L.

FIGURE 2.10. Illustration of water movement through soil.

$$\eta = \frac{\text{Pore volume}}{\text{Bulk volume}} \tag{2.2}$$

The open area of cross section that conveys water is cross-sectional area $\times$ η. This suggests that fine-textured, highly porous soil should favor greater water movement than coarse-textured soils. However, small pores provide more surface attraction for water molecules, and frictional forces opposing flow are greater in fine-grained soils than in coarse-grained ones. Hydraulic conductivity must be used instead of porosity in assessing water movement in soil. *Hydraulic conductivity* (k) is the velocity of water movement in response to a hydraulic gradient of 1.00 in a porous material. The hydraulic gradient is the difference in hydraulic head divided by the length of the flow path. Values of hydraulic conductivity have been determined for soil particles of different sizes (Table 2.6). Values of k decrease rapidly with decreasing particle size; coarse sand has a k about 10^6 times greater than that of clay.

Darcy's equation relates the seepage velocity in soil or other porous media to the hydraulic conductivity and the *hydraulic gradient* (i):

$$v = ki \tag{2.3}$$

where v = seepage velocity (m sec^{-1}). If the volume of seepage is needed, the relation between velocity of flow, area of flow (A), and discharge (Q) may be substituted into Darcy's equation:

$$Q = Av \tag{2.4}$$

$$\frac{Q}{A} = ki \quad \text{and} \quad Q = kiA \tag{2.5}$$

Because we are using hydraulic conductivity, porosity does not have to be considered, and A is the cross-sectional area of the porous material instead of the pore space.

Table 2.6. Hydraulic Conductivity (k) of Soil Separates

Soil Separate	k (cm sec^{-1})
Clean gravel	2.5–4.0
Fine gravel	1.0–3.5
Coarse, clean sand	0.01–1.0
Mixed sand	0.005–0.01
Fine sand	0.001–0.05
Silty sand	0.0001–0.002
Silt	0.00001–0.0005
Clay	10^{-9}–10^{-6}

Example 2.3: Use of Darcy's Equation to Estimate Seepage Velocity
Compare seepage velocity for 1-m-thick layers of coarse sand and silt above
which stands 0.1 m of water.

Solution: A median value of k for each soil material will be estimated from
data in Table 2.6; k is 0.055 cm sec^{-1} and 0.000255 cm sec^{-1} for coarse sand
and silt, respectively. The hydraulic head is the vertical distance from the
water surface to the end of the path of flow. This vertical distance represents
gravitation energy available to overcome frictional losses between water and
soil particles and cause flow through the soil. It is conventional to express
available energy in terms of height of water. Thus, hydraulic head is 0.1 m
of water depth above the soil plus 1 m depth of saturated soil, or 1.1 m.
The hydraulic gradient is

$$i = \frac{\text{Hydraulic head}}{\text{Flow path}} = \frac{1.1}{1.0} = 1.1$$

Seepage velocities are

$$V_{\text{coarse sand}} = 0.055 \text{ cm sec}^{-1} \times 1.1 = 0.06 \text{ cm sec}^{-1}$$
$$V_{\text{silt}} = 0.000255 \text{ cm sec}^{-1} \times 1.1 = 0.00028 \text{ cm sec}^{-1}$$

In computing seepage by the Darcy equation, we assume that the water can
flow away at the lower surfaces of soil layers at a velocity equal to or greater
than the flow through the soil layers. If this assumption is false, the Darcy
equation will not provide a reliable estimate, because seepage will be impeded
at the lower surface. An example is a pond with a highly permeable sandy bottom
constructed in an area with a water table just below the pond bottom. Although
the pond bottom will seep rapidly, the high water table will block the seepage
and reduce its rate.

Capillarity

The rise of water in small-bore tubes or soil pores is called *capillary action.*
To explain capillary action, one must consider cohesion, adhesion, and surface
tension. *Cohesive* forces result from attraction between like molecules. Water
molecules exhibit cohesion because they form hydrogen bonds with each other.
Adhesive forces result from attraction between unlike molecules. Water adheres
to a solid surface if the surface molecules form hydrogen bonds with water. Such
a surface that wets easily is called a *hydrophilic* surface. Adhesive forces between
water and the hydrophilic surface are greater than cohesive forces among water
molecules. A *hydrophobic* surface is one that does not attract water molecules
by adhesion. Water will bead on a hydrophobic surface and run off, for cohesion
is stronger than adhesion. For example, raw wood wets readily, but a coat of
paint causes wood to shed water. The net cohesive force on molecules in a body

of water is zero, because cohesive forces cannot act above the surface, and molecules of the surface layer are subjected to an inward cohesive force from molecules below the surface. Surface molecules act as a skin over the surface to provide *surface tension,* which permits insects and spiders to walk over the surface of water, and needles and razor blades to float when gently placed on a water surface. The strength of the surface film decreases with increasing tempera-ture and when impurities are dissolved in water. Soap and most other organic substances decrease surface tension when dissolved in water.

Water rises to considerable heights in a thin glass tube or in clay or fine-silt soils. Capillary action is the combined effect of surface tension, adhesion, and cohesion. In a thin tube, water adheres to the walls and spreads upward over as much surface as possible. Water moving up the wall is attached to the surface film, and molecules in the surface film are joined by cohesion to molecules below. As adhesion drags the surface film upward, it pulls a column of water up the tube against the force of gravity. The column of water below the surface film is under tension because the water pressure is less than the atmospheric pressure.

The height to which water will rise by capillary action is limited by the weight of water that can be supported by the film of water adhering to the tube or pore (Fig. 2.11). In a very small diameter tube or pore the upward surface tension force is

$$F = \gamma 2\pi r \qquad (2.6)$$

where F = upward force
γ = viscosity of water (0.073 N m^{-1} at 20°C)
π = 3.14
r = radius of pore or tube (m)

In large pores or tubes the force is directed according to the angle θ between the meniscus and the wall of the pore or tube (Fig. 2.11). The upward force is

$$F = \gamma 2\pi r \cos \theta \qquad (2.7)$$

We have no way of knowing θ in most situations, so we assume that the force is directed parallel to the pore or tube.

The downward force or weight of water being lifted by capillarity is

$$w = mg \qquad (2.8)$$

where w = weight
m = mass
g = gravitational acceleration (9.8 m sec^{-2})

Equation 2.8 may be rearranged as follows:

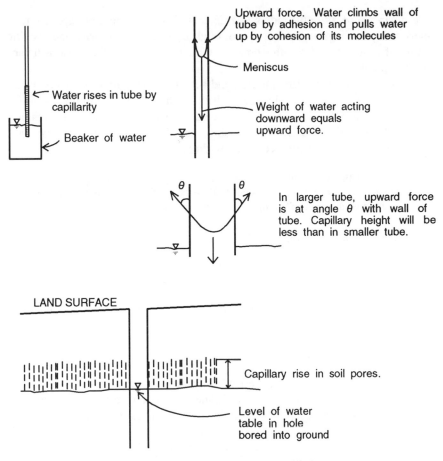

FIGURE 2.11. Illustration of capillarity.

$$w = (\rho V)g = \rho(\pi r^2 h)g$$

where V = volume of column

ρ = specific weight of water $(1{,}000 \text{ kg m}^{-3})$

h = height of water column (m)

By setting upward and downward forces equal,

$$\gamma 2\pi r = \rho(\pi r^2 h)g$$

we obtain a convenient equation for estimating the height of water in a capillary:

$$h = \frac{2\gamma}{\rho g r} \qquad (2.9)$$

At a given temperature the radius (diameter $= 2r$) of the pore or capillary tube determines how high water will rise through capillarity. Viscosity and specific weight of water decreases with increasing temperature, so capillary rise will vary with temperature.

In soil the pores are interconnected and act like a capillary tube. Thus, water can rise from the water table into dry soil through capillarity. Capillarity in soil increases in the following order: sand < silt < silty clay < clay. If capillary tube walls or soil pores are hydrophobic, capillary action can be downward instead of upward, or it can occur in any direction.

Example 2.4: Height of Capillary Rise in Soil
    The capillary fringe in a soil extends 0.35 m above the water table. Calculate the effective diameter of the interconnected pore space.

Solution: Rearranging Equation 2.9 gives

$$r = \frac{2\gamma}{h\rho g} = \frac{2(0.073\ \mathrm{N\ m^{-1}})}{(0.35\ \mathrm{m})(1{,}000\ \mathrm{kg\ m^{-3}})(9.8\ \mathrm{m\ sec^{-2}})} = 4.3 \times 10^{-5}\ \mathrm{m}$$

Diameter is $2r$, so the effective pore diameter is 8.6×10^{-5} m, or 0.086 mm.

2.6.3 Moisture Concentration

Soil moisture concentration in terrestrial soils varies greatly with weather conditions, and it is an important variable influencing plant growth. The moisture concentration of soils is reported on a dry-mass basis as follows:

$$\text{Moisture (\%)} = \left[\frac{\text{Weight wet soil}}{\text{Weight dry soil}} - 1\right] \times 100 \tag{2.10}$$

Soil moisture concentrations at maximum retentive capacity, field capacity, and permanent wilting percentage depend on soil texture, organic matter content, and type of clay minerals present (Fig. 2.9). In general, a soil with a high specific surface area holds more water and retains the water more tightly than a soil with a lower specific surface area. Expandable clay minerals have a greater surface area per unit weight than do nonexpandable ones. For example, Green found the specific surface area of pond soils at Comayagua, Honduras, to be 439 $\mathrm{m^2 \cdot g}$ clay^{-1}, and pond soils at Auburn, Alabama, had specific surface areas around 100 $\mathrm{m^2 \cdot g}$ clay^{-1}.[9] Clay in Honduran soil was largely smectite, and clay in Alabama soil was kaolinite, mica, and hydrous oxides. Boyd and Teichert–Coddington found that biologically available water in dry pond bottoms became deficient for microbial activity at a higher soil moisture concentration in Honduran ponds than in Alabama ponds.[10]

Table 2.7. *Soil Moisture Content (% dry weight) for Three Fish Ponds of Different Ages at Auburn, Alabama*

Soil Layer (cm)	Pond[a]		
	F	M	G
0–5	288	358	294
5–10	216	266	115
10–15	216	213	39
15–20	206	216	31
20–25	172	164	28
25–30	141	51	
30–35	117	34	
35–40	102	33	
40–45	90	33	
45–50	58	35	
50–55	32	—	
55–60	28	—	

Source: Munsiri and Boyd (unpublished data).

[a]Pond ages: F = 52 years; M = 23 years; G = 2 years.

In a pond filled with water, soil moisture concentration decreases with soil depth. Soil at all depths is saturated, and the ratio of solids to water is larger near the soil surface than at greater depths (Table 2.7). Old ponds with a thick sediment layer have a deeper layer of high-water-content soil than do newer ponds.

2.7 Gases

The movement of oxygen and carbon dioxide in soils is extremely important, because exchange of these two gases with the atmosphere is essential for maintaining aerobic conditions within a soil. Evaporation from unvegetated soils, such as an empty pond bottom, depends on the outward movement of water vapor. In an unsaturated soil, gases move through the pore spaces. Oxygen is consumed in respiration by plant roots, animals, and soil microorganisms, so the oxygen concentration, or partial pressure, in the soil is lower than in the air. The partial-pressure gradient causes oxygen to move into soil. Oxygen must move by diffusion, a slow process. Oxygen cannot diffuse through soil fast enough to meet biological requirements for oxygen and still maintain the same oxygen concentration (20.95%) found in the atmosphere. Soil air in deeper horizons may contain only 10–15% oxygen. Water vapor is generated from soil moisture and carbon dioxide produced by the respiration of the soil organisms.

These two gases have greater partial pressures in soil air than in atmospheric air, and their net diffusion is from soil to air.

Because gases move within the pore spaces of a soil, soils with large pores (coarse-textured soils) are better aerated than those with small pores (fine-textured soils). Water also can enter pore spaces, and when pore spaces are filled with water, movement of gases by diffusion is markedly impeded by the greater molecular density of water as compared to air. A saturated soil usually becomes depleted of oxygen except at its surface. Pond soils are saturated with water except when they are dried between aquaculture crops.

2.8 Soil Colloids

Colloidal particles are the most reactive fraction of soils. *Colloids* are particles comprising aggregates of molecules that are larger than ordinary molecules but small enough to act differently from coarse dispersions. Colloids are formed simply by grinding coarse materials into finer particles. Colloidal particles are 1–100 mμ in diameter. They do not dissolve in water, but they are so small they remain dispersed. Colloidals are electrically charged, exhibit Brownian movement because they are bombarded by molecules of the dispersion medium, have particle diameters greater than the average wavelength of white light and interfere with the passage of light by scattering it (Tyndall effect). They also have very large surface areas and can strongly adsorb ions and other dissolved substances, and do not pass through ordinary semipermeable membranes.[11] Clay and humus are the predominant colloidal particles in soil.

2.8.1 Clay

Clays are the smallest mineral particles formed by weathering. Clay particles are 0.002 mm or smaller in diameter, and the smallest ones are colloidal. Basic clay types are layer silicate, iron and aluminum oxide, and amorphous clays.[12]

Silicate Clays

Layer silicate clays are crystalline substances composed of layers. A layer is made of sheets, of which there are two types. One type is a tetrahedral sheet made of basic units of a tetrahedron of one silicon atom surrounded by four oxygen atoms (Fig. 2.12). The basic units are connected by shared oxygen atoms to form the sheet. Sheets comprising basic octahedron units (Fig. 2.13) made of aluminum and magnesium ions each surrounded by six oxygen atoms or hydroxyl groups also occur. The units are joined by shared oxygens or hydroxyls to form the octahedral sheet. Octahedral sheets dominated by aluminum are called dioctahedral sheets; those dominated by magnesium are called trioctahedral sheets. The sheets are connected by shared oxygen atoms to form layers. Crystals

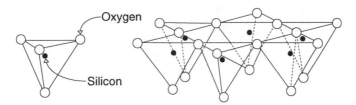

Basic Tetrahedral Unit   Attachment of Basic Units into a Sheet

FIGURE 2.12.   Basic tetrahedral unit of silicon and oxygen.

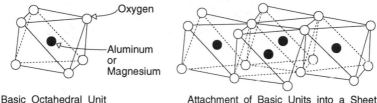

Basic Octahedral Unit   Attachment of Basic Units into a Sheet

FIGURE 2.13.   Basic octahedral unit of either aluminum or magnesium and oxygen.

or *clay micelles* contain many layers connected by hydrogen bonding (Fig. 2.14).
Spaces between layers are called interlayers.

Isomorphous substitution can occur in tetrahedral and octahedral sheets. Aluminum often substitutes for silicon in the tetrahedral sheet, and iron or zinc can substitute for aluminum in the octahedral sheet. Substitution of aluminum (Al^{3+}) for silicon (Si^{4+}) leaves an excess negative charge. Iron (Fe^{2+}) or zinc (Zn^{2+}) substitution for aluminum leaves an excess positive charge. The usual overall effect of isomorphous substitution is to provide a negative charge, and most clays are negative charged.

Layers constituting clay micelles may contain one tetrahedral sheet and one

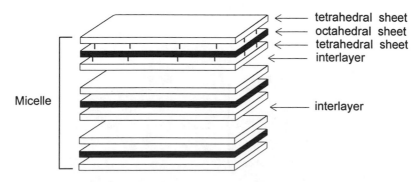

FIGURE 2.14.   Simple representation of a clay micelle.

octahedral sheet (1:1 clay), two tetrahedral sheets and one octahedral sheet (2:1 clay), and 2:1 clay interspaced with an octahedral sheet (2:1:1 clay). These forms are illustrated in Figure 2.15.

Kaolinite is a common 1:1 clay. Sheets and layers of 1:1 clays fit together very tightly, and there is little isomorphic substitution in the sheets. Cations and water cannot easily enter the interlayers. The surface area of 1:1 clays is restricted to outside surfaces, they have comparatively small electrical charges and do not swell when wetted.

There are three basic groups of 2:1 clays: smectite, vermiculite, and illite. Montmorillonite is the most well-known member of the smectite group. Layers are weakly joined, and the surface area exposed by the faces of the interlayer is greater than the external surface area. Cations and associated water molecules are attracted by negative charges in the interlayer. When wetted, clays of the smectite group swell. The high negative charge of smectites results primarily from substitution of Mg^{2+} for Al^{3+} ions in the dioctahedral sheet. In vermiculites, there is a very high negative charge resulting from substitution of Al^{3+} for Si^{4+} in the tetrahedral sheets. Because of the more highly charged faces exposed in the interlayers, water and cations entering the interface tend to hold the layers together rather than pulling them apart as in smectite. Thus, vermiculites do not swell as much as smectites, but they swell more than kaolinite.

Illite has a high degree of substitution of Al^{3+} for Si^{4+} in the octahedral sheets, and it has a very great charge. Potassium ions fit snugly between the layers and

```
Tetrahedral sheet  ⎫
Octahedral  sheet  ⎬   Kaolinite  ( 1 : 1  type )
                   ⎭


Tetrahedral  sheet ⎫
Octahedral   sheet ⎪
Tetrahedral  sheet ⎪
 (Interlayer)      ⎬   Smectite  ( 2 : 1  type )
Tetrahedral  sheet ⎪
Octahedral   sheet ⎪
Tetrahedral  sheet ⎭


Tetrahedral  sheet ⎫
Octahedral   sheet ⎪
Tetrahedral  sheet ⎪
 (Interlayer)      ⎪
Octahedral   sheet ⎬   Chlorite  ( 2 : 1 : 1  type )
 (Interlayer)      ⎪
Tetrahedral  sheet ⎪
Octahedral   sheet ⎪
Tetrahedral  sheet ⎭
```

FIGURE 2.15. Representations of different types of silicate layer clays.

bind them tightly. This neutralizes much of the negative charge, makes the interlayer inaccessible to water, and as a result, illite does not expand upon wetting.

Chlorite is an example of a 2:1:1 clay. The negative charge on chlorite is small compared to smectites and vermiculites, and it does not swell upon wetting.

Several types of clay may be mixed in native soils. There also is considerable variation in the typical micelles described here, which result from a micelle having individual layers of more than one type. Clays described here simply serve as models to assist the reader in understanding the nature of silicate clays.

Iron and Aluminum Oxides

Highly weathered soil, especially in the tropics and subtropics, contains iron and aluminum oxides such as [1]$Al(OH)_3$ (gibbsite) and [1]$FeOOH$ (goethite), which exist as fine particles and are considered part of the clay fraction.* Some iron and aluminum oxides have a definite crystalline structure, and others are amorphous. They have no strong electrical charges and do not expand when wet. In many tropical soils, iron and aluminum oxides dominate the clay fraction.

Amorphous Clays

Many soils contain colloidal mineral matter that is not crystalline. The most important amorphous colloid is allophane, which is $Al_2O_3 \cdot 2SiO_3 \cdot H_2O$ or some similar composition. Allophane is common in soils developed from volcanic ash.

Important Properties of Clays

Characteristics of clays have been mentioned several times, but will be summarized here. A clay content of 15–20% and above makes a soil plastic. That is, the wet soil can be readily molded into various shapes such as balls or cylinders. In a soil with a very high clay content, there is a specific moisture content where the clay changes from plastic to a semiliquid that will flow. This moisture content is the *plastic limit*. Clay particles are sticky because of adhesion of clay particles to water molecules between them. Where water can enter the interlayers of the micelles of clays, clays will swell upon wetting and shrink upon drying. The particles of clay are negatively charged and repel each other, causing clays to disperse in water. If the charges on dispersed clay particles are neutralized by cations or by changing the pH, clay particles will flocculate and settle out of water. Soils can be dispersed if the primary cation in the soil is sodium, which does not effectively reduce the negative charge on the clay particles. Adding calcium ion to the soil can reduce the charge and cause flocculation in the soil. Applications of gypsum ($CaSO_4$) to soils of a high sodium content flocculate the

---

*Gibbsite and goethite sometimes are written as $Al_2O_3 \cdot 3H_2O$ and $Fe_2O_3 \cdot H_2O$, respectively.

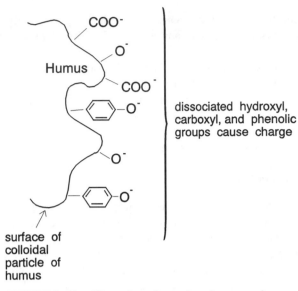

FIGURE 2.16. Illustration of negative charges on humus.

clay particles and improve the soil texture. The negative charges on soil and other soil colloids are responsible for the cation-exchange property of soils.

2.8.2 Humus

The formation of humus was discussed earlier. It consists of a complex mixture of various organic compounds such as polyphenols, polyquinones, polyuronides, and polysaccharides. Humus particles are closely associated with mineral particles in soil. Humus is colloidal, so it has a high surface area and is absorptive. It is particularly important in causing soil aggregation and in increasing the ability of soil to hold water. The surfaces of humus are even more strongly charged than vermiculites and smectites. The origin of this negative charge is dissociated hydroxy, carboxylic, and phenolic groups on the surface of humus (Fig. 2.16).

2.9 Color

Color is a diagnostic property of soil for use in soil taxonomy, because certain substances can impart a characteristic color to soil. Iron oxides can produce yellow to bright red colors. Organic matter causes a darkening of the soil, and some organic soils are almost black in color. Sandy soils often are light brown. Waterlogged soils develop gray or black colors because of reducing conditions and the presence of ferrous compounds. Pond soils typically have an oxidized surface layer. This layer is brown, but the color of the native soil may also be obvious. Beneath the oxidized surface layer, the typical gray or black color of waterlogged soils occurs.

Soil colors are most conveniently measured by comparison with a color chart. The Munsell Soil Color Charts[13] are widely used for measuring soil color. Three variables—hue, value, and chroma—are combined to describe soil colors with a designation such as 5YR 5.5/6. In this designation YR refers to yellow–red and 5 indicates the position of the color in the yellow–red hue. In the notation 5.5/6, 5.5 indicates the color value and 6 indicates the chroma. The chart gives this complete color designation the name yellowish red.

Recent studies by the author have revealed that below the thin oxidized surface layer, pond sediment has a dark color indicative of highly reduced conditions. Lighter colors are found in the original pond bottom soil. This suggests that the original soil is not as reduced as the sediment.

2.10 Ion Exchange

2.10.1 Source of Charges

It has been shown that substitution of Al^{3+} for Si^{4+} in the tetrahedral sheet or substitution of Mg^{2+} for Al^{3+} in the dioctahedral sheet of a clay micelle produces an unsatisfied negative charge. Substitution of Al^{3+} for Mg^{2+} in the trioctahedral sheet of a micelle provides an unsatisfied positive charge. It is possible for clays to be negatively or positively charged, but most clays carry a net negative charge.

Some charges on clay particles are not associated with isomorphous substitution in the sheets. These charges are caused by hydroxy groups on surfaces of mineral colloids. The charge resulting from hydroxy groups increase as pH increases:

$$\text{Micelle--OH} + \text{OH}^- = \text{Micelle--O}^- + H_2O \tag{2.11}$$

At low pH the hydroxy groups can cause a positive charge:

$$\text{Micelle--OH} + H^+ = \text{Micelle--OH}_2^+ \tag{2.12}$$

There is a pH-dependent charge on many clays, and as pH increases the charge increases.

Humus is strongly charged because of hydroxy, carboxy, and phenolic groups. These groups in humus can be thought of as organic acids, and a small percentage of the groups will be dissociated to provide a charge even at low pH. However, as pH rises, the charge on humus increases because of a greater percentage of dissociation of the groups:

$$\text{(hydroxy)} \quad \text{--OH} + \text{OH}^- = \text{O}^- + H_2O \tag{2.13}$$

$$\text{(carboxyl)} \quad \text{--COOH} + \text{OH}^- = \text{COO}^- + H_2O \tag{2.14}$$

$$\text{(phenolic)} \quad \text{--OH} + \text{OH}^- = \text{--}\langle\underline{}\rangle\text{--O}^- + H_2O \tag{2.15}$$

2.10.2 Attraction of Ions

Negative charges on soil colloids attract cations, and positive charges attract anions:

$$\equiv Colloid \equiv + 2Ca^{2+} + Na^+ + K^+ + Mg^{2+} = \begin{matrix} Ca^{2+} \\ Na^+ \\ K^+ \\ Mg^{2+} \end{matrix} \equiv Colloid \equiv Ca^{2+} \quad (2.16)$$

$$Colloid + NO_3^- + Cl^- = \begin{matrix} NO_3^- \\ Colloid \\ Cl^- \end{matrix} \quad (2.17)$$

Most colloids have a net negative charge, but some sites on the surface are negatively charged and others are positively charged. It is more descriptive to show ion exchange as follows:

$$\equiv Colloid \equiv + Ca^{2+} + Mg^{2+} + K^+ + Na^+ + Cl^- + H_2PO_4^- = \quad (2.18)$$

$$\begin{matrix} K^+ & Cl^- \\ Mg^{2+} \equiv & Colloid & \equiv Ca^{2+} \\ Na^+ & H_2PO_4^- \end{matrix}$$

Cation adsorption is a more important process than anion adsorption in soils, and our discussion focuses on cation adsorption.

All species of cations are not attracted to negatively charged colloids with equal tenacity. The order of attraction is

$$Al^{3+} > Ca^{2+} > Mg^{2+} > K^+ = NH_4^+ > Na^+$$

Attraction decreases as valence decreases, but different ions of the same valence are not always attracted equally; for example, Ca^{2+} is held more tightly than Mg^{2+}. Hydrogen ion can be attracted to colloids, but there is little H^+ in soil solutions compared to the other cations. Copper, zinc, iron, and other metals can occur as cations, and they are adsorbed by electrical attraction to negatively charged colloids. Their concentrations usually are very low, and they make up an insignificant proportion of the adsorbed cations. It is important to remember that cations are dissolved in pore water and in water attracted to soil particles, and reactions indicated in Equations 2.11 through 2.18 occur in aqueous solution.

2.10.3 Cation Exchange

An equilibrium between cations in soil solution and cations adsorbed on colloids exists as

$$Ca^{2+} \quad \begin{array}{c} Al^{3+} \ Mg^{2+} \\ Colloid \quad K^+ \\ Na^+ \ NH_4{}^+ \end{array} = Al^{3+} + Ca^{2+} + Mg^{2+} + K^+ + Na^+ + NH_4{}^+ \quad (2.19)$$

If the concentration of one of the cation species in the soil solution (e.g. Ca^{2+}) is increased, the equilibrium is disrupted and a new equilibrium is established between cations on colloids and cations in solution. The new equilibrium has more Ca^{2+} on colloids. To make room for more Ca^{2+} on colloids, a proportion of all the other adsorbed cations must be released into solution. At the new equilibrium, concentrations of all ions other than Ca^{2+} decrease on colloids, and concentrations of all cations increase in solution. This process is called *cation exchange*. Cations can be held in the soil by adsorption onto colloids. They can be released by exchange with other cations in solution. Adsorbed ions are available to plants and soil microorganisms through exchange processes.

Cations strongly attracted to colloids have greater replacing power than those weakly attracted. Calcium can readily replace sodium or potassium, but high concentration can overcome strength of attraction. If a very high concentration of sodium is added to a soil solution, sodium will replace most of the other adsorbed cations.

2.10.4 Cation-Exchange Capacity

The ability of colloids in a soil to adsorb cations is called the *cation-exchange capacity (CEC)*. The CEC can be measured by using a strong concentration of some cation to leach all of the adsorbed cations from a soil sample. One popular method employs ammonium chloride. All cations on cation-exchange sites are replaced with $NH_4{}^+$. Excess $NH_4{}^+$ in soil solution must be displaced with an electrolyte-free solution. Finally, a strong concentration of another cation such as K^+ in potassium chloride solution is used to leach all of the $NH_4{}^+$ from the sample. The $NH_4{}^+$ concentration in the leachate is measured. The amount of $NH_4{}^+$ adsorbed by the soil and then displaced by K^+ can be calculated from the volume and $NH_4{}^+$ concentration of leachate. The cation-exchange capacity is calculated as milliequivalents of $NH_4{}^+$ adsorbed by 100 g soil (dry weight basis).

Example 2.5: Calculation of Cation-Exchange Capacity

A 10-g sample of soil is leached with 1 N NH_4Cl. The adsorbed $NH_4{}^+$ is displaced in 200 mL of 1 N KCl solution. The NH_4 concentration in the leachate is 40 mg L^{-1}.

Solution: The weight of $NH_4{}^+$ in the leachate is

$$40 \text{ mg L}^{-1} \times \frac{200 \text{ mL}}{1,000 \text{ mL L}^{-1}} = 8 \text{ mg}$$

The weight of $NH_4{}^+$ can be changed to milliequivalents as follows:

$$8 \text{ mg} \div 18 \text{ mg NH}_4 \text{ meq}^{-1} = 0.444 \text{ meq}$$

The 10-g soil sample had the capacity to adsorb 0.444 meq of cations. The cation-exchange capacity is

$$0.444 \text{ meq} \times \frac{100}{10 \text{ g soil}} = 4.44 \text{ meq } 100 \text{ g}^{-1}$$

The CEC of soils depends on the quantity of colloids in the soil and on the type of colloids. The CEC of humus may be as high as 200 meq 100 g^{-1}. Values of CEC (meq 100 g^{-1}) for typical clay minerals[12] are as follows: vermiculite, 150; smectite, 100; chlorite, 30; kaolinite, 8; gibbsite and geothite, 4. A soil containing 30% smectite clay and 5% organic matter has a CEC of 35–40 meq 100 g^{-1}. A soil containing 10% kaolinite, 20% iron and aluminum oxide clay, and 1% organic matter has a CEC of 3–4 meq 100 g^{-1}. Soils vary greatly in cation-exchange capacity; lowest values are <1 meq 100 g^{-1}, and highest values are >100 meq 100 g^{-1}. Bottom soils from 28 public fishing impoundments in Alabama[14] had CEC values of 0.5–26 meq 100 g^{-1}. The CEC of pond soils also varies with location in the pond and with depth below the soil surface.

2.11 Acidity

An acid is a substance that releases H$^+$ in solution, and a base is one that provides OH$^-$ in solution. The pH is an index of H$^+$ concentration and indicates the ratio of H$^+$ to OH$^-$ in solution (Fig. 2.17). The pH can be expressed mathematically as

$$pH = -\log(H^+) \tag{2.20}$$

The pH normally is said to range from 0 to 14, and a pH of 7 is considered neutral. The basis for pH is the ionization of water:

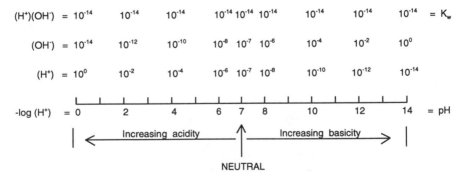

FIGURE 2.17. The pH scale.

$$H_2O = H^+ + OH^-$$

The product of H^+ and OH^- is the ionization constant of water, and it is equal to 10^{-14} at 25°C. Water is considered neutral (neither acidic or basic) because $(H^+) = (OH^-)$. We can write

$$(H^+)(OH^-) = K_w = 10^{-14} \tag{2.21}$$

and, because $(H^+) = (OH^-)$, we may substitute (H^+) for (OH^-):

$$(H^+)(H^+) = 10^{-14}, \qquad (H^+) = 10^{-7}$$

The pH of pure water at 25°C is

$$pH = -\log(10^{-7}) = 7$$

A pH value below 7 indicates an acidic condition because $(H^+) > (OH^-)$; a pH above 7 indicates a basic solution for $(OH^-) > (H^+)$.

The pH is a measure of the intensity of the H^+ concentration, but acidity is the capacity of a substance to supply hydrogen ions. Both pH and acidity are a concern in soils. The pH regulates the solubility of minerals in soils and is a master variable in soil chemical equilibria. The acidity of the soil must be considered in determining the amount of lime necessary to change the pH of soil by a given amount.

2.11.1 Exchange Acidity

The pH and acidity of most soils results from exchangeable aluminum ions in the soil as illustrated in the reaction

$$\text{Colloid—Al}^{3+} = \text{Al}^{3+} + 3H_2O = \text{Al(OH)}_3 + 3H^+ \tag{2.22}$$

The H^+ concentration and pH depends on the equilibrium concentration of Al^{3+} in solution which is related to the proportion of adsorbed Al^{3+} to the CEC of the soil colloids. The acidity is governed by the amount of Al^{3+} adsorbed on soil colloids. Actually, H^+ and Fe^{3+} adsorbed on soil colloids also are acidic. Ferric iron is acidic because it can hydrolyze in the same manner as Al^{3+} (Eq. 2.22). There is little H^+ or Fe^{3+} adsorbed on soil colloids, and it is necessary only to consider Al^{3+} in most cases. The concentration of acidic cations given in milliequivalents per 100 g is called the *exchange acidity*. All other cations are considered to be basic, and the proportion or percentage of the CEC occupied by acidic cations is known as the *base unsaturation*:

$$\text{Base unsaturation} = \frac{\text{Exchange acidity}}{\text{CEC}} \qquad (2.23)$$

Some authors refer to base saturation $(1.0 -$ base unsaturation); base unsaturation or base saturation may be presented as a percentage instead of a decimal fraction.

Example 2.6: Calculation of Base Unsaturation of a Soil
    The soil has a CEC of 2.65 meq 100 g^{-1} and 100 ppm Ca, 25 ppm Mg, 40 ppm Na, and 12 ppm K.

Solution: Convert the cation concentrations from parts per million to milliequivalents per 100 g. Since 1 ppm = 1 mg kg^{-1}, 1 ppm = 0.1 mg 100 g^{-1}:

$$
\begin{aligned}
&10 \text{ mg Ca } 100 \text{ g}^{-1} &\div\ &20.04 \text{ mg Ca meq}^{-1} &=\ &0.50 \text{ meq } 100 \text{ g}^{-1}\\
&2.5 \text{ mg Mg } 100 \text{ g}^{-1} &\div\ &12.16 \text{ mg Mg meq}^{-1} &=\ &0.21 \text{ meq } 100 \text{ g}^{-1}\\
&4 \text{ mg Na } 100 \text{ g}^{-1} &\div\ &23.0 \text{ mg Na meq}^{-1} &=\ &0.17 \text{ meq } 100 \text{ g}^{-1}\\
&1.2 \text{ mg K } 100 \text{ g}^{-1} &\div\ &39.1 \text{ mg K meq}^{-1} &=\ &0.03 \text{ meq } 100 \text{ g}^{-1}
\end{aligned}
$$

The sum of the basic ions is 0.91 meq 100 g^{-1}. Other ions adsorbed on the soil can be considered acidic. The base unsaturation is

$$\frac{0.91}{2.65} = 0.34 \text{ or } 34\%$$

The pH of an acidic, highly-weathered soil that does not contain iron pyrite is closely related to its degree of base unsaturation (Fig. 2.18). The data in Fig. 2.18 are for pond soils and agricultural soils in Alabama. The shape of the regression line relating pH to base unsaturation may differ for sets of soil samples taken from other locations. Pond soils had a lower pH than agricultural soils at the same base unsaturation. For example, at a base unsaturation of 0.2, agricultural soils had a pH of about 7, whereas the pH of pond soils was about 6. For Figure 2.18 the pond soils contained considerably more organic matter than the agricultural soils. This apparently caused the difference in the relationship between pH and base unsaturation in the two sets of soil samples. Soils that are nearly base saturated will be neutral or alkaline, but as base unsaturation increases, pH decreases. Soils with a base unsaturation of 0.5 and above normally have a pH of 5 or less.

The exchange acidity of a soil is greatly influenced by its CEC. Two Alabama pond soils of the same pH will have the same base unsaturation (Fig. 2.18), but if one of the soils has a greater CEC than the other, it will have larger exchange acidity.

Example 2.7: Illustration of Relationship of Exchange Acidity to pH
    Two soil samples from Alabama ponds have pH values of 5.1. One is a loam

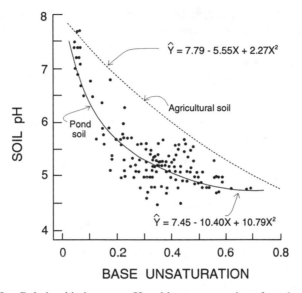

FIGURE 2.18. Relationship between pH and base unsaturation of pond soils (dots and solid line) and agricultural soils (dashed line) in Alabama. (*Sources: Pond soil data from Boyd*[28]; *agricultural soil data from Adams and Evans.*[29])

of low-organic-matter concentration with a CEC of 1.10 meq 100 g^{-1}. The other is a clay with a moderate amount of organic matter. It has a CEC of 5.15 meq 100 g^{-1}. The exchange acidity of each soil will be calculated.

Solution: From Figure 2.18 the base unsaturation of both soils is about 0.40. The exchange acidity may be obtained by rearranging Equation 2.23:

$$\text{Exchange acidity} = (\text{Base unsaturation}) (\text{CEC})$$
$$\text{Exchange acidity}_{loam} = (0.4)(1.10) = 0.44 \text{ meq } 100 \text{ g}^{-1}$$
$$\text{Exchange acidity}_{clay} = (0.4)(5.15) = 2.06 \text{ meq } 100 \text{ g}^{-1}$$

In the example the soil with the higher CEC had an exchange acidity almost five times greater than the soil with the lower CEC even though both soils had the same pH. It would take about five times as much lime to neutralize the loam as the clay.

2.11.2 Sulfide Acidity

Some soils may contain iron pyrite (FeS_2) or closely related sulfide compounds. Iron pyrite accumulation is especially common in soils of coastal areas. Sulfide can form in sediment in coastal wetlands. Freshwater streams bring soil particles into wetlands, and wetland vegetation provides a source of organic matter that mixes with settling soil particles. Decomposition of organic matter in sediment causes oxygen depletion, and, under anaerobic conditions, iron oxides associated with the soil are reduced to provide ferrous iron (Fe^{2+}). Brackish water in coastal wetlands has a high sulfate concentration, and, under anaerobic conditions, microbial activity produces sulfides and elemental S. Iron, sulfides, and sulfur can react by several pathways to form iron pyrite (FeS_2), which precipitates into the sediment. A summary equation for iron pyrite formation[15] is

$$Fe_2O_3 \quad + 4SO_4^{2-} \quad +8CH_2O \quad + \frac{1}{2}O_2 \rightarrow 2FeS_2 + 8HCO_3^- + 4H_2O$$

Iron III   Sulfate in   Organic matter
oxide from   seawater   from wetland
the sediment   vegetation

$$(2.24)$$

Iron pyrite is stable as long as it is in an anaerobic environment. It is not uncommon for some coastal soils to contain up to 5% total sulfur that is mostly in sulfide form.

When exposed to the atmosphere, sulfide in soils oxidizes to form sulfuric acid. Soils that contain more than about 0.75% total sulfur are known as *potential acid-sulfate soils*. When the sulfide oxidizes from exposure to atmospheric oxygen, sulfide-bearing soils are termed *acid-sulfate soils*. Of course, there are pyrite-bearing soils that formed in sedimentary environments where calcium carbonate precipitation also occurred. If enough calcium carbonate is present to neutralize acidity formed through pyrite oxidation, acid-sulfate soils may not develop in spite of the presence of pyrite.

The oxidation of iron pyrite involves a number of reactions as follows:

$$FeS_2 + H_2O + 3.5O_2 \rightarrow FeSO_4 + H_2SO_4 \tag{2.25}$$

$$2FeSO_4 + 0.5O_2 + H_2SO_4 \rightarrow Fe_2(SO_4)_3 + H_2O \tag{2.26}$$

$$FeS_2 + 7Fe_2(SO_4)_3 + 8H_2O \rightarrow 15FeSO_4 + 8H_2SO_4 \tag{2.27}$$

According to Sorensen et al., the production of ferric sulfate from ferrous sulfate is greatly accelerated by the activity of bacteria of the genus *Thiobacillus,* and, under acidic conditions, the oxidation of pyrite by ferric sulfate is very rapid.[16] In addition, ferric sulfate may hydrolyze according to the following reactions:

$$Fe_2(SO_4)_3 + 6H_2O \rightarrow 2Fe(OH)_3 + 3H_2SO_4 \tag{2.28}$$

$$Fe_2(SO_4)_3 + 2H_2O \rightarrow 2Fe(OH)(SO_4) + H_2SO_4 \tag{2.29}$$

Ferric sulfate may react with iron pyrite to form elemental sulfur, and the sulfur may be oxidized to sulfuric acid by microorganisms:

$$Fe_2(SO_4)_3 + FeS_2 \rightarrow 3FeSO_4 + 2S \tag{2.30}$$

$$S + 1.5O_2 + H_2O \rightarrow H_2SO_4 \tag{2.31}$$

Ferric hydroxide can react with adsorbed bases, such as potassium in acid-sulfate soils, to form jarosite, a basic iron sulfate:[17]

$$3Fe(OH)_3 + 2SO_4^{2-} + K^+ + 3H^+ \rightarrow KFe_3(SO_4)_2(OH)_6 \cdot 2H_2O + H_2O \tag{2.32}$$

Jarosite is relatively stable, but in older acid-sulfate soils where acidity has been neutralized it tends to hydrolyze:

$$KFe_3(SO_4)_2(OH)_6 \cdot 2H_2O + 3H_2O \rightarrow 3Fe(OH)_3 + K^+ + 2SO_4^{2-} + 3H^+ + 2H_2O \tag{2.33}$$

Sulfuric acid dissolves aluminum, manganese, zinc, and copper from soil, and acidic runoff from acid-sulfate soils or mine spoils may contain potentially toxic metallic ions. The amount of acid produced in acid-sulfate soils depends on the amount and particle size of iron pyrite, the presence or absence of exchangeable bases and carbonates within the pyrite-bearing material, the exchange of oxygen and solutes with the pyrite-bearing material, and the abundance of *Thiobacillus*. Because the exchange of oxygen and solutes and the abundance of *Thiobacillus* are restricted with depth, acid-sulfate acidity is primarily a problem in the upper few centimeters of the soil profile.

Most sulfuric acid production in ponds will occur in above-water portions of levees or in the bottom when it is allowed to dry after fish or shrimp harvest. For example, fish ponds were constructed on potential acid-sulfate soils at Illo-Illo, Philippines. During the dry season, pyrite in dry surface soils of the levees oxidized to form sulfuric acid. In the wet season, runoff from levees carried sulfuric acid into ponds, and pond water pH was depressed enough to kill fish.[18] Bottoms of these ponds become extremely acidic when dried after fish harvest. Waters at the soil–water interface and in the surface layer of pond soil usually contain dissolved oxygen. Contact with oxygenated water allows pyrite oxidation to progress at a slow rate even in the bottom soil of ponds filled with water.

It is usually easy to identify potential acid-sulfate soils and acid-sulfate soils. A potential acid-sulfate soil will have a total sulfur content of 0.75% or higher, and when dried its pH will drop to 3.5 or less. Fleming and Alexander reported that pH was 2–3 in dry soil-distilled water mixtures made with soils from a tidal marsh area in South Carolina.[19] These soils contained up to 5.5% total sulfur. Another test for potential acid-sulfate soil is to mix equal parts of soil and 10–30% hydrogen peroxide. After 10 min, the pH of the mixture should be 2.5 or less. Universal pH paper with a pH range of 0–6 is suitable for field identification of potential acid-sulfate soils.

An acid-sulfate soil will have a pH less than 3.5 (1:1 mixture of soil and distilled water). The yellowish mottling of jarosite can sometimes be seen in surface layers. In strongly acid-sulfate soils, yellow deposits of elemental sulfur are often visible and the odor of sulfur is apparent. In Australia and Indonesia,

I have seen 0.5- to 1-cm-diameter by 3- to 5-cm-long cylindrical, rust-colored particles of iron oxides and hydroxides on levee surfaces where soils were highly sulfuric.

2.12 Soil pH

Although soil pH ranges from <2 in active acid sulfate soil to >9 in alkali soils, most highly leached mineral soils do not have pH < 4. Calcareous and alkaline soils have pH < 8.5, and even saline–alkaline soils seldom have pH > 8.5. Alkali soils have a pH range of 8.5–10. Some highly organic soils may have pH values as low as 3, and others may have pH of 8 or more.[20]

A wide range in pH was found in soils of aquaculture ponds.[21] The pH of 358 freshwater pond soils ranged from 3.9 to 8.0 and from 1.2 to 9.8 for 346 brackish-water pond soils (Fig. 2.19). Average pH was 6.69 for freshwater pond soils and 6.50 for brackish-water pond soils. When acid-sulfate soils were removed, brackish-water pond soils had an average pH of 7.23.

The pH of waterlogged agricultural soils usually is near neutral, and the pH of freshly collected, water-saturated samples of soil from the bottom of ponds is consistently between 6.0 and 7.0. The pH in such samples reflects the pH of the pore water. In water-saturated soil the pore water at a depth of 1–2 cm below the soil–air or soil–water interfaces usually is anaerobic. The solubility of soil iron compounds is enhanced by anaerobic conditions, but the solubility of soil aluminum compounds is not increased. High Fe^{2+} concentrations occur in anaerobic interstitial water, and the mineral siderite ($FeCO_3$) controls the solubility of Fe^{2+} according to the equilibrium:

$$FeCO_3(s) + H^+ = Fe^{2+} + HCO_3^-; \qquad K = 10^{-0.3} \tag{2.34}$$

Assuming that HCO_3^- and Fe^{2+} concentrations in the interstitial water of an aquaculture pond are 60 mg L^{-1} ($10^{-3} M$) and 20 mg L^{-1} ($10^{-3.74} M$), respectively, we may calculate the pH of the interstitial water by substituting Fe^{2+} and HCO_3^- concentrations into the mass-action form of Equation 2.34:

$$\frac{(Fe^{2+})(HCO_3^-)}{(H^+)} = 10^{-0.3} \tag{2.35}$$

The result is $(H^+) = 10^{-6.44} M$ or pH = 6.44. At the ranges of Fe^{2+} and HCO_3^- found in interstitial water of pond sediment, pH is between 6 and 7. Of course, in the oxidized surface layer of sediment, the pH is controlled by exchangeable acidity or iron pyrite oxidation, and the range of pH values in the oxidized surface layer can be much wider than that of anaerobic soils.

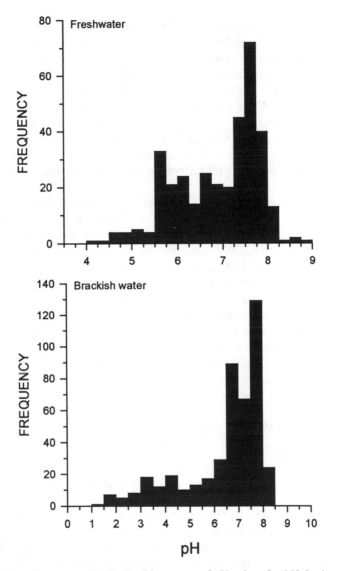

FIGURE 2.19. Frequency distribution histograms of pH values for 358 freshwater fish ponds and 346 brackish-water shrimp ponds. (*Source: Boyd et al.*[21])

2.13 Soil Classification

Soil classification methods are based on the notion that the soil is comprised of many individual and recognizable units each of which is called a soil. Many soil classification systems have been proposed, but the *soil taxonomy*[2] system is the most widely used. Soil taxonomy is based on examination of specific soil properties in natural, in-place soils. Properties can be evaluated by field observation of soil profiles and by laboratory analyses of physical and chemical properties. Data from the examination of soils are used with a system of nomenclature related to major soil characteristics to provide a name for each soil unit that occurs in the soil. When a soil is classified in the soil taxonomy system it is assigned to six categories. The most general category is order, and the other categories, in order of increasing specificity, are suborder, great group, subgroup, family, and series. The most basic characteristics used in soil taxonomy are features of the surface and subsurface horizons and moisture and temperature regimes.

In the United States and many other countries, soils have been classified over most of the land surface, and the locations of individual soil units plotted on maps. Soil maps are discussed later in this chapter after a method for classifying soils has been described. Soil maps provide useful information about soils in a particular area, but for specific projects, such as pond construction, soil properties must be evaluated in situ for the specific site.

Soil classification data are useful in planning aquaculture projects, for soil maps based on soil taxonomy can help in the initial assessment of the suitability of soil properties at a particular pond site. The soil profile is greatly altered in pond bottoms by construction, erosion, and deposition. No attempt has been made to develop a general classification system of pond soils, but many of the soil properties used in soil taxonomy are properties important to pond management operations.

2.13.1 Diagnostic Horizons and Climatic Conditions

Surface Horizons

Diagnostic surface horizons are known as *epipedons,* of which there are seven.

1. *Mollic horizons.* These epipedons typically develop in grasslands. When dry, mollic epipedons are soft and not massive. Soil structure is well developed because of the permeation of the soil by grass roots. Mollic epipedons are at least 18 cm thick and dark because the organic carbon concentration is 1% or greater. Base saturation is 50% or greater, and soil pH is usually above 7.0.

2. *Umbric horizons.* In very humid forests, umbric epipedons, which are quite similar to mollic epipedons, develop. However, because of high rainfall, umbric epipedons are more highly leached than mollic epipedons. Base saturation is below 50%, and pH is lower in umbric horizons than in mollic horizons.

3. *Ochric horizons.* These soil layers are found in forests where rainfall is not sufficient to cause umbric epipedons. Ochric horizons are thinner, lower in organic carbon concentration, and lighter in color than umbric epipedons.

4. *Histic horizons.* A histic horizon comprises largely organic matter. It is an O horizon, but can be up to several meters thick. Histic horizons are classified as fibric, hemic, and sapric. In a *fibric* horizon, the organic matter is not well decomposed, and recognizable leaves and other plant parts are present. When the plant parts are in an intermediate stage of decomposition, the term *hemic* is used. *Sapric* is used for highly decomposed plant material. The symbols *i, e,* and *a* are used to indicate fibric, hemic, and sapric, respectively. Histic soils develop in wetlands, and a histic horizon must be saturated with water for at least 30 days per year.

5. *Melanic horizons.* These epipedons develop from lava or volcanic ash and are said to have andic soil properties. They are very thick and black, and organic carbon concentrations range from 6% to 25%. Melanic epipedons usually develop in soils with large amounts of grass root residues, but they are thicker and higher in organic carbon concentration than mollic epipedons. They differ from histic epipedons in that they do not develop in wetlands, and they contain appreciable mineral matter.

6. *Anthropic horizons.* Color, organic carbon concentrations, and soil structure in anthropic horizons are similar to mollic horizons. However, anthropic epipedons are the result of agriculture. Fertilizer application has raised the concentration of phosphorus extractable in 1% citric acid to at least 250 ppm.

7. *Plaggen horizons.* Long-term application of manure and plowing gives rise to these horizons. Plaggen epipedons are similar to mollic horizons.

Subsurface Horizons

Diagnostic subsurface horizons are E or B horizons resulting from illuviation processes in the soil profile. In soil formation, residues from plant growth add organic matter to the upper portion of the parent material, and weathering results in formation of clay. Humus formed from plant residues gives the upper soil layer the dark color that defines it as the A horizon. The layer just beneath the A horizon is more weathered than the parent material and is called the C horizon. Over time, another horizon—the B horizon—may begin to develop between the A and C horizons. This layer has a lighter color than the A horizon but is darker than the C horizon. Some enrichment of the C horizon with fine particles also occurs. A weakly developed B horizon is called a *Bw horizon.* In some cases, clay particles from the A horizon are translocated downward, and a marked accumulation of clay is found in the B horizon. A clay-enriched B horizon is termed a *Bt horizon.* In sandy parent materials, clay accumulation does not occur in the B horizon, but humus and iron and aluminum oxides accumulate to form a *Bhs horizon.* In dry climates, salts leached from the topsoil may accumulate in the B horizon as calcium carbonate (Bk horizon), gypsum (By horizon), or more-soluble salts (Bz horizon). In soils with thin A horizons, leaching of clay and soluble minerals out of the upper part of the soil profile may cause the formation of a "washed-out" grayish or whitish E horizon between the A and B horizons. The common diagnostic subsurface horizons are described next.

1. *Cambric horizons*. These soil strata are weakly developed B horizons (Bw). They are not highly weathered, and the parent material is usually sand or even finer particles.

2. *Agrillic horizons*. In some soils the clay concentration in the B horizon increases as a result of illuviations, and Bw horizons evolve into Bt horizons. When the clay concentration in the Bt horizon increases to 1.2 times the overlying horizons, the Bt horizon is called an agrillic horizon.

3. *Natic horizons*. This horizon is an agrillic horizon that has accumulated salts dissolved from upper horizons. A natic horizon must have a sodium absorption ratio of 13 or more and 15% or more exchangeable sodium. It is called a Btn horizon.

4. *Kandic horizons*. A kandic horizon is like an agrillic horizon, but the clay has a low cation-exchange capacity (<16 meq/100 g). Clay minerals are typically kaolinite or hydrous oxides. Many tropical and subtropical soils have kandic horizons.

5. *Spodic horizons*. In soils formed from sand-textured parent material in areas of high precipitation, iron and aluminum oxides leached from surface soils may accumulate in the B horizon to form spodic horizons. Sometimes the iron and aluminum may be chelated by finely divided organic matter, and spodic horizons result from the chelated compounds.

6. *Albic horizons*. These soil layers are white-colored horizons that develop between A and B horizons as a result of leaching of clay or iron and aluminum oxides. Albic horizons usually are found above agrillic, kandic, and spodic horizons; they are E horizons.

7. *Oxic horizons*. In high-rainfall climates, soils can become highly leached, and a B horizon comprised of iron and aluminum oxides, quartz sand, and kaolinite may develop. This stage in soil development represents the final stages of weathering, and the B horizon is called an oxic or Bo horizon. An oxic horizon is very low in cation-exchange capacity and contains small concentrations of plant nutrients.

8. *Calcic, gypsic, and salic horizons*. These horizons result from the accumulation of calcium carbonate or calcium–magnesium carbonate, gypsum, or salts more highly soluble than gypsum in the B horizon. The horizons are termed Bk, By, or Bz for calcic, gypsic, and salic horizons, respectively.

Soil Moisture Regimes

Three soil moisture conditions occur in soils. *Wet soils* are saturated with water; *moist soils* have water contents lower than saturation but greater than the permanent wilting percentage; *dry soils* are at or below the permanent wilting percentage. In determining the soil moisture regime for soil classification, one usually considers the moisture content in the 10- to 90-cm depth layer. Soils with an *aquic moisture regime* occur in areas where natural drainage is slow. The subsoil, or even the entire soil profile, is saturated part of the year. *Aridic moisture regimes* are found in areas where the subsoil (10–90-cm depth) is dry in some or all parts for more than half of the growing season. Also, no part of the subsoil is moist for more than 90 consecutive days during the growing season in years with normal precipitation. A soil that is not dry for as long as 90 cumulative days in years with normal precipitation has a *udic moisture regime*. Such a moisture regime is common in well-drained soils in a humid climate with well-distributed precipitation during the warm months. If heavy rainfall (precipitation is greater than evapotranspiration every month) occurs and udic conditions still exist, the soil moisture condition is called *perudic*. An *ustic soil moisture regime* is said to exist in places where the subsoil is dry at some point in the 10–90-cm layer for as long as 90 cumulative days in years with normal rainfall. Ustic conditions are common in areas with a pronounced dry season. A special case of the ustic soil moisture regime occurs in places with a cool, moist winter and a hot, dry summer. Such moisture regimes are termed *xeric*.

Soil Temperature Regimes

Temperature regimes are based on soil temperature in the roots zone between 5 and 100 cm deep. The names and temperature ranges for the soil temperature regimes are *pergelic*, <0°C; *frigid*, 0–8°C; *mesic*, 8–15°C; *thermic*, 15–22°C; *hyperthermic*, >22°C.

2.13.2 Categories of Soil Classification

The most general or highest category in soil taxonomy is order. There are 11 soil orders (Table 2.8), and they are divided into suborders (Table 2.9). Names for suborders contain two syllables. For example, an Ultisol from an udic moisture regime is given the suborder name Udults. The *ud* in Udults is the prefix for the udic moisture regime and *ults* is the formative syllable for the order Ultisols. Soils of the suborder Torrents are Vertisols with an aridic moisture regime. The suborder is frequently used in preparing general soil maps.

Soil suborders are divided into great groups. The great group name consists of the suborder name plus a prefix (Table 2.10). To illustrate, a Tropaquepts is an Inceptisol with an aquic moisture regime located in a continually warm and humid climate. The subgroup follows the great group and indicates if a soil is typical for the great group to which it is assigned. For example, a soil near Auburn might be a Typic Hapludults. This soil is typical for the great group consisting of Ultisols from an udic moisture regime that have weakly developed

Table 2.8.   Soil Orders and Their Description

Order	Formative Syllable	Description
Histosols	ist	Soils with a high concentration of organic matter (histic epipedons).
Andisols	and	Soils developed primarily from weathering of volcanic materials.
Spodosols	od	Soils with a spodic horizon. Aluminum, iron, and humus have been transported into the B horizon.
Oxisols	ox	Soils with an oxic horizon. Oxisols are common in the tropics. They are old, deep red, and contain much iron.
Vertisols	ert	Soils with high concentrations of 2:1 type clays, which swell when wetted and shrink when dried. They develop deep cracks when dry. Swelling and shrinking often mix the horizons.
Aridisols	id	Soils with an aridic soil moisture regime. There is not enough moisture to cause sufficient plant growth to provide a surface layer of organic matter. Bases are not leached and accumulate in the A horizon.
Ultisols	ult	Soils with a kandic or argillic horizon. Base saturation is less than 35%. Ultisols usually are found in warm, humid climates.
Alfisols	alf	Soils with a kandic or argillic horizon. Base saturation is greater than 35%.
Mollisols	oll	Soils with a mollic horizon. They are usually found in grasslands.
Inceptisols	ept	Soils with a cambic horizon. These are recent soils in which the profiles have begun to develop.
Entisols	ent	Soils that cannot be fit into any of the other 10 orders. There also are recent soils with little evidence of profile development. Entisols are common in alluvial areas or on steep slopes.

Source: Soil Survey Staff[5]

horizons. In Table 2.10 the prefix *hapl* means minimum horizon. The category below subgroups is family, which gives information on soil texture, mineralogy, temperature, and thickness. For example, a Typic Kandiudult at Auburn, Alabama, might be clayey, kaolinitic, thermic, and shallow. The lowest or most specific category in soil classification is the soil series. The soil series is delineated on detailed soil maps, and most interpretations of land use are based on it. The name of a soil series usually reflects a geographic or physiographic feature of the locality where the series was first described. For example, near the community of Marvyn in Lee County, Alabama, a soil series has been delineated, which is a fine-loamy, siliceous, thermic Typic Hapludults. This soil series was given the name *Marvyn*.

In ponds, the upper part of the original soil profile may be removed and the remaining soil compacted, or the original soil may be covered with compacted fill excavated from the pond bottom area or from a site outside the pond. A pond soil profile develops as already discussed, but the characteristics of the horizons in pond soil profiles are different from those of diagnostic horizons used in soil taxonomy. The author currently is working towards the development of a scheme for classifying pond soils based on characteristics of pond soil profile horizons.

Table 2.9.   Soil Suborders with Meaning of Formative Element[a]

Suborder	Formative Element and Meaning	
Aqualfs, Aquands, Aquents, Aquepts, Aquolls, Aquox, Aquods, Aquults	Aqu	Aquatic moisture regime
Boralfs, Barolls	Bor	Northern, cool
Udalfs, Udands, Udolls, Udox, Udults, Uderts	Ud	Udic moisture regime
Ustalfs, Ustands, Ustolls, Ustox, Ustults, Usters	Ust	Ustic moisture regime
Xeralfs, Xerands, Xerolls, Xerults, Xererts	Xer	Xeric moisture regime
Cryands	Cry	Cryic or cold
Torrands, Torrox, Torrerts	Torr	Aridic moisture regime
Vitrands	Vitr	Presence of glass
Argids	Arg	Argillic horizon
Orthids, Orthods, Orthents	Orth	True or common
Arents	Ar	Mixed by plowing
Fluvents	Fluv	Floodplain parent material
Psamments	Psamm	Sand textured
Fibrists	Fibr	Slightly decomposed organic matter
Folists	Fol	Mass of leaves
Hemists	Hem	Moderately decomposed organic matter
Saparists	Sapr	Highly decomposed organic matter
Ochrepts	Ochr	Ochric epipedon
Plaggepts	Plagg	Plaggen epipedon
Tropepts	Trop	Tropical
Umbrepts	Umbr	Umbric epipedon
Rendolls	Rend	High carbonate content
Perox	Per	Perudic moisture regime
Ferrods	Ferr	Presence of iron
Humods, Humults	Hum	Presence of organic matter
Albolls	Alb	Albic horizon

[a]Suborder name made by adding the formative syllable for the order (see Table 2.8) after the formative element of the suborder.

Source: Foth[22]

2.14 Soil Maps

Soil maps are based on soil survey data and contain a wealth of information on soil characteristics. They are important initial sources of information on the suitability of soils at a given site for pond aquaculture. In the United States, soil maps can be obtained from the United States Department of Agriculture, Soil Conservation Service. There is a USDA/SCS office in most counties. In other nations, soil maps usually are available from federal agricultural agencies.

Soil maps are made to many different scales. The scales of world soil maps and soil maps of large nations, such as those in Foth, are too small to provide useful information for specific areas.[22] Soil maps of regions, provinces, or states

Table 2.10. Prefixes and Meanings for Great Groups of Soils[a]

Prefix	Meaning	Prefix	Meaning
acr	Highly weathered	luv	Illuvial
agr	Agric horizon	med	Temperate climates
alb	Albic horizon	natr	Natric horizon
and	Andic properties	ochr	Ochric epipedon
anthr	Anthropic epipedon	pale	Old
arg	Argillic horizon	pell	Low chroma
bor	Northern, cool	plac	Thin cemented layer
calc	Calcic horizon	plag	Plaggen horizon
camb	Cambic horizon	plinth	Plinthite (laterite)
chrom	High chroma	psamm	Sandy
cry	Cold	quartz	Quartz
dur	Hardpan	rhod	Dark red color
dystr, dys	Low base saturation	sal	Salic horizon
eutr, eu	High base saturation	sider	Free iron oxides
ferr	High iron content	sombr	Dark horizon
flub	Floodplain	sphagn	Sphagnum moss
frag	Fragipan	sulf	Sulfides
gibbs	Gibbsite	torr	Aridic moisture regime
gloss	Tongued	trop	Warm and humid (tropic)
gyps	Gypsic horizon	ud	Udic moisture regime
hal	High salt content	umbr	Umbric epipedon
halp	Minimum horizon development	ust	Ustic moisture regime
hum	High humus content	verm	Mixed by worms
hydr	Presence of water	vitr	Glass
kand	Low activity clay	xer	Xeric moisture regime

[a]Great group name made by adding the descriptive prefix in front of the suborder name.

Source: Foth[22]

(Fig. 2.20) also have small scales and only provide general information.[23] However, in most nations, soil maps with larger scales are available. For example, I have obtained excellent soil maps for specific areas in Thailand from the Thailand Department of Land Development, and in Indonesia I obtained very good detailed soil maps from the Ministry of Agriculture. In the United States, soil survey reports are available for most counties. These reports usually have a general soil map of the county (Fig. 2.21), and a series of detailed soil map sheets provide information of smaller areas. The USDA/SCS Soil Survey Report for Lee County, Alabama,[24] contains 51 detailed soil maps to supplement the general soil maps so that soils data can be obtained for specific localities (Fig. 2.21). Soils of the Auburn University Fisheries Research Unit are mapped on map sheet number 12 (see Figs. 2.21 and 2.22). The numbers on individual map sheets, such as Figure 2.22, denote map units. Map units define specific areas on the landscape and contain one or more soils for which the map unit is named. Map Unit 31 is for Pacolet sandy loam with 1–6% slopes.

General information on the Pacolet sandy loam is provided in the soil survey report as follows:

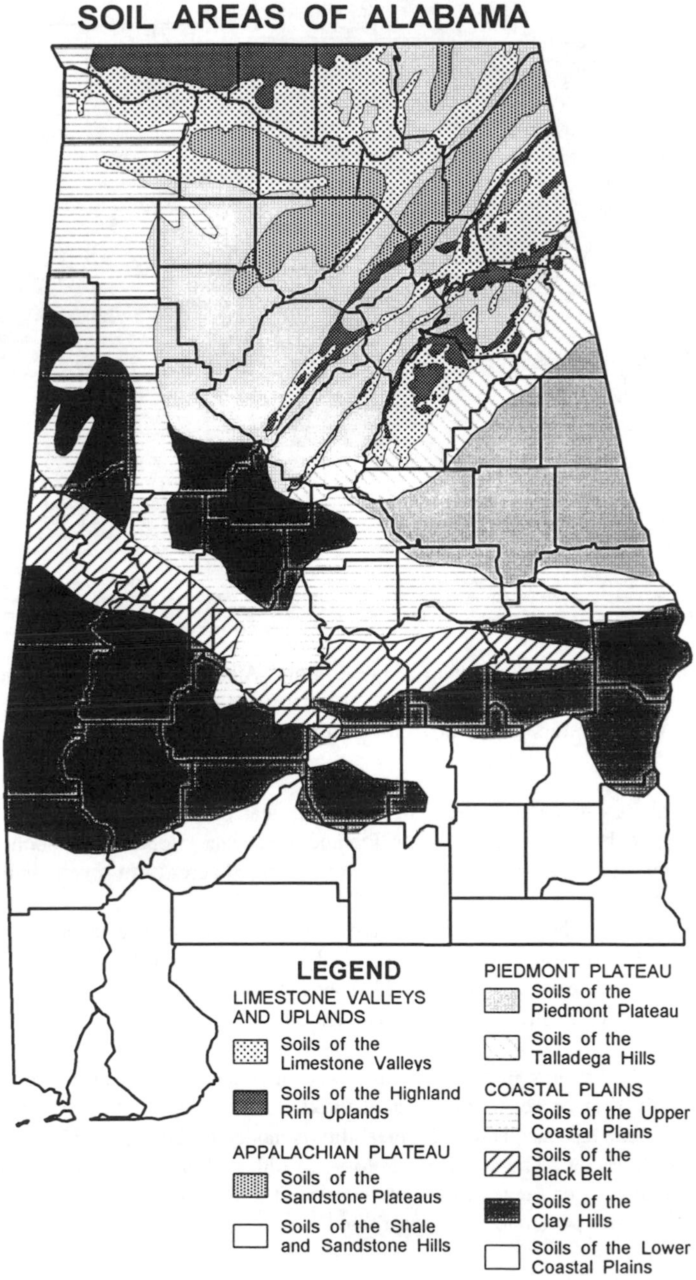

SOIL AREAS OF ALABAMA

LEGEND

LIMESTONE VALLEYS AND UPLANDS

Soils of the Limestone Valleys

Soils of the Highland Rim Uplands

APPALACHIAN PLATEAU

Soils of the Sandstone Plateaus

Soils of the Shale and Sandstone Hills

PIEDMONT PLATEAU

Soils of the Piedmont Plateau

Soils of the Talladega Hills

COASTAL PLAINS

Soils of the Upper Coastal Plains

Soils of the Black Belt

Soils of the Clay Hills

Soils of the Lower Coastal Plains

FIGURE 2.20. General soil map of Alabama. The heaviest lines indicate county boundaries, and soil areas are depicted by shading. (*Source: Hajek et al.*[23])

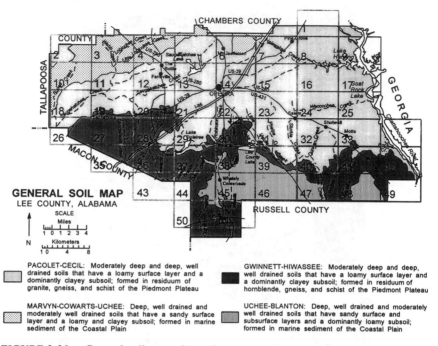

PACOLET-CECIL: Moderately deep and deep, well drained soils that have a loamy surface layer and a dominantly clayey subsoil; formed in residuum of granite, gneiss, and schist of the Piedmont Plateau

GWINNETT-HIWASSEE: Moderately deep and deep, well drained soils that have a loamy surface layer and a dominantly clayey subsoil; formed in residuum of hornblende, gneiss, and schist of the Piedmont Plateau

MARVYN-COWARTS-UCHEE: Deep, well drained and moderately well drained soils that have a sandy surface layer and a loamy and clayey subsoil; formed in marine sediment of the Coastal Plain

UCHEE-BLANTON: Deep, well drained and moderately well drained soils that have sandy surface and subsurface layers and a dominantly loamy subsoil; formed in marine sediment of the Coastal Plain

FIGURE 2.21.   General soil map of Lee County, Alabama. Soil areas are depicted by shading, and the grids indicate areas covered by individual, detailed soil maps. For example, grid 12 refers to the area shown in map sheet 12. (*Source: McNutt.*[24])

This is a moderately deep, well drained, gently sloping soil on moderately broad to broad ridgetops of the Piedmont Plateau. Slopes are smooth and convex. Individual areas are 10 to several hundred acres (4 to several hundred hectares) in size.

Typically, the surface layer is brown sandy loam about 6 inches (15 cm) thick. The subsoil is yellowish red sandy clay loam to a depth of 11 inches (28 cm), red clay to a depth of 23 inches (58 cm), and red sandy clay loam to a depth of 33 inches (84 cm). The underlying material is multicolored soft saprolite.

This soil is low in natural fertility and in content of organic matter. It is strongly acid or very strongly acid throughout except for the surface layer where lime has been added. Permeability is moderate, and the available water capacity is low. The soil has fair to good tilth and can be worked within a moderately wide range of moisture content. The root zone is moderately deep and is easily penetrated by plant roots.[24]

The soil survey report[24] also contains descriptions of soil profiles and tabular data on engineering properties, physical and chemical properties, chemical analyses, and suitability for various uses. A few facts that can be obtained about soils

LEE COUNTY, ALABAMA - GRID 12

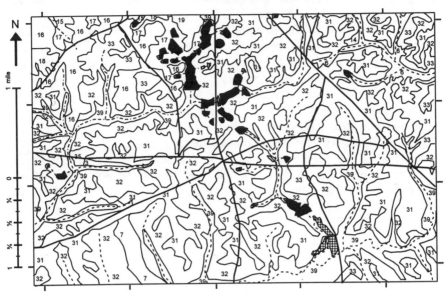

FIGURE 2.22. Detailed soil map sheet 12 for grid 12 of the soil survey report for Lee County, Alabama. Heaviest solid lines are roads, lighter solid lines are soil areas, dashed lines are streams, and hatched and shaded areas represent ponds on the Auburn University Fisheries Research Unit. The number or map unit refers to different soils: 6 = Cartecay; 7 = Cecil; 14 = Durham; 15 = Enoree; 16, 17, and 18 = Gwinnett; 31, 32, and 33 = Pacolet; 39 = Sacul. Descriptions, properties, land use, and other information about each map unit are provided in the soil survey report. (*Source: McNutt.*[24])

in Map Unit 31 are as follows: 8–20% clay in the 0–15-cm layer and 35–60% clay in the 15–90 cm layer; pH 4.5–6.0; base unsaturation 70–86%; calcium, magnesium, and potassium concentrations of 154, 41, and 51 ppm, respectively, in the 0–15-cm layer; bulk density of 1.62 g cm^{-3}. Of course, the maps are based on sampling, and one cannot expect the data to be highly accurate for a given site. Soil survey reports provide a first approximation of soil conditions at a site, and site-specific data should be collected for verification.

2.14.1 Limitation Ratings

Data on engineering properties of soils reported in soil survey reports are placed in three classes according to their limitations for specific uses. The rating class is given in terms of limitations and restrictive feature(s). Only the most restrictive feature should be listed when a limitation class is given; for example, severe, excesses humus, forms a limitation class name. Restrictive features need to be treated to overcome limitations imposed by soil characteristics. If the rating class is slight, no restrictive feature is indicated in the limitation class name and no special treatment will be necessary.

Soils should be rated in-place and assigned a slight, moderate, or severe limitation class for property. A moderate or severe limitation does not mean that a soil cannot be used. Developers can modify soil features, adjust plans, and redesign to compensate for many moderate and severe soil limitations. However, the initial cost of pond and dike construction and maintenance cost must be considered when on-site soils have a restrictive feature. Limitation ratings are for single properties; consequently, efforts to overcome limitations are different, depending on the property and local conditions.

Many properties and limitation classes for soils and soil survey map units can be obtained from soil taxonomic limits[2] and from limits in Part 603 of the *National Soil Handbook*.[25] The following limitation rating definitions are essentially the same as those used by USDA-SCS:

1. *Slight:* This rating indicates that on-site soils have properties favorable for use. No unusual construction, design, management, or maintenance are required for the designated use.

2. *Moderate:* This rating indicates that on-site soils have one or more properties that require special attention for the designated use. This degree of limitation can be overcome or modified by special planning, design, management, or maintenance. Examples of moderate limitations are the need for lime on highly acidic soils and the requirement for additional compaction and treatment of soils to reduce seepage.

3. *Severe:* The severe rating is given when one or more properties of on-site soils are unfavorable for the rated use. Examples are steep slopes, potential soil acidity because of sulfides, and very high hydraulic conductivity. The severe rating usually means that major reclamation, and modifications in design, management, or maintenance are required for the designated use. Some soils with a severe rating can be improved by removing the limiting feature, replacing it with other soil material, or installing artificial barriers. In most cases, significant cost and effort are required to compensate for a severe limitation. Sometimes, it may not be economically feasible to use a site with one or more severe limitations.

2.14.2 Limitation Ratings for Pond Construction

Excavated Fish Ponds

Interpretation of soil limitations for excavated fish ponds in Table 2.11 considers soil properties to a depth of 150 cm. The properties are rated as a final (after compaction of sides and bottom) barrier against seepage. Installation of synthetic

Table 2.11. Soil Limitation Ratings for Excavated Fish Ponds

Property	Limitation Rating			Restrictive Feature
	Slight	Moderate	Severe	
Depth to sulfidic or sulfuric layer (cm)	>100	50–100	<50	Potential acidity or toxicity
Thickness of organic soil material[a] (cm)	<50	50–80	>80	Seepage; hard to compact
Exchangeable acidity[b] (%)	<20	20–35	>35	Exchangeable acidity
Lime requirement (ton ha^{-1})	<2	2–10	>10	Mineral acidity
pH of 50–100-cm layer of pond bottom	>5.5	4.5–5.5	<4.5	Too acid
Clay content (%)	>35	18–35	<18	Too sandy/silty; excessive seepage
	Clayey	Loamy	Sandy/silty	
Slope of terrain (%)	<2	2–5	>5	Slope
Depth to water table[c] (%)	>75	25–75	<25	Hard to drain; dilution
Frequency of flooding[d]	None	Occasional	Frequent	Flooding
Small stones[e] (%)	<50	50–75	>75	Small stones
Large stones[e] (%)	<25	25–50	>50	Large stones
Decomposed OM[f] (%)				
Low-clay-content soil (<60% clay)	<4	4–12	>12	Excessive humus
High-clay-content soil (>60% clay)	<8	8–18	>18	Reducing environment
Depth to rock (cm)	>150	100–150	<100	Shallow; seepage

[a]Unified classes PT, OL, OH; sapric fibric material. Hemic.

[b]For soils with CEC $\leq$ 24 meq/100 g clay; if clay activity is >24 meq/100 g clay, increase values by 10%.

[c]Rate slight to moderate if management is not affected. Difficult to construct.

[d]None: no possibility of flooding. Rate: 0–5 times in 100 years. Occasional: 5–50 times in 100 years. Frequent: >50 times in 100 years. Common: occasional and frequent combined.

[e]In 50–100-cm layer or at design bottom; small stones = material >2 mm, large stones >7 cm.

[f]Highly decomposed organic matter.

or clay liners is not considered in the evaluation, but they may be needed to overcome a severe limitation.

Fish Pond Embankments, Dikes, and Levees

Embankments, dikes, and levees are raised structures of soil material constructed to impound water. The soil material is considered as being mixed and compacted to medium density. Soil used for these applications must resist seepage, piping, and erosion. The final material should be able to support plant cover and should not cause toxic leachate to enter ponds. Erodibility factors needed for use in Table 2.12 often are included in soil survey reports, or they may be found in Chapter 7.

Plant growth can be restricted by deficiencies or excessive amounts of substances. For example, excessive sodium, salt, sulfur, copper, and nickel inhibit vegetable growth. Other minerals such as selenium, boron, and arsenic can be taken up by plants and can be toxic to animals that consume this vegetation. Substances or factors not considered in this report should be evaluated if local experience indicates the need. Excavated soil material often is low in plant-available phosphorus and nitrogen, and liming and fertilization may be necessary. Soil materials that are extremely acidic or have the potential to become acidic have been rated as severe because of toxicity to fish, crustaceans, and plants.

Table 2.12. Limitation Ratings for Fish Pond Embankments, Dikes, and Levees

	Limitation Rating			
Property	Slight	Moderate	Severe	Restrictive Feature
Clay content[a] (%)	>35	18–35	<18	Too sandy
	Clayey	Loamy	Sandy	
Depth to sulfidic or sulfuric material (cm)	>100	50–100	<50	Toxicity; potential acidity
Engineering classes[b,c]			All "G" and	Too stony
			"S" classes	Too sandy
Slope (%)	<8	8–15	>15	Slope
Thickness of organic material[c,d] (cm)	<15	15–50	>50	Subsides; excess humus; difficult to compact
Depth to water table (cm)	>100	50–100	<50	Wetness
Fraction >8 cm diameter[e] (%)	<25	25–50	>50	Large stones
Depth to bedrock (cm)	>100	50–100	<50	Depth to rock
Shrink-swell potential[f]	Low	Medium to high	Very high	Shrink-swell
Erodibility[g] (K)	<0.1	0.1–0.3	>0.3	Erosion

[a]Weighted mean from surface to 100-cm depth.

[b]Weighted average to 100-cm depth.

[c]Unified classes are G = gravelly; >50% of soil is more than 0.5 cm in diameter; S = sandy, soils are loamy sands or sands; PT = high organic material, fibric materials; OL = hemic or sapric materials with less than 35% clay; OH = hemic or sapric materials with more than 35% clay; SW = well-graded sandy soil material, free-draining; SM = silty sands, normally have little dry strength. For SM and SW classes rate moderate if >20% passes a number 200 sieve, and slight if >30% passes the sieve.

[d]Cumulative depth (cm) within the surface 100 cm; fibric or sapric material; unified classes, PT, OH, OL.

[e]Percent by weight; weighted average to 100 cm.

[f]Defined as the change in length of an unconfined clod or core and called the coefficient of linear expansion (COLE). Collect a core from a wet or moist soil, measure its wet length, set it upright in a dry place. Let the core air-dry and measure the dry length. COLE = 100 × (moist length − dry length)/dry length. COLE: <3% = low; 3–6% = medium; 6–9% = high; >9% = very high.

[g]Erodibility factor (see Chap. 7).

Example 2.8: Determination of Limitation Ratings from a Soil Survey Report
   Soil limitation ratings for construction of excavated fish ponds on a proposed
   site about 8 km north of Auburn, Alabama, will be obtained through use of
   a recent soil survey of Lee County.[24]

Solution:

Step 1. Locate the proposed site for the ponds on the Index of Map Sheets,
which precedes the individual soil map sheets (Fig. 2.21). The site occurs in
sheet 12 (Fig. 2.22) of the soil survey report.
Step 2. On the map sheet find the area where the ponds are to be located and
locate the map unit symbols. In this case, the symbols are 31 and 32.
Step 3. Locate the Index of Map Units from the Table of Contents. This index
shows that both map unit symbols represent soils of the Pacolet Series. Unit
31 is a Pacolet sandy loam of 1–6% slopes, and unit 32 is a Pacolet sandy
loam of 6–10% slopes. The Index of Map Units also gives the pages of the
section Detailed Soil Map Units, where the soils designated by map units 31
and 32 are described.
Step 4. To determine limitation ratings, one must find data on those soil
properties listed in Table 2.11 for construction of excavated ponds. In most
soil survey reports prepared after 1965, required data can be found in the
section on detailed soil map units and the following tables: Engineering Index
Properties, Physical and Chemical Properties of Soils, and Physical Analyses
of Selected Soils. The locations of these tables in the soil survey report are
indicated in the Summary of Tables.
Step 5. Because of the steep slope, only land indicated by map unit 31 is
suitable. The properties of soils under this map unit are given in the following
table:

Property	Range of Property	
	Value	Rating
Depth to sulfidic or sulfuric layer (cm)	>100	Slight
Thickness of organic soil material (cm)	0	Slight
Exchangeable acidity (%)	20–35	Moderate
Lime requirement (ton ha^{-1})	2–10	Moderate
pH of 50–100 cm layer of pond bottom	5.0–5.5	Moderate
Clay content (%)	36	Slight
Slope of terrain (%)	2–6	Moderate
Depth to water table (cm)	>75	Slight
Frequency of flooding	None	Slight
Small stones (%)	0	Slight
Large stones (%)	0	Slight
Decomposed OM (%)	<2	Slight
Depth to rock (cm)	83	Severe

The rating of the proposed pond sites is severe because of the shallow soil depth above rock. Also, acidity and slope impose a moderate limitation. If a soil survey report is not available, one must visit the site and collect the necessary data for making the limitation ratings.

2.15 Pond Soils

Pond soils do not differ greatly from terrestrial soils in their major physical, chemical, and mineralogical features. The major difference in pond soils and agricultural soils is that pond soils are flooded almost continuously. Pond soils are similar to wetland soils, but the dense higher aquatic vegetation typical of wetlands normally is absent from aquaculture ponds. Nutrient inputs to aquaculture ponds are much greater than for wetlands.

The surface layer of pond soils usually is exposed to oxygenated pond water, and oxidizing conditions exist at the soil surface. Deeper layers of pond soils are anaerobic, and reducing conditions prevail. Some ponds are constructed in organic soil, and bottom soils contain high concentrations of organic matter. The majority of ponds are built on mineral soils containing no more than 5–10%

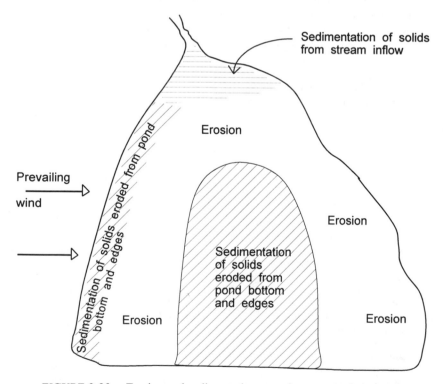

FIGURE 2.23. Erosion and sedimentation areas in an aquaculture pond.

organic matter. There is a tendency for organic matter concentrations to increase gradually over time in pond soils, but organic soils normally do not develop in pond bottoms.

The topography of fish pond bottoms changes over time as a result of erosion and sedimentation. According to Hejný and Kvĕt, in the shallow water regions of large ponds some areas are subjected to erosion and others receive sedimentation.[26] In deeper, central parts of ponds, sedimentation prevails (Fig. 2.23). Organic matter tends to accumulate where sedimentation dominates. Erosion and sedimentation patterns shown in Figure 2.23 have been described even in small research ponds.[27] The surface 10–15-cm layer of pond soils tends to have a lower bulk density than found in the surface of agricultural soils, and this difference is especially pronounced for soils from sedimentation areas of ponds.

References

1. Yoo, K. H., and C. E. Boyd. *Hydrology and Water Supply for Aquaculture.* 1994. Chapman & Hall, New York.

2. Soil Survey Staff. *Soil Taxonomy: A Basic System of Soil Classification for Making and Interpreting Soil Surveys.* 1975. USDA-SCS Agricultural Handbook 436, U.S. Government Printing Office, Washington, D.C.

3. Hajek, B. F., and C. E. Boyd. 1994. Rating soil and water information for aquaculture. *Aquacultural Eng.* **13**:115–128.

4. Alexander, M. *Introduction to Soil Microbiology.* 1977. John Wiley & Sons, New York.

5. Soil Survey Staff. *Keys to Soil Taxonomy,* 4th ed. 1990. Soil Management Support Monograph No. 19, Virginia Polytechnical Institute and State University, Blacksburg, Va.

6. Nelson, D. W., and L. E. Sommers. 1982. Total carbon, organic carbon, and organic matter. In *Methods of Soil Analysis, Part 2, Chemical and Microbiological Properties,* A. L. Page, R. H. Miller, and D. R. Keeney, eds., pp. 539–579. American Society of Agronomy and Soil Science Society of America, Madison, Wisc.

7. Boyd, C. E., and M. L. Cuenco. 1980. Refinement of the lime requirement procedures for fish ponds. *Aquaculture* **21**:293–299.

8. Kramer, P. J. *Plant and Soil Water Relationships: A Modern Synthesis.* 1969. McGraw-Hill, New York.

9. Green, B. 1993. Water and chemical budgets for organically fertilized fish ponds in the dry tropics. Ph.D. dissertation, Auburn University, Ala.

10. Boyd, C. E., and D. Teichert-Coddington. 1994. Pond bottom soil respiration during fallow and culture periods in heavily-fertilized tropical fish ponds. *J. World Aquaculture Soc.* **25**:417–423.

11. Sawyer, C. N., and P. L. McCarty. *Chemistry for Sanitary Engineers.* 1967. McGraw-Hill, New York.

12. Brady, N. C. *The Nature and Properties of Soils*. 1990. Macmillan, New York.

13. Kollmorgen Instruments Corporation. *Munsell Soil Color Charts*. 1992. Macbeth, Division of Killmorgen Instruments Corporation, Newburgh, N.Y.

14. Boyd, C. E. 1970. Influence of organic matter on some characteristics of aquatic soils. *Hydrobiologia* **36**:17–21.

15. Dent, D. *Acid Sulfate Soils: A baseline for Research and Development*. 1986. Publication 39, International Institute of Land Reclamation and Improvement, Wageningen, The Netherlands.

16. Sorensen, D. L., W. A. Knieb, D. B. Porcella, and B. Z. Richardson. 1980. Determining the lime requirement for the Blackbird Mine spoil. *J. Environ. Qual.* **9**:162–166.

17. Gaviria, J. I., H. R. Schmittou, and J. H. Grover. 1986. Acid sulfate soils: Identification, formation, and implications for aquaculture. *J. Aquaculture Tropics* **1**:99–109.

18. Singh, V. P. 1980. Management of fishponds with acid-sulfate soils. *Asian Aquaculture* **5**:4–6.

19. Fleming, J. F., and L. T. Alexander. 1961. Sulfur acidity in South Carolina tidal marsh soils. *Soil Sci. Soc. Amer. Proc.* **25**:94–95.

20. Millar, C. E. *Soil Fertility*. 1955. John Wiley & Sons, New York.

21. Boyd, C. E., M. E. Tanner, M. Madkour, and K. Masuda. 1994. Chemical characteristics of bottom soils from freshwater and brackishwater aquaculture ponds. *J. World Aquaculture Soc.* **25**: 517–534.

22. Foth, H. D. *Fundamentals of Soil Science*. 1990. John Wiley & Sons, New York.

23. Hajek, B. F., F. L. Gilbert, and C. A. Steers. *Soil Associations of Alabama*. 1975. Dept. Agronomy and Soils Departmental Series No. 24, Alabama Agricultural Experiment Station, Auburn University, Ala.

24. McNutt, R. B. *Soil Survey of Lee County, Alabama*. 1981. USDA Soil Conservation Service, U.S. Governmment Printing Office, Washington, D.C.

25. Soil Survey Staff. *National Soils Handbook*. 1983. USDA-Soil Conservation Staff, Washington, D.C.

26. Hejný, S., and J. Květ. 1978. Introduction to the ecology of fishpond littorals. In *Ecological Studies*, vol. 28, D. Dykyjova and J. Květ, eds., pp. 1–9. Springer-Verlag, Berlin, Germany.

27. Masuda, K. 1993. Phosphorus in fishpond mud. Ph.D. dissertation, Auburn University, Alabama.

28. Boyd, C. E. 1974. *Lime Requirements of Alabama Fish Ponds*. Bulletin 459, Alabama Agricultural Experiment Station, Auburn University, Ala.

29. Adams, F., and C. E. Evans. 1962. A rapid method for measuring lime requirement of red-yellow podzolic soils. *Soil Sci. Sci. Amer. Proc.* **26**:355–357.

3

Soil Nutrients

3.1 Introduction

Plants obtain carbon from carbon dioxide in the atmosphere and hydrogen from water. Oxygen is produced in photosynthesis or acquired from the atmosphere. Essential mineral nutrients (Table 3.1) are supplied by the soil. Most vascular plants absorb carbon dioxide through their leaves and take in water and mineral nutrients through their roots. The root system extends into the soil and provides a large surface area for absorbing nutrients and water. Soil minerals dissolve in pore water to make the *soil solution*. Plant roots absorb dissolved mineral nutrients from the soil solution.

Aquatic plants need the same nutrients as terrestrial plants. Other than vascular aquatic plants rooted in shallow water with leaves above the water surface, aquatic plants absorb most of their nutrients directly from the water. Phytoplankton, the

Table 3.1. Essential Mineral Nutrients for Plants

Macronutrients	Micronutrients
Nitrogen	Iron
Phosphorus	Manganese
Sulfur	Copper
Calcium	Zinc
Magnesium	Boron[a]
Potassium	Molybdenum
Sodium[a]	Cobalt[a]
Silicon[a]	Chlorine[a]

[a]Not all required by all species. There is evidence that certain algae and some higher plants need one or more of these elements.

PHYTOPLANKTON   LAND PLANT

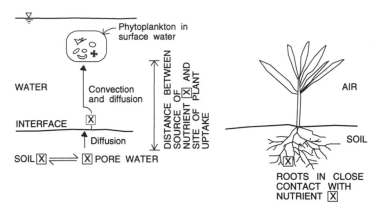

FIGURE 3.1. Illustration of pathway of nutrient (x) in soil to site of absorption by plants in a pond and in a terrestrial soil.

dominant flora of aquaculture ponds, absorb all of their nutrients from water. Phytoplankton and other algae do not have root systems for extracting nutrients directly from the soil solution. Mineral nutrients in pond soil dissolve in pore water, and nutrients in pore water can diffuse into the overlaying pond water where they can be absorbed by algae. Uptake of mineral nutrients by land plants and phytoplankton is contrasted in Figure 3.1.

This chapter discusses mineral nutrients and the soil solution. The mineral nutrient composition of pond soils also is presented and contrasted with that of terrestrial soils.

3.2 Soil Solution

Plants usually absorb ionic forms of mineral nutrients. Ion species in soil solution are in equilibrium with solid-phase minerals, which controls their solubilities. Soil solutions are dilute salt solutions, and the behavior of electrolytes in soil solutions obeys the laws of thermodynamics.[1] The complexity of soil solution chemistry results from the many simultaneous equilibria that exist between solution ions and solid-phase minerals, among solution-phase ions, and between dissolved and atmospheric gases. Chemical equilibria in soil solutions are illustrated for Ca^{2+} in a calcareous soil containing gypsum (Fig. 3.2) and for Cu^{2+} in a slightly basic soil containing calcium carbonate (Fig. 3.3). Solution Ca^{2+} is simultaneously in equilibrium with solid-phase minerals and with exchangeable calcium (Fig. 3.2). Solution Cu^{2+} (Fig. 3.3) is simultaneously in equilibrium with solid-phase copper mineral, with inorganic and organic complexes of copper,

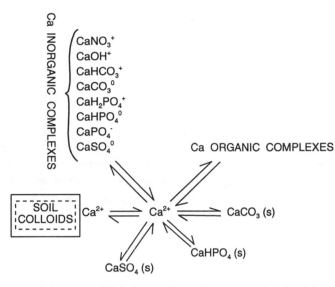

FIGURE 3.2. Calcium equilibria in a soil containing gypsum and calcium carbonate.

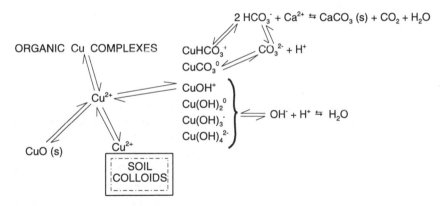

FIGURE 3.3. Copper equilibria in a slightly basic soil containing free calcium carbonate.

and with exchangeable copper. Solubility of calcium carbonate in both illustrations is controlled by atmospheric carbon dioxide.

3.3 Chemical Equilibrium

Many chemical reactions in natural environments stop before all reactants are converted to products. An equilibrium state is reached in which there is a definite ratio of reactants and products:

$$aA + bB = cC + dD \qquad (3.1)$$

If A and B are combined, C and D are formed. A back reaction also is possible in which C and D combine to form A and B. At equilibrium, there is a mixture of products and reactants, and if the equilibrium is disrupted by adding any one or more of substances A, B, C, or D, the reaction will proceed in the direction necessary to reestablish equilibrium.

The *law of mass action* states that at equilibrium a definite mathematical relationship exists between reactants and products. The product of the activities of the products, each raised to the power of its numerical coefficient, divided by the product of the activities of the reactants, each also raised to the power of its numerical coefficient, equals a constant at a given temperature:

$$\frac{(C)^c(D)^d}{(A)^a(B)^b} = K \qquad (3.2)$$

The constant, K, is the *equilibrium constant*. It also may be called the *solubility product* for a salt or the *dissociation constant* or *ionization constant* for an acid or base. Activities of dissolved substances in Equation 3.2 are expressed in moles per liter; gases are expressed in atmospheres of pressure, and solids and water have unit activity.

3.3.1 Relationship to Thermodynamics

Equilibrium is a thermodynamic property of chemical reactions, because the driving force causing a reaction to proceed is the difference in free energy of products relative to the free energy of reactants. A reaction will progress in the direction of lower free energy, and at equilibrium the free energy of products equals the free energy of reactants. In Equation 3.1, when A and B are brought together, they contain a certain amount of free energy. They react to form C and D, and the free energy of the "left-hand side" of the equation declines as C and D are formed on the "right-hand side." When the free energies on both sides of the equation are equal, a state of equilibrium exists.

The free energy of the reaction in which one mole of a substance in its standard state is formed from stable elements at standard conditions of 1 atm pressure and 25°C is the *standard free energy of formation*, ΔF_f° (Table 3.2). A more complete list of ΔF_f° values may be found in Garrels and Christ.[2] The *standard free energy of reaction*, ΔF_r°, is estimated as follows:

$$\Delta F_r^\circ = \sum \Delta F_f^\circ \text{ products} - \sum \Delta F_f^\circ \text{ reactants} \qquad (3.3)$$

Example 3.1: Calculation of ΔF_r°
    ΔF_r° will be calculated for the reaction

Table 3.2. Standard Free Energies of Formation (ΔF_f°) of Selected Substances[a]

Substance	State[b]	ΔF_f° (kcal)	Substance	State[b]	ΔF_f° (kcal)
Al^{3+}	aq	-115.0	$FeCO_3$	c	-161.06
$Al(OH)_3$	am	-271.9	$Fe(OH)_3$	c	-166.0
Ca^{2+}	aq	-132.18	Fe_2O_3	c	-177.1
$CaCO_3$	c	-269.78	Mn^{2+}	aq	-54.4
$CaSO_4 \cdot 2H_2O$	c	-429.19	MnO	c	-86.8
$Ca_3(PO_4)_2$	c	-429.19	MnO_2	c	-111.1
$CaHPO_4 \cdot 2H_2O$	c	-514.60	MnO_4^{2-}	aq	-120.4
CO_2	g	-94.26	$Mn(OH)_4$	c	-149.34
CO_2	aq	-92.31	Mg^{2+}	aq	-136.13
H_2CO_3	aq	-149.00	$Mg(OH)_2$	c	-199.27
HCO_3^-	aq	-140.31	$MgCO_3$	c	-246.0
CO_3^{2-}	aq	-126.22	NO_2^-	aq	-8.25
Cl^-	aq	-31.35	NO_3^-	aq	-26.43
Cl_2	g	0	NH_3	aq	-6.37
Cl_2	aq	1.65	NH_4^+	aq	-19.00
HCl	aq	-31.35	OH^-	aq	-37.59
Cu^{2+}	aq	15.53	H_2O	l	-56.69
$CuSO_4 \cdot 5H_2O$	c	-449.3	PO_4^{3-}	aq	-245.1
H^+	aq	0	HPO_4^{2-}	aq	-261.5
H_2	g	0	$H_2PO_4^-$	aq	-271.3
Fe^{2+}	aq	-20.30	H_3PO_4	aq	-274.2
Fe^{3+}	aq	-2.52	SO_4^{2-}	aq	-177.34
$Fe(OH)_3$	c	-166.0	H_2S	aq	-6.54
FeS_2	c	-36.0	H_2SO_4	aq	-177.34

[a]The elemental forms (i.e., Ca, Mg, Al, O_2, H_2, etc.) have $\Delta F_f^\circ = 0$ kcal.

[b]aq = aqueous, am = amorphous, c = crystalline, g = gas, l = liquid.

$$HCO_3^- = H^+ + CO_3^{2-}$$

Solution: From Table 3.2, ΔF_f° values are $HCO_3^- = -140.31$ kcal; $H^+ = 0$ kcal, $CO_3^{2-} = -126.22$ kcal. By substitution, we obtain

$$\Delta F_r^\circ = \Delta F_f^\circ CO_3^{2-} + \Delta F_f^\circ H^+ - \Delta F_f^\circ HCO_3^-$$
$$= (-126.22) + (0) - (-140.31)$$
$$= 14.09 \, kcal$$

In the calculation of ΔF_r°, activities of all products and reactants must be taken as unity, for ΔF_f° is the free energy of reaction for the formation of one mole of a substance under standard conditions. The standard free energy of reaction also can be expressed as

$$\Delta F_r^\circ = -RT \ln K \qquad (3.4)$$

where R = universal gas law constant (0.001987 kcal/°A),

T = absolute temperature (°A),

°A = degree absolute (273.15 + °C)

The term $-RT$ ln has the value -1.364 at 25°C:

$$(-0.001987 \text{ kcal/°A}) (298.15°A)(2.303) = -1.364$$

and

$$\Delta F_r^° = -1.364 \log K \tag{3.5}$$

Equation 3.5 provides a technique for calculating the equilibrium constant of any reaction for which we know $\Delta F_r^°$.

Example 3.2: Estimation of K from $\Delta F_r^°$
    The K for the reaction $HCO_3^- = CO_3^{2-} + H^+$ will be calculated.

Solution: The $\Delta F_r^°$ for the equation was found to be 14.09 kcal in Example 3.1.

$$\Delta F_r^° = -1.364 \log K$$

$$\log K = -\frac{\Delta F_r^°}{1.364} = -\frac{14.09}{1.364} = -10.33$$

Equations 3.4 and 3.5 apply only to the standard free-energy change with the reaction at equilibrium. A general equation for the free energy of reaction under nonstandard or nonequilibrium conditions, ΔF_r, is

$$\Delta F_r = \Delta F_r^° + RT \ln \frac{(C)^c(D)^d}{(A)^a(B)^b} \tag{3.6}$$

The term $(C)^c(D)^d/(A)^a(B)^b$ is equal to K only when a reaction is at equilibrium. Remember that K is the equilibrium constant, and when a chemical reaction is at equilibrium, $\Delta F_r = 0$ and $\Delta F_r^° = -RT \ln K$. It is conventional to call the term $(C)^c(D)^d/(A)^a(B)^b$ the *reaction quotient, Q,* when the reaction is not at equilibrium (i.e., $\Delta F_r \neq 0$).

3.3.2 Activities

Activities only equal measured molar concentrations in very dilute solutions. In all other solutions, *activity* is defined as

$$a = \gamma m \tag{3.7}$$

where a = activity (M)
γ = activity coefficient (dimensionless)
m = measured molar concentration

In dilute solutions such as freshwaters and soil solutions, activities of single ions may be estimated with the Debye–Hückel equation:

$$-\log \gamma_i = \frac{(A)(Z_i)^2(I)^{\frac{1}{2}}}{1 + (B)(S_i)(I)^{\frac{1}{2}}} \tag{3.8}$$

where γ_i = activity coefficient of ion i
$A = 0.5085$ at 25°C
Z_i = charge of ion i
$B = 0.3281$ at 25°C
S_i = effective size of ion i (Table 3.3)
I = ionic strength (M)

The ionic strength of a solution is a measure of the strength of the electrostatic field caused by its ions. Ionic strength varies with concentrations and charges of ions and may be calculated as follows:

$$I = \frac{1}{2}\sum_{i}^{m}(m_i)(Z_i)^2 + \ldots + (M_n)(Z_n)^2 \tag{3.9}$$

Example 3.3: Calculation of Activities
Activities will be calculated for a solution with $HCO_3^- = 63$ mg L^{-1}, $SO_4^{2-} = 4.2$ mg L^{-1}, $Cl^- = 2.6$ mg L^{-1}, $Ca^{2+} = 15$ mg L^{-1}, $Mg^{2+} = 4.3$ mg L^{-1}, $Na^+ = 2.6$ mg L^{-1}, $K^+ = 1.0$ mg L^{-1}.

Solution: Divide analytical concentrations (mg L^{-1}) of ions by their molecular (ionic) weight to obtain molar concentrations. For HCO_3^-, this calculation is

$$0.063 \text{ g } L^{-1} \div 61 \text{ g mole}^{-1} = 0.00103 \ M$$

Concentrations of other ions follow: $SO_4^{2-} = 0.000044 \ M$, $Cl^- = 0.000073 \ M$, $Ca^{2+} = 0.00037 \ M$, $Mg^{2+} = 0.00018 \ M$, $Na^+ = 0.00011 \ M$, and $K^+ = 0.000026 \ M$. The ionic strength is

Table 3.3.   Values for Ion Sizes (S_i) for Use in the Debye-Hückel Equation

Ions	S_i
OH^-, HS^-, K^+, Cl^-, NO_2^-, NO_3^-, NH_4^+	3
Na^+, PO_4^{3-}, SO_4^{2-}, HPO_4^{2-}, HCO_3^-, $H_2PO_4^-$	4
CO_3^{2-}	5
Ca^{2+}, Cu^{2+}, Zn^{2+}, Mn^{2+}, Fe^{2+}	6
Mg^{2+}	8
Al^{3+}, Fe^{3+}, H^+	9

Source: Modified from Adams.[4]

$$I = \frac{1}{2}\sum[0.00103(1)^2 + 0.000044(2)^2 + 0.000073(1)^2 + 0.00037(2)^2$$
$$+ 0.00018(2)^2 + 0.00011(1)^2 + 0.000026(1)^2]$$
$$= 0.0018\ M$$

The activity coefficient for HCO_3^- is

$$-\log \gamma_{HCO_3^-} = \frac{(0.5085)(1)^2(0.0018)^2}{1 + (0.3281)(4)(0.0018)^{\frac{1}{2}}}$$

$$= \frac{0.0216}{1.0557} = 0.0205$$

$$\gamma_{HCO_3^-} = 0.954$$

The activity of HCO_3^- is

$$a_{HCO_3^-} = (0.954)(0.00103) = 0.00098\ M$$

By similar computations, the activities of the other ions are as follows: $a_{SO_4^{2-}} = 0.000036\ M$, $a_{Cl^-} = 0.000070\ M$, $a_{Ca^{2+}} = 0.00031\ M$, $a_{Mg^{2+}} = 0.00015\ M$, $a_{Na^+} = 0.00010\ M$, and $a_{K^+} = 0.000025\ M$.

Calculations of ion activities are necessary because electrostatic effects among ions reduce their effective concentrations.[3] At infinite dilution, the activity coefficient is 1.0, and measured molarity equals activity. At greater solute concentrations, electrostatic interactions increase and reduce the effective concentrations of ions. An analytical concentration of $0.0003\ M\ Ca^{2+}$ has an activity of $0.00026\ M$ at $I = 0.001\ M$, but only $0.00020\ M$ at $I = 0.01\ M$.

3.3.3 Ion pairs and Complexes

A significant proportion of the ions in soil solution are strongly attracted to ions of opposite charge. Ions attracted to each other by electrical charge either behave as uncharged particles or as ions of a smaller charge than expected. Ions attracted by mutual charges are called complex ions or ion pairs. In a solution containing Ca^{2+}, K^+, SO_4^{2-}, and HCO_3^-, the ion pairs $CaSO_4^0$, $CaHCO_3^-$, KSO_4^-, and $KHCO_3^0$ will form. Ion pairs are soluble, but they are not charged as strongly as unpaired ions. Values of K for formation of important ion pairs for major ions are listed in Table 3.4. Activity coefficients for ion pairs are considered unity.

Example 3.4: Ion-Pair Calculation
  The concentration of $CaHCO_3^+$ will be calculated for a solution $10^{-3}\ M$ with respect to Ca^{2+} and HCO_3^-.

Solution: From Table 3.4, find the expression

Table 3.4. *Equilibrium Constants at 25°C and Zero Ionic Strength for Ion Pairs in Natural Waters*

Reaction	Equilibrium Constant (K)
$CaSO_4^0 = Ca^{2+} + SO_4^{2-}$	$10^{-2.28}$
$CaCO_3^0 = Ca^{2+} + CO_3^{2-}$	$10^{-3.20}$
$CaHCO_3^+ = Ca^{2+} + HCO_3^-$	$10^{-1.26}$
$MgSO_4^0 = Mg^{2+} + SO_4^{2-}$	$10^{-2.23}$
$MgCO_3^0 = Mg^{2+} + CO_3^{2-}$	$10^{-3.40}$
$MgHCO_3^+ = Mg^{2+} + HCO_3^-$	$10^{-1.16}$
$NaSO_4^- = Na^+ + SO_4^{2-}$	$10^{-0.62}$
$NaCO_3^- = Na^+ + CO_3^{2-}$	$10^{-1.27}$
$NaHCO_3^0 = Na^+ + HCO_3^-$	$10^{0.25}$
$KSO_4^- = K^+ + SO_4^{2-}$	$10^{-0.96}$

Source: Modified from Adams.[4]

$$CaHCO_3^+ = Ca^{2+} + HCO_3^- \qquad K = 10^{-1.26}$$

Rearrange the expression and calculate $CaHCO_3^+$ as follows:

$$\frac{(Ca^{2+})(HCO_3^-)}{(CaHCO_3^+)} = 10^{-1.26}$$

$$(CaHCO_3^+) = \frac{(10^{-3})(10^{-3})}{10^{-1.26}} = 10^{-4.74} M$$

Analytical methods (specific ion electrodes excluded) do not distinguish free ions and ion pairs. Sulfate in natural water is distributed among the following species: SO_4^{2-}, $CaSO_4^0$, $MgSO_4^0$, KSO_4^-, and $NaSO_4^-$. Conventional analytical techniques do not distinguish among free SO_4^{2-} and SO_4 in ion pairs. The SO_4^{2-} concentration must be calculated by subtracting sulfate in ion pairs from the measured sulfate concentration. Actual ionic concentrations are less than measured ionic concentrations in solutions containing ion pairs, and ionic activities calculated directly from analytical data as done in Example 3.3 are not exact. Adams demonstrated that ion pairs affect ionic strength, ionic concentrations, and ionic activities in soil solutions.[4] Adams developed a method to correct for ion pairing:

1. Use measured ionic concentrations to calculate ionic strength.

2. Calculate ionic activities.

3. Calculate ion-pair concentrations with respective ion-pair equations, equilibrium constants, and initial estimates of ionic activities.

4. Revise ionic concentrations and ionic strength by subtracting from them the calculated ion-pair concentrations.

5. Repeat steps 2–4 until all ionic concentrations and activities remain unchanged with succeeding calculations.

Adams discusses all aspects of the calculations and gives examples.[1,4] The iterative procedure is tedious and slow unless it is programmed into a computer.

Boyd calculated activities of major ions in natural water with and without corrections for ion pairs.[5] Ion pairs had little effect on ionic activity in weakly mineralized water ($I < 0.002$ M). In more strongly mineralized waters, ionic activities corrected for ion pairs were appreciably smaller than uncorrected ones. As a general rule, if waters contain less than 500 mg L^{-1} of total dissolved ions, ion pairs may be ignored in calculating ionic activities unless highly accurate activities are required.

Ion pairs are not restricted to major ions. Equilibrium constants for ion pairs of aluminum, boron, phosphorus, iron, manganese, ammonium, and zinc are presented by Adams.[4] Schindler presented equilibrium constants on the formation of soluble inorganic complexes (ion pairs) or iron, manganese, zinc, and copper with carbonate and hydroxide.[6] Organic compounds also can form complexes with inorganic compounds. Many natural organic substances in soils are negatively charged and can complex cations. Synthetic organic compounds sometimes are added to soils or water to form complexes with metals. An organic substance that forms complexes with cations is called a *chelating agent* or a *ligand*. Ethylenediamine tetraacetic acid is a common synthetic chelating agent that can form complexes with cations as in Equation 3.10 for the calcium ion:

$$(3.10)$$

$$
\underset{\substack{HOOCH_2 \\ HOOCH_2}}{} \overset{H\ H}{\underset{H\ H}{N-C-C-N}} \underset{\substack{CH_2COOH \\ CH_2COOH}}{} + 2Ca^{2+} = Ca \underset{\substack{OOCH_2 \\ OOCH_2}}{} \overset{H\ H}{\underset{H\ H}{N-C-C-N}} \underset{\substack{CH_2COO \\ CH_2COO}}{} Ca + 4H^+
$$

Sillen and Martell provided equilibrium constants for complexes of selected ligands and metal ions.[7]

Only small proportions of major cations in soil solutions are bound in ion pairs or organic complexes. A larger percentage of dissolved iron, manganese, zinc, and copper exists in iron pairs; concentrations of micronutrient cations normally are <10% of total concentrations of micronutrient cations.

3.3.4 Equilibrium Calculations

Accounting for electrostatic effects of ionic solutions with activity coefficients, ion pairs, and metal complexes result in tedious computations that are seldom feasible or necessary in practical applications. It is usually permissible to assume that measured molar concentrations of major ions are equal to activities. Ion pairing and complex formation by micronutrients often must be considered. The following examples provide typical calculations that may be needed in investigations of soil and water chemistry.

Example 3.5: Solubility of Al(OH)$_3$
   The solubility of Al(OH)$_3$ will be calculated for pH 4, 5, and 6.

Solution: The appropriate chemical equation is

$$Al(OH)_3 + 3H^+ = Al^{3+} + 3H_2O$$

and $K = 10^9$. In a solubility problem involving a sparingly soluble mineral, the mineral and water may be omitted from the equilibrium expression:

$$\frac{(Al^{3+})}{(H^+)^3} = 10^9$$

$$(Al^{3+}) = (H^+)^3 \, 10^9$$

At pH 4

$$(Al^{3+}) = (10^{-4})^3 10^9 = 10^{-3} \, M$$

At pH 5

$$(Al^{3+}) = (10^{-5})^3 10^9 = 10^{-6} \, M$$

At pH 6

$$(Al^{3+}) = (10^{-6})^3 10^9 = 10^{-9} \, M$$

The molecular weight of Al^{3+} is 26.98. At pH 4, 0.001 M × 26.98 g Al mole^{-1} = 0.027 g L^{-1} = 27 mg Al^{3+} L^{-1}. Concentrations at pH 5 and pH 6 are 0.027 mg Al^{3+} L^{-1} and 0.000027 mg Al^{3+} L^{-1}, respectively.

Example 3.6: Solubility of Phosphorus
   The equilibrium concentration between dicalcium phosphate (CaHPO$_4$) and dissolved phosphate will be calculated.

Solution: The dissolution of CaHPO$_4$ is

$$CaHPO_4 = Ca^{2+} + HPO_4^{2-}$$

for which $K = 10^{-6.56}$. Because CaHPO$_4$ is only slightly soluble, its concentration changes little upon dissociation, and CaHPO$_4$ may be taken as unit activity. This gives

$$(Ca^{2+})(HPO_4^{2-}) = 10^{-6.56}$$

Concentrations of Ca^{2+} and HPO$_4^{2-}$ are equal at equilibrium, so let $x = (Ca^{2+}) = (HPO_4^{2-})$:

$$(x)(x) = 10^{-6.56}$$

$$x^2 = 10^{-6.56}$$

$$x = 10^{-3.28} M \quad \text{or} \quad 0.00052 \, M$$

Phosphorus has a molecular weight of 30.98. The concentration of HPO_4^{2-} expressed in terms of phosphorus is $0.00052 \, M \times 30.98 \, \text{g mole}^{-1} = 0.016$ $\text{g P L}^{-1} = 16 \text{ mg P L}^{-1}$.

Example 3.7: Regulation of Solubility by Ions from Another Equilibrium
The solubility of calcium phosphate ($CaHPO_4$) will be calculated in a solution at equilibrium with gibbsite.

Solution: Suppose that a few milligrams per liter of $CaHPO_4$ are dissolved in a water at equilibrium with gibbsite [$Al(OH)_3$] at pH 5. Gibbsite provides Al^{3+} to solution, and Al^{3+} can react with phosphorus to precipitate $AlPO_4 \cdot 2H_2O$ (variscite). The solubility of gibbsite in water of pH 5 is $10^{-6} \, M$ (see Example 3.5). Dissolution of variscite may be written as

$$AlPO_4 \cdot 2H_2O + 2H^+ = Al^3 + H_2PO_4^- + H_2O$$

for which $K = 10^{-2.5}$. Thus,

$$\frac{(Al^{3+})(H_2PO_4^-)}{(H^+)^2} = 10^{-2.5}$$

$$(H_2PO_4^-) = \frac{(H^+)^2 \, 10^{-2.5}}{(Al^{3+})}$$

$$= \frac{(10^{-5})^2 \, 10^{-2.5}}{10^{-6}} = 10^{-6.5} \, M$$

Expressed in terms of phosphorus, $10^{-6.5} \, M$ is $0.0000032 \, M \times 30.98$ g P $\text{mole}^{-1} = 0.000001$ g P L^{-1} or 0.001 mg P L^{-1}. In this example, phosphate solubility is controlled by variscite, because variscite is less soluble than $CaHPO_4$.

Example 3.8: Ion-Pair Calculation
The ion pair $CuCO_3^0$ forms in dilute, slightly alkaline solutions of Cu^{2+}. The stable mineral form of copper in such solutions is CuO (tenorite). The concentration of $CuCO_3^0$ will be computed for water containing $10^{-5.3} \, M \, CO_3^{2-}$ at pH 8.3.

Solution: The dissolution of tenorite may be written as

$$CuO + 2H^+ = Cu^{2+} + H_2O$$

for which $K = 10^{7.65}$. Therefore,

$$\frac{(Cu^{2+})}{(H^+)^2} = 10^{7.65}$$

$$(Cu^{2+}) = (10^{-8.3})^2 \, 10^{7.65} = 10^{-8.95} \, M$$

The formation of $CuCO_3^0$ is

$$Cu^{2+} + CO_3^{2-} = CuCO_3^0$$

for which $K = 10^{6.77}$. Hence,

$$\frac{CuCO_3^0}{(Cu^{2+})(CO_3^{2-})} = 10^{6.77}$$

$$CuCO_3^0 = (10^{-8.95})(10^{-5.3})(10^{6.77}) = 10^{-7.48} \, M$$

Copper has a molecular weight of 63.54. Thus, $(Cu^{2+}) = 0.071 \, \mu g \, L^{-1}$ and $(CuCO_3^0) = 2.1 \, \mu g \, Cu \, L^{-1}$. There is about 30 times as much copper in $CuCO_3^0$ as in cupric ion.

Equilibrium calculations can be much more complex than illustrated here, but if the reader understands the examples little difficulty will be encountered with equilibria discussed in this book.

3.4 Macronutrient Cations

Basic exchangeable cations, Ca^{2+}, Mg^{2+}, Na^+, and K^+, are plant macronutrients. They originate from various minerals from which soils were formed. These ions are dissolved from minerals by water flowing over the land surface or infiltrating the soil. Limestone is a common source of calcium and magnesium in soils. Limestones may be calcitic ($CaCO_3$), dolomitic [$MgCa(CO_3)_2$], or carbonates with calcium and magnesium in other proportions. Limestones are relatively insoluble in water, but their solubility is enhanced by carbon dioxide:

$$CaCO_3 + CO_2 + H_2O = Ca^{2+} + 2HCO_3^- \qquad (3.11)$$

$$CaMg(CO_3)_2 + 2CO_2 + 2H_2O = Ca^{2+} + Mg^{2+} + 4HCO_3^- \qquad (3.12)$$

Feldspars dissolve to provide potassium and sodium as shown in Equations 3.13 and 3.14 for representative compounds:

$$KAlSi_3O_8 + H_2O \rightarrow HAlSi_3O_8 + K^+ + OH^- \qquad (3.13)$$

$$NaAlSi_3O_8 + H_2O \rightarrow HAlSi_3O_8 + Na^+ + OH^- \qquad (3.14)$$

Water in contact with soil normally contains carbon dioxide, and the reaction of feldspar with water is described more accurately by the reaction

$$NaAlSi_3O_8 + CO_2 + 5\frac{1}{2}H_2O \rightarrow Na^+ + HCO_3^- + 2H_4SiO_4 + \frac{1}{2}Al_2Si_2O_5(OH)_4$$

$$(3.15)$$

Dissolved ions are removed from the soil profile by surface runoff or downward infiltration. Leaching of basic cations is more extensive in humid climates than in arid ones. In high-rainfall areas, dissolved cations are leached downward into the ground water, and soil in all horizons may be acidic because few basic cations are left in the soil and soil colloids are highly base unsaturated. There are some soils in humid areas that formed from limestone deposits. These soils often contain limestone and they are neutral or basic in reaction.

Leaching is less intense in drier climates. Calcium carbonate and other bases accumulate near the surface and soils are neutral or slightly alkaline. Soils containing limestone usually have a pH near 8. In some places there may be enough rainfall to leach basic cations from surface soil but not enough rainfall to leach these ions from the solum. Calcium carbonate, and sometimes calcium sulfate, may be deposited in the subsoil.

In extremely arid climates there is not enough rainfall to flush soluble salts out of surface soil. Such soils are called *saline soils,* and they have a high salinity and pH. The *exchangeable sodium percentage* (ESP) is the percentage of the CEC filled with sodium ions. If the ESP exceeds 15%, pH normally will be above 8.5. The pH of saline soils can rise to 10 if the ESP is very high. Another index of soil salinity is the *sodium adsorption ratio* (SAR):

$$SAR = \frac{Na}{[0.5(Ca + Mg)]^{0.5}} \qquad (3.16)$$

where Na, Ca, and Mg are cation concentrations in soil solution (meq L^{-1}). At SAR above 13, a soil is classified as saline. When ESP and SAR exceed 15% and 13, respectively, most plants will not grow well because of osmotic effects of high salt concentration.

3.5 Bicarbonate and Carbonate

Dissolution of limestone (Eq. 3.11 and 3.12) is a primary source of bicarbonate and carbonate, but bicarbonate can originate from dissolution of feldspars (Eq. 3.15). Carbonate only occurs in measurable concentration when the pH of water exceeds 8.3. The relationship between bicarbonate and carbonate is established by the expression

$$HCO_3^- = CO_3^{2-} + H^+; \qquad K = 10^{-10.33} \qquad (3.17)$$

Bicarbonate and carbonate are readily lost from soil by leaching. In humid areas most soil contains little or no limestone, and bicarbonate concentrations in soil solution and runoff are low. However, where soils developed from deposits of limestone, even soils in humid climates may be calcareous. For example, soils in the Black-Belt Prairie region of Alabama and Mississippi developed from Selma chalk. Bottom soils of ponds in this region contain calcium carbonate. Surface water from soils in semiarid climates contain more bicarbonate than those of humid regions. Precipitation of carbonates occurs in the subsoil in semiarid climates and in surface soils of areas with arid climates.

3.6 Nitrogen

Nitrogen occurs in soil as a constituent of both organic and inorganic substances. Nitrogen gas in the atmosphere is the major source of this element, and there is a well-defined nitrogen cycle in ecosystems (Fig. 3.4) in which nitrogen is transformed from the gaseous state to organic form (nitrogen fixation) to inorganic compounds (decomposition and mineralization) and back to a gas (denitrification).

In the nitrogen cycle, bacteria and blue–green algae reduce nitrogen gas (N_2) to ammonia and use ammonia in amino acid synthesis. Amino acids are used to make protein. When nitrogen-fixing microorganisms die, their remains becomes soil organic matter containing organic nitrogen. Nitrogen fixation is not the only source of nitrogen in ecosystems. Electrical activity in the atmosphere forms nitric acid that enters the ecosystem in rainfall. In agricultural and aquacultural

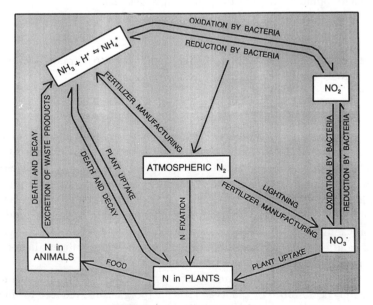

FIGURE 3.4. Nitrogen cycle.

ecosystems, organic nitrogen is applied in manures and feeds, and inorganic nitrogen is added in chemical fertilizers. Nitrogen in feeds and manures originated from plant material, and nitrogen in chemical fertilizer was made by industrial fixation of atmospheric nitrogen. Sodium nitrate also occurs in deposits of the mineral caliche in Chile. Caliche is mined and sodium nitrate is extracted for use as a fertilizer.

Decomposition of soil organic matter by soil microbes transforms organic nitrogen to ammonia (NH_3) or ammonium (NH_4^+). This process is known as *mineralization* or organic nitrogen. When organic nitrogen is mineralized to inorganic nitrogen or ammonium fertilizers are applied, a pH- and temperature-dependent equilibrium is established between NH_3 and NH_4^+:

$$NH_3 + H^+ = NH_4^+ \tag{3.18}$$

In the pH range of most soils (5–8), 90% or more of the mineralized nitrogen will be NH_4^+. The term *ammonia N* refers to NH_3 and NH_4^+ collectively. When it is necessary to indicate the ammonia nitrogen species, NH_3 and NH_4^+ are used. A large amount of the ammonia N in soils is absorbed by soil microbes or plants and transformed to organic nitrogen. Ammonium is a cation and can be adsorbed by negatively charged soil colloids. Ammonia N is soluble and is readily lost by leaching; it also can be oxidized to nitrate by nitrification. Ammonia (NH_3) can be lost from the soil or water by diffusion into the air when pH is high.

The first step in nitrification involves oxidation of ammonium to nitrite by chemoautotrophic bacteria of the genus *Nitrosomonas*:

$$NH_4^+ + 1\tfrac{1}{2}O_2 \rightarrow NO_2^- + 2H^+ + H_2O \tag{3.19}$$

The second step of nitrification, oxidation of nitrite to nitrate, is effected by chemoautotrophic bacteria of the genus *Nitrobacter*:

$$NO_2^- + \tfrac{1}{2}O_2 \rightarrow NO_3^- \tag{3.20}$$

Nitrite produced by *Nitrosomonas* normally does not accumulate in soils, for it is rapidly changed to nitrate by *Nitrobacter*. Chemoautrophic organisms use the energy derived from the oxidation of inorganic compounds to synthesize organic matter from carbon dioxide instead of relying on light energy to transform carbon dioxide photosynthetically into organic matter. Nitrification is an important process in the nitrogen cycle, but the amount of organic matter it produces is insignificant in comparison to photosynthesis. The nitrification process consumes oxygen and produces acidity. Both of these effects are significant factors in the water-quality dynamics of aquaculture ponds.[8]

Nitrate in soils originates from rainfall, nitrate fertilizers, and nitrification. Nitrate may be absorbed by plants and soil microorganisms, reduced to ammonia, and incorporated into protein. Nitrate is soluble and easily lost through leaching.

Nitrate also can be lost through denitrification. Several pathways of denitrification are:

$$HNO_3 + 2H^+ \rightarrow HNO_2 + H_2O \tag{3.21}$$

$$2HNO_2 + 4H^+ \rightarrow \underset{\text{(hyponitrite)}}{H_2N_2O_2} + 2H_2O \tag{3.22}$$

$$H_2N_2O_2 + 4H^+ \rightarrow \underset{\text{(hydroxylamine)}}{2HN_2OH} \tag{3.23}$$

$$NH_2OH + 2H^+ \rightarrow NH_3 + 2H_2O \tag{3.24}$$

$$H_2N_2O_2 \rightarrow N_2O + H_2O \tag{3.25}$$

$$H_2N_2O_2 + 2H^+ \rightarrow N_2 + 2H_2O \tag{3.26}$$

$$\underset{\text{(nitrous oxide)}}{N_2O} + 2H^+ \rightarrow N_2 + H_2O \tag{3.27}$$

The reactions occur under anaerobic conditions.[9] Microorganisms that carry out these transformations use oxidized inorganic compounds as terminal electron and hydrogen acceptors in respiration instead of molecular oxygen. Nitrate can be reduced to nitrite, which can then be reduced to hyponitrite (Eqs. 3.21 and 3.22). At this point several possibilities exist. Hyponitrite may be reduced to hydroxylamine, which can be further reduced to ammonia (Eqs. 3.23 and 3.24), or hyponitrite may be reduced to N_2O or N_2 (Eqs. 3.25 and 3.26). Nitrous oxide may be reduced to N_2 (Eq. 3.27). Nitrogen and nitrous oxide are lost from the soil by volatilization. Ammonia also may be lost by volatilization if the pH is high, but in most soils where denitrification is proceeding, pH is low because of high-carbon-dioxide concentrations.

In highly acidic soils, denitrification may occur through purely chemical pathways:

$$\underset{\text{(amino acids)}}{RNH_2} + HNO_2 \rightarrow N_2 + ROH + H_2O \tag{3.28}$$

$$3HNO_2 \rightarrow 2NO + HNO_3 + H_2O \tag{3.29}$$

Microbially mediated denitrification is of much greater magnitude than chemical denitrification in most soils.

The nitrogen stored in soil is contained primarily in soil organic matter. Nitrate and ammonium usually are absorbed by microbes and plants or lost through leaching because of their high solubility. Nevertheless, detectable concentrations of ammonium and nitrate occur in most soils.

The ratio of carbon to nitrogen in natural ecosystems is usually established by the C:N ratio of humus, which is about 10:1. In agricultural systems the C:N ratio may be quite variable because of nitrogen inputs in management. The C:N ratio is an important soil property, because organic matter with a low or narrow C:N ratio decomposes much faster than organic matter with a high or wide C:N ratio. If a soil contains 5% organic carbon, it usually has around 0.5% organic nitrogen. The total nitrogen concentration may be higher because of the presence of ammonium and nitrate.

3.7 Sulfur

Sulfur is present in soil in various minerals, it is present in atmospheric gases, and it is contained in organic matter. Its inorganic forms are highly soluble, and it can be transformed from one valence state to another by biochemical and chemical processes. The sulfur cycle in an ecosystem (Fig. 3.5) is similar to that of nitrogen, but sulfur is less abundant than nitrogen in organic matter.

Atmospheric sulfur originates from volcanic activity, and in modern times, much sulfur has been added to the atmosphere through combustion of sulfur-containing fossil fuels. Sulfur can be absorbed directly from the atmosphere by plants and incorporated into organic matter. It also can be oxidized to sulfuric acid in the atmosphere and reach the land surface in rain. Sulfuric acid in rainfall is the major cause of acid rain.

Organic matter in soil decomposes and sulfur is released (mineralized) as sulfide, which in the presence of oxygen is oxidized to sulfate. The oxidation

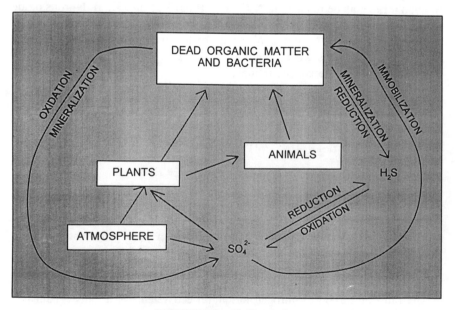

FIGURE 3.5. Sulfur cycle.

of sulfide to sulfate is illustrated in Equations 2.25 to 2.31. Sulfide oxidation can occur through simple chemical reactions, but it occurs faster when mediated by soil microorganisms. Sulfate in soil is highly soluble. It is absorbed by plants or microorganisms, and it may be leached from the soil, but there usually is measurable sulfate in the soil solution.

In anaerobic soils, sulfate may be used instead of oxygen as a hydrogen acceptor in microbial metabolism. Some representative reactions follow:

$$SO_4^{2-} + 8H^+ \rightarrow S^{2-} + 4H_2O \tag{3.30}$$

$$SO_3^{2-} + 6H^+ \rightarrow S^{2-} + 3H_2O \tag{3.31}$$

$$S_2O_3^{2-} + 8H^+ \rightarrow 2SH^- + 3H_2O \tag{3.32}$$

Alexander provides a pathway for the microbial production of hydrogen sulfide, which contains the following steps:[9]

$$H_2SO_4 + 2H^+ \rightarrow \underset{\text{(sulfite)}}{H_2SO_3} + H_2O \tag{3.33}$$

$$H_2SO_3 + 2H^+ \rightarrow \underset{\text{(sulfoxylate)}}{H_2SO_2} + H_2O \tag{3.34}$$

$$H_2SO_2 + 2H^+ \rightarrow \underset{\text{(sulfur hydrate)}}{H_4SO_2} \tag{3.35}$$

$$H_4SO_2 + 2H^+ \rightarrow H_2S + 2H_2O \tag{3.36}$$

Except for the special case of potential acid-sulfate soils that contain iron pyrite (FeS_2), most of the sulfur in soils usually is contained in organic matter. According to Brady, the ratio of C:N:S is normally around 13:10:0.13.[10] A soil containing 5% organic carbon would have about 0.05% total sulfur. Potential acid-sulfate soils may contain up to 5% total sulfur.

3.8 Phosphorus

Phosphorus is a key nutrient for plant growth. In aquatic systems, phosphorus is usually the most important single factor regulating phytoplankton productivity.[8] Soil phosphorus is present in minerals and organic matter, and there is an equilibrium between the concentration of phosphorus in the soil solution and phosphorus contained on soil solids. The equilibrium concentration of phosphorus in the soil solution or in water bodies is a major factor affecting productivity of natural ecosystems, but in agricultural ecosystems natural concentrations of phosphorus generally are too low for optimal rates of plant productivity. Phosphate fertilizers are applied to overcome phosphorus limitations. Phosphorus not removed in the crop is largely bound in the soil in forms unavailable to plants. Phosphorus must be applied on a regular basis to maintain adequate concentrations of available phosphorus for optimal plant productivity.

3.8.1 Inorganic Forms of Phosphorus

Inorganic phosphorus in soils normally can be considered an ionization product of orthophosphoric acid:

$$H_3PO_4 = H^+ + H_2PO_4^- \qquad K = 10^{-2.13} \qquad (3.37)$$

$$H_2PO_4^- = H^+ + HPO_4^{2-}, \qquad K = 10^{-7.21} \qquad (3.38)$$

$$HPO_4^{2-} = H^+ + PO_4^{3-}, \qquad K = 10^{-12.36} \qquad (3.39)$$

At a pH of 7.21, concentrations of $H_2PO_4^-$ and HPO_4^{2-} are equal. There is little H_3PO_4 except in highly acidic solutions, and PO_4^{3-} predominates only in highly basic solutions. Effects of pH on proportions of H_3PO_4, $H_2PO_4^-$, HPO_4^{2-}, and PO_4^{3-} are provided in Figure 3.6. Acidic soils contain primarily $H_2PO_4^-$, and basic soils contain more HPO_4^{2-} than $H_2PO_4^-$.

Inorganic phosphorus reacts with iron and aluminum in acidic soils to form highly insoluble compounds. Representative aluminum and iron phosphate compounds that have been identified in soil are

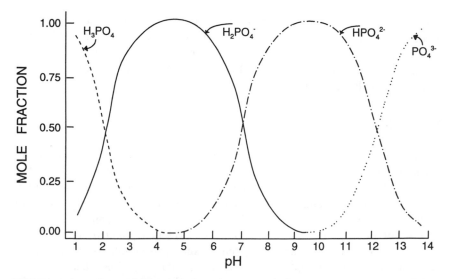

FIGURE 3.6. Effects of pH on relative proportions (mole fractions) of H_3PO_4, $H_2PO_4^-$, HPO_4^{2-}, and PO_4^{3-} in an orthophosphate solution.

$$Al^{3+} + H_2PO_4^- + H_2O = \underset{(variscite)}{AlPO_4 \cdot 2H_2O} + 2H^+ \tag{3.40}$$

$$Fe^{3+} + H_2PO_4^- + 2H_2O = \underset{(strengite)}{FePO_4 \cdot 2H_2O} + 2H^+ \tag{3.41}$$

Concentrations of Al^{3+} and Fe^{3+} available to react with $H_2PO_4^-$ increase as soil pH decreases. Two representative iron and aluminum compounds that may control Al^{3+} and Fe^{3+} concentrations in acidic soils are

$$\underset{(gibbsite)}{Al(OH)_3} + 3H^+ = Al^{3+} + 3H_2O \tag{3.42}$$

$$\underset{(amorphorus\ iron\ III\ oxide)}{Fe(OH)_3} + 3H^+ = Fe^{3+} + 3H_2O \tag{3.43}$$

The solubility of gibbsite often is expressed in another manner as:

$$Al(OH)_3 = Al^{3+} + 3OH^-, \qquad K = 10^{-33} \tag{3.44}$$

Because of greater concentrations of Fe^{3+} and Al^{3+} in soil solution at lower pH, the availability of native soil phosphate and phosphate applied in fertilizer tends to decrease as pH decreases.

There is often a lot of colloidal iron and aluminum oxides, such as gibbsite and goethite, in acidic soils. Phosphate can be adsorbed by iron and aluminum oxides as follows:

$$H_2PO_4^- + Al(OH)_3 = Al(OH)_2H_2PO_4 + OH^- \tag{3.45}$$

$$H_2PO_4^- + FeOOH = FeOH_2PO_4 + OH^- \tag{3.46}$$

Silicate clays can fix phosphorus. Phosphorus is substituted for silicate in the clay structure. Clays also have some ability to adsorb anions, because they have a small number of positive charges on their surfaces. The anion-exchange process is

$$-Colloid+ + Ca^{2+} + Mg^+ + K^+ + H_2PO_4^- = K^+ - \underset{Ca^{2+}}{\overset{Mg^{2+}}{Colloid+}} H_2PO_4^- \tag{3.47}$$

The adsorption of phosphate by colloids is favored by decreasing pH.

Primary phosphate compounds in neutral and basic soils are calcium phosphates. Common calcium phosphates are listed in Table 3.5. The most soluble compound is monocalcium phosphate $Ca(H_2PO_4)_2$. This is the form of phosphorus normally applied in fertilizer. In neutral or basic soils, $Ca(H_2PO_4)_2$ is transformed

Table 3.5   Solubility Expressions and Equilibrium Constants (K) for Some Selected Compounds Found in Soils

Name	Reaction	K
Calcite	$CaCO_3 = Ca^{2+} + CO_3^{2-}$	$10^{-8.35}$
Dolomite	$CaMg(CO_3)_2 = Ca^{2+} + Mg^{2+} + 2CO_3^{2-}$	$10^{-16.7}$
Calcium hydroxide	$Ca(OH)_2 = Ca^{2+} + 2OH^-$	$10^{-5.43}$
Albite	$NaAlSi_3O_8 + CO_2 + 5\frac{1}{2}H_2O = Na^+ + HCO_3^- +$	$10^{-1.9}$
	$2H_4SiO_4 + \frac{1}{2}Al_2Si_2O_5(OH)_4$	
Variscite	$AlPO_4 \cdot 2H_2O + 2H^+ = Al^{3+} + H_2PO_4^- + 2H_2O$	$10^{-2.5}$
Strengite	$FePO_4 \cdot 2H_2O + 2H^+ = Fe^{3+} + H_2PO_4^- + 2H_2O$	$10^{-6.85}$
Vivianite	$Fe(PO_4)_2 \cdot 8H_2O + 4H^+ = 3Fe^{2+} + 2H_2PO_4^- +$	$10^{3.11}$
	$8H_2O$ $(Fe^{3+} + e^- = Fe^{2+})$	$10^{13.01}$
Gibbsite	$Al(OH)_3 + 3H^+ = Al^{3+} + 3H_2O$	10^9
Fe (III) oxide	$Fe(OH)_3 + 3H^+ = Fe^{3+} + 3H_2O$	$10^{3.54}$
Goethite	$FeOOH + 2H^+ = Fe^{3+} + H_2O$	$10^{3.55}$
Hematite	$Fe_2O_3 + 6H^+ = 2Fe^{3+} + 3H_2O$	$10^{-1.45}$
Hydroxyapatite	$Ca_5(PO_4)_3OH + 7H^+ = 5Ca^{2+} + 3H_2PO_4^- + H_2O$	$10^{14.46}$
Monocalcium phosphate	$Ca(H_2PO_4)_2 = Ca^{2+} + 2H_2PO_4^-$	
Dicalcium phosphate	$CaHPO_4 \cdot 2H_2O = Ca^{2+} + HPO_4^{2-} + 2H_2O$	10^{-7}
Tricalcium phosphate	$Ca_3(PO_4)_2 = 3Ca^{2+} + 2PO_4^{3-}$	10^{-25}
Siderite	$FeCO_3 = Fe^{2+} + CO_3^{2-}$	$10^{-10.4}$
Iron (II) hydroxide	$Fe(OH)_2 = Fe^{2+} + 2OH^-$	10^{-14}
Manganese (IV) oxide	$MnO_2 + 4H^+ = Mn^{4+} + 2H_2O$	—
Manganite	$MnOOH + 2H^+ = Mn^{3+} + 2H_2O$	$10^{0.3}$
Manganese (II) hydroxide	$Mn(OH)_2 = Mn^{2+} + 2OH^-$	$10^{-12.8}$
Manganese carbonate	$MnCO_3 = Mn^{2+} + CO_3^{2-}$	$10^{-9.3}$
Malachite	$Cu_2(OH)_2CO_3 + 4H^+ = 2Cu^{2+} + 3H_2O + CO_2$	$10^{14.16}$
Tenorite	$CuO + 2H^+ = Cu^{2+} + H_2O$	$10^{7.65}$
Zinc oxide	$ZnO + 2H^+ = Zn^{2+} + H_2O$	$10^{11.18}$
Zinc carbonate	$ZnCO_3 + 2H^+ = Zn^{2+} + H_2O + CO_2$	$10^{7.95}$
Cobalt carbonate	$CoCO_3 = Co^{2+} + CO_3^{2-}$	$10^{-12.8}$
Gypsum	$CaSO_4 \cdot 2H_2O = Ca^{2+} + SO_4^{2-} + 2H_2O$	$10^{-4.63}$
Silicon dioxide	$SiO_2 + 2H_2O = H_4SiO_4$	$10^{-2.74}$

Sources: Hem;[3] Adams;[4] Schindler.[6]

through dicalcium, octacalcium, and tricalcium phosphates to apatite. Apatite is extremely insoluble. A representative apatite, hydroxyapatite, dissolves as follows:

$$Ca_5(PO_4)_3OH + 7H^+ = 5Ca^{2+} + 3H_2PO_4^- + H_2O \qquad (3.48)$$

A high concentration of Ca^{2+} of a high pH favors formation of hydroxyapatite from phosphate applied in fertilizer.

The maximum availability of phosphorus in mineral soil occurs between pH 6 and 7 (Fig. 3.7). In this pH range there is less Fe^{3+} and Al^{3+} to react with phosphorus and a smaller tendency of iron and aluminum oxides to adsorb phosphorus than at lower pH. Also, at a pH of 6 to 7 the activity of Ca^{2+} is

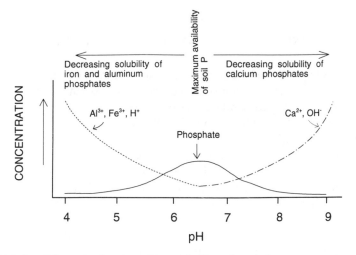

FIGURE 3.7. Schematic showing effects of pH on the relative concentrations of dissolved phosphate in an aerobic soil.

normally lower than at higher pH. Nevertheless, even in the pH range of 6 to 7, most of the phosphorus added to soils normally is rendered insoluble through adsorption by soil colloids or precipitation as insoluble compounds.

3.8.2 Organic Phosphorus

Phosphorus contained in organic matter is released by microbial activity. A portion of the phosphorus released in decomposition is absorbed by plants, but the remainder reacts with Fe^{3+}, Al^{3+}, Ca^{2+}, and soil colloids and is fixed in the soil. In a soil with a high concentration of organic matter, there is less mineral matter to react with phosphorus, and Fe^{3+}, Al^{3+}, and Ca^{2+} ions in soil solution are complexed by organic compounds and less available to react with phosphorus.

3.8.3 Phosphorus Concentrations in Soil

Many techniques are used to measure phosphorus concentrations in soils. Total phosphorus concentration is measured by perchloric acid treatment to dissolve all types of soil phosphorus. Phosphorus soluble in water or dilute sodium bicarbonate solution is an estimate of reactive or plant available soil phosphorus.[11] Hieltjes and Liklema used a three-step extraction to fractionate soil phosphorus: (1) 1 *M* NH$_4$Cl to remove loosely bound phosphorus; (2) 0.1 *N* NaOH to remove iron- and aluminum-bound phosphorus; (3) 0.5 *N* HCl to remove calcium-bound phosphorus.[12] Extraction with 0.03 *N* NH$_4$F and 0.025 *N* HCl[13] is thought to remove soil phosphorus held by Al^{3+} and Fe^{3+} ions as follows:

$$3NH_4F + 3HF + AlPO_4 \rightarrow H_3PO_4 + (NH_4)_3AlF_6 \qquad (3.49)$$

$$3NH_4F + 3HF + FePO_4 \rightarrow H_3PO_4 + (NH_4)_3AlF_6 \qquad (3.50)$$

Dilute hydrochloric or sulfuric acids are probably the most widely used extracting solutions. These solutions remove a phosphorus fraction that is closely related to the plant-available phosphorus in noncalcareous soils. An extracting solution commonly used in the southeastern United States is a 1:1 mixture of 0.05 N HCl and 0.025 N H_2SO_4.[14]

Different extracting solutions provide different estimates of phosphorus concentrations for the same sample. However, the amount of phosphorus extracted from a soil by a solution often is related to the potential for plant growth in the soil. This is illustrated in Table 3.6 for the growth of phytoplankton in nutrient solutions where soil was the only source of phosphorus.[15] Different extracting solutions removed different concentrations of phosphorus, but, for any extracting solution, algal growth increased with increasing phosphorus concentration.

A soil's ability to remove phosphorus from solution or to release phosphorus may be measured. The *phosphorus adsorption capacity* (PAC) is the amount of phosphorus that a soil sample can remove from a concentrated solution of phos-

Table 3.6 Growth of Scenedesmus dimorphus *(individuals per mL* $\times$ *10^3) in Cultures Where Soils Were the Only Source of Phosphorus. The amounts of phosphorus extracted from soils with four solutions and the correlations coefficients between extractable phosphorus and growth are also presented.*

Soil Series	Growth	Solution I[a]	Solution II[b]	Solution III[c]	Solution IV[d]
Baxter	21	2.0	9.5	14.3	78
Decatur	148	4.6	28.0	39.9	283
Anniston	12	1.0	3.0	7.2	60
Colbert	2	1.0	5.0	7.2	42
Holstun	146	21.0	180.0	113.0	462
Dewey	300	31.5	312.5	579.0	1250
Cecil	6	1.0	5.5	7.9	82
Davidson	37	1.2	12.5	15.0	134
Appling	0	0.8	2.5	6.8	63
Madison	11	0.8	5.0	17.2	143
Lucedale	63	3.0	23.0	31.1	242
Amite	164	5.0	60.0	77.5	372
Correlation coefficients		0.875	0.891	0.951	0.972

[a]Dilute nutrient solution without phosphorus.

[b]0.05 HCl plus 0.025 N H_2SO_4.

[c]0.002 N H_2SO_4 plus 3 gm L^{-1} K_2SO_4.

[d]0.1 N HCl plus 0.03 N NH_4F.

phorus during a 17-h period of shaking.[16] The *net phosphorus sorption capacity* (NPS) is an amount of phosphorus that can be removed from a soil saturated with phosphorus during the PAC test by further shaking in phosphorus-free water for 40 h.[17]

The *equilibrium phosphorus concentration* (EPC) and the *phosphorus buffering capacity* (PBC) are useful indices of soil phosphorus status.[16] The EPC is the phosphorus concentration in soil solution at which there is no net adsorption or release of phosphorus by the soil. The PBC is an index of the amount of adsorbed phosphorus required to provide a given phosphorus concentration in the soil solution. To determine the EPC and PBC, one puts subsamples of a soil in a series of phosphorus solutions with concentrations ranging from less than the EPC to well above it and agitates until equilibrium is attained. The phosphorus concentration in each solution is measured, and the amount of adsorbed phosphorus is calculated. Amounts of phosphorus adsorbed from solutions are plotted in Figure 3.8. The EPC is the *x*-intercept, and the PBC is the tangent to the regression line at the *x*-intercept. A soil with a large EPC value will release phosphorus to solution more easily than one with a small EPC. A soil with a large PBC needs a higher phosphorus concentration to provide a given phosphorus concentration in solution than a soil with a small PBC. When a given amount of phosphorus is removed from the soil by a crop, the decrease in solution phosphorus is greater for a soil with a small PBC than for a soil with a high PBC.

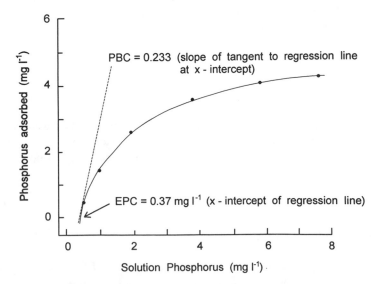

FIGURE 3.8. Calculation of the equilibrium phosphorus concentration (EPC) and the phosphorus buffering capacity (PBC).

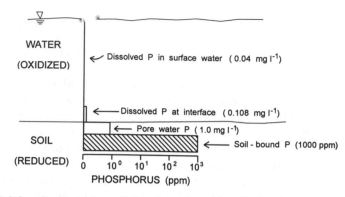
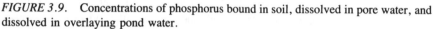

FIGURE 3.9. Concentrations of phosphorus bound in soil, dissolved in pore water, and dissolved in overlaying pond water.

3.8.4 Phosphorus Solubility under Reducing Conditions

The preceding discussion concerned phosphorus in aerobic soils. In flooded soils microbial decomposition of organic matter can cause oxygen depletion and low redox potential. Iron from iron(III) oxides and iron(III) phosphate compounds is converted to iron(II) oxides and compounds that are more soluble than the trivalent forms. This causes an increase in soluble iron and phosphorus in the soil solution. An example of this phenomenon is shown in Figure 3.9, with unpublished data from the author's laboratory. Concentrations of phosphorus in aerobic pond water are much greater than those in anaerobic pore water of pond soils.

3.9 Silicon

Silicon is a constituent of soil minerals and a basic component of layered silicate clays. Silica is not an essential element for vascular plants, but it is required by diatoms, which are common algae in aquatic ecosystems. Silica is derived from silicon dioxide in soil:

$$SiO_2 + 2H_2O = H_4SiO_4 \tag{3.51}$$

The silicic acid dissociates as follows:

$$H_4SiO_4 = H^+ + H_3SiO_4^-, \qquad K = 10^{-9.46} \tag{3.52}$$

$$H_3SiO_4^- = H^+ + H_2SiO_4^{2-}, \qquad K = 10^{-12.56} \tag{3.53}$$

The ratio of $H_3SiO_4^-$:H_4SiO_4 is calculated as

$$\frac{(H_3SiO_4^-)}{(H_4SiO_4)} = \frac{10^{-9.46}}{(H^+)}$$

Percentages of $H_3SiO_4^-$ at different pH values are: pH 6, 0.03%; pH 7, 0.34%; pH 8, 3.35%; pH 9, 25.7%. At the pH of most soils, dissolved silica is present primarily as undissociated silicic acid.

3.10 Micronutrient Cations

Iron, manganese, zinc, copper, and cobalt are metallic cations that occur in the soil solution. Small concentrations of these metals are needed for normal growth and development for plants and animals, but slightly higher concentrations may be toxic. A wide array of compounds of these five metals occurs in soils. Some representative compounds and equations showing their dissociations are provided in Table 3.5. Their solubilities are controlled by pH, and low pH favors high solubility. Considering $FeOOH$, MnO_2, CuO, and $ZnCO_3$ as stable mineral forms of iron, manganese, copper, and zinc, respectively, in soils, Figure 3.10 shows that the equilibrium concentrations of the free cations of these micronutrients decrease markedly with increasing pH. The solubility of micronutrient cations in the soil solution is so low that equilibrium concentrations of ionic forms is inadequate to meet plant requirements. However, the organic matter and certain anions in soils form complexes with micronutrient cations and greatly increase the total concentration of dissolved micronutrient cations for use by plants.

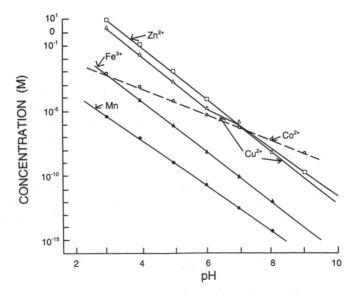

FIGURE 3.10. Effects of pH on concentrations of ionic forms of trace metal in soils.

Table 3.7   Equations and Equilibrium Constants for Selected Ion Pair of Trace Metals with Major Anions in Water

Reaction	K	Reaction	K
$Fe^{2+} + SO_4^{2-} = FeSO_4^0$	$10^{2.7}$	$Cu^{2+} + 4OH^- = Cu(OH)_4^{2-}$	$10^{16.1}$
$Fe^{3+} + SO_4^{2-} = FeSO_4^-$	$10^{4.15}$	$Cu^{2+} + SO_4^{2-} = CuSO_4^0$	$10^{2.3}$
$Fe^{3+} + Cl^- = FeCl^{2+}$	$10^{1.42}$	$MnSO_4^0 = Mn^{2+} + SO_4^{2-}$	$10^{-2.27}$
$FeCl^{2+} + Cl^- = FeCl_2^+$	$10^{0.66}$	$Zn^{2+} + OH^- = ZnOH^+$	$10^{5.04}$
$FeCl_2^+ + Cl^- = FeCl_3^0$	10^1	$Zn^{2+} + 3OH^- = Zn(OH)_3^-$	$10^{13.9}$
$Cu^{2+} + HCO_3^- = CuHCO_3^+$	$10^{2.7}$	$Zn^{2+} + 4OH^- = Zn(OH)_4^{2-}$	$10^{15.1}$
$Cu^{2+} + CO_3^{2-} = CuCO_3^0$	$10^{6.77}$	$Zn^{2+} + Cl^- = ZnCl^+$	$10^{0.43}$
$Cu^{2+} + 2CO_3^{2-} = Cu(CO_3)_2^{2-}$	$10^{10.01}$	$Zn^{2+} + 2Cl^- = ZnCl_2^0$	$10^{0.61}$
$Cu^{2+} + OH^- = CuOH^+$	$10^{6.0}$	$Zn^{2+} + 3Cl^- = ZnCl_3^-$	$10^{0.53}$
$2Cu^{2+} + 2OH^- = Cu_2(OH)_2^{2+}$	$10^{17.0}$	$Zn^{2+} + 4Cl^- = ZnCl_4^{2-}$	$10^{0.20}$
$Cu^{2+} + 3OH^- = Cu(OH)_3^-$	$10^{15.2}$	$Zn^{2+} + SO_4^{2-} = ZnSO_4^0$	$10^{2.38}$

Source: Schindler,[6] Adams.[4]

Some ion pairs that can form in soil solution between micronutrient cations and major anions are listed in Table 3.7. According to Pagenkopf, humic and fulvic acids are the common naturally occurring ligands in natural systems.[18] Pagenkopf used salicylic acid as a model of natural, organic ligands. The complexation of manganese ion by salicylic acid is:

$$\tag{3.54}$$

The bidentate coordination by the carboxylic oxygen and the phenolic oxygen results in a six-membered ring. This type of binding of a metal by an organic compound is called *chelation*. This word is derived from a Greek word meaning claw. The analogy to a claw of an organic ligand binding a metallic ion is obvious from Equation 3.54.

Where the identity of the ligand is known, equilibrium constants can be determined and concentrations of complexed cations may be computed. Triethanolamine is a chelating agent often used to form chelated micronutrient cation preparations for use as fertilizers or algicides. According to Sillen and Martell, triethanolamine (HTEA) dissociates as[7]

$$HTEA = H^+ + TEA^-, \qquad K = 10^{-8.08} \tag{3.55}$$

and TEA^- can form the following complexes with Cu^{2+}:

$$Cu^{2+} + TEA^- = CuTEA^+, \qquad K = 10^{4.44} \tag{3.56}$$

$$Cu^{2+} + TEA^- + OH^- = CuTEAOH^0, \qquad K = 10^{11.9} \tag{3.57}$$

$$Cu^{2+} + TEA^- + 2OH^- = CuTEA(OH)_2{}^-, \qquad K = 10^{18.2} \tag{3.58}$$

Example 3.9: Ion Pairs and Organic Complexes of a Micronutrient Cation
A solution has a pH of 8, 10^{-3} M HCO_3^-, $10^{-5.33}$ M CO_3^{2-}, 10^{-4} M SO_4^{2-}, and 10^{-4} M triethanolamine. The equilibrium concentrations of Cu^{2+}, Cu ion pairs, and Cu-triethanolamine complexes will be computed.

Solution: The Cu^{2+} ion concentration is regulated by solubility of tenorite (Table 3.5):

$$CuO + 2H^+ = Cu^{2+} + H_2O$$

$$\frac{(Cu^{2+})}{(H^+)^2} = 10^{7.65}$$

$$Cu^{2+} = (10^{-8})^2 \, 10^{7.65} = 10^{-8.35} \, M$$

The different ion pairs can be found in Table 3.7:

$Cu^{2+} + HCO_3^- = CuHCO_3^+$

$CuHCO_3^+ = (Cu^{2+})(HCO_3^-) \, K = (10^{-8.35}) \, (10^{-3}) \, (10^{2.7}) = 10^{-8.65} \, M$

$Cu^{2+} + CO_3^{2-} = CuCO_3^0$

$CuCO_3^0 = (Cu^{2+}) \, (CO_3^{2-}) \, K = (10^{-8.35}) \, (10^{-5.33}) \, (10^{6.77}) = 10^{-6.91} \, M$

$Cu^{2+} + 2CO_3^{2-} = Cu(CO_3)_2{}^{2-}$

$Cu(CO_3)_2{}^{2-} = (Cu^{2+}) \, (CO_3^{2-})^2 \, K = (10^{-8.35}) \, (10^{-5.33})^2 \, (10^{10.01}) = 10^{-9} \, M$

$Cu^{2+} + OH^- = CuOH^+$

$CuOH^+ = (Cu^{2+}) \, (OH^-) \, K = (10^{-8.35}) \, (10^{-6}) \, (10^{6.0}) = 10^{-8.35} \, M$

$2Cu^{2+} + 2OH^- = Cu_2 (OH_2)^{2+}$

$Cu_2(OH)_2{}^+ = (Cu^{2+})^2 \, (OH^-)^2 \, K = (10^{-8.35})^2 \, (10^{-6})^2 \, (10^{17.0}) = 10^{-11.7} \, M$

$Cu^{2+} + 3OH^- = Cu(OH)_3{}^-$

$Cu(OH)_3{}^- = (Cu^{2+}) \, (OH^-)^3 \, K = (10^{-8.35}) \, (10^{-6})^3 \, (10^{15.2}) = 10^{-11.15} \, M$

$Cu^{2+} + 4OH^- = Cu(OH)_4{}^{2-}$

$Cu(OH)_4{}^{2-} = (Cu^{2+}) \, (OH^-)^4 \, K = (10^{-8.35}) \, (10^{-6})^4 \, (10^{16.1}) = 10^{-16.25} \, M$

$Cu^{2+} + SO_4^{2-} = CuSO_4^0$

$CuSO_4^0 = (Cu^{2+}) \, (SO_4^{2-}) \, K = (10^{-8.35}) \, (10^{-4}) \, (10^{2.3}) = 10^{-10.05} \, M$

The total concentration of copper in ion pairs is obtained by adding the molar concentrations, $10^{-6.89}$ M.

Only the triethanolamine concentration is given, so the TEA$^-$ concentration available to chelate copper must be computed. From Equation 3.55, we obtain

$$\frac{(H^+)\,(TEA^-)}{(HTEA)} = 10^{-8.08}$$

$$\frac{(TEA^-)}{(HTEA)} = \frac{10^{-8.08}}{10^{-8}} = 10^{-0.08} = 0.83$$

$$\%TEA^- = \frac{0.83}{1.83} \times 100 = 45.4\%$$

The HTEA concentration is 10^{-4} M = 0.0001 M, and TEA$^-$ = (0.0001) (0.454) = 0.0000454 M = $10^{-4.34}$ M.

Equations 3.56 to 3.58 allow computations of the triethanolamine-chelated copper:

$Cu^{2+} + TEA^- = CuTEA^+$
$CuTEA^+ = (Cu^{2+})\,(TEA^-)\,K = (10^{-8.35})\,(10^{-4.34})\,(10^{4.44}) = 10^{-8.25}\,M$
$Cu^{2+} + TEA^- + OH^- = CuTEAOH^0$
$CuTEAOH^0 = (Cu^{2+})\,(TEA^-)\,(OH^-)\,K = (10^{-8.35})\,(10^{-4.34})\,(10^{-6})\,(10^{11.9}) = 10^{-6}\,M$
$Cu^{2+} + TEA^- + 2OH^- = CuTEA(OH)_2^-$
$CuTEA(OH)_2^- = (Cu^{2+})\,(TEA^-)\,(OH^-)^2\,K = (10^{-8.35})\,(10^{-4.34})\,(10^{-6})^2\,(10^{18.2}) = 10^{-6.49}\,M$

The total amount of copper chelated by triethanolamine is $10^{-6.49}\,M$. The total copper concentration is $10^{-6.32}\,M$, compared with $10^{-8.35}\,M$ for Cu^{2+}. The concentration of total copper in solution is increased by two orders of magnitude through ion pairs and organic chelates.

In natural systems the exact compositions of the micronutrient cation-organic complexes are not known, and equilibrium constants are not available. All micronutrient cations form complexes with organic ligands, and total concentrations of these cations in soil solution or surface water greatly exceeds concentrations expected from equilibria between free cations and minerals controlling their solubility. For example, in a pond water of pH 7, the solubility of iron from $Fe(OH)_3$ (Eq. 3.42) is only $10^{-17.7}$ M, but actual concentrations usually are 0.5–1.0 mg Fe L^{-1} ($10^{-5.04}$ to $10^{-4.74}$ M).

Solubilities of iron and manganese in particular are affected by the redox potential. When the redox potential declines in response to poor soil aeration, solubilities of iron and manganese increase, as seen in the E_h–pH diagram (Fig. 3.11).

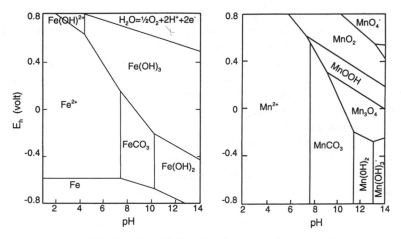

FIGURE 3.11. E_h–pH diagrams for iron and manganese.

3.11 Micronutrient Anions

Chloride occurs in soil in ionic form and is readily leached. Soils in humid regions have low concentrations of chloride, but chloride concentrations in soils increase as precipitation declines.

Boron also is easily leached from soils, but it can be bound by soil colloids. Concentrations of boron are greater in soils of semiarid or arid climates than in humid climates. Dissolved boron in soil solutions exists as boric acid and its dissociation products:

$$H_3BO_3 = H^+ + H_2BO_3^-, \qquad K = 10^{-9.24} \tag{3.59}$$

One-half of the acid is dissociated at pH 9.24. At the pH of most soils, boron exists in solution primarily as undissociated boric acid.

Molybdenum is a rather rare element in the earth's crust, but small amounts occur in soil. According to Hem, concentrations observed in solutions are related directly to the abundance of molybdenum in the soil.[3] The most available form of molybdenum to plants is MoO_4^{2-}, which is dominant at higher pH. This form of molybdenum behaves much like phosphate ion and can be adsorbed by iron and aluminum oxides.

3.12 Nutrient Concentrations in Pond Soils

Soil samples for nutrient analyses were obtained from 358 freshwater fish ponds in the United States, Honduras, Rwanda, and Bhutan, and from 346 brackish-water shrimp ponds in Mexico, Honduras, Colombia, Ecuador, Thailand, and

Philippines.[19] Approximately one-third of the freshwater ponds were used for sportfishing (sunfish, largemouth bass) or for producing sportfish fingerlings. These ponds were fertilized with chemical fertilizers at 2–10 kg N and P_2O_5 ha^{-1} at three- to four-week intervals during warm months. Other freshwater ponds were used for production of tilapia, carp, and channel catfish. Tilapia and carp ponds were treated with 100–500 kg dry weight manure ha^{-1} at one- to two-week intervals. Channel catfish ponds received feed at 5,000–10,000 kg ha^{-1} per year. Ponds ranged in age from 2 to 50 years, but exact ages of many ponds could not be obtained. Brackish-water shrimp ponds in Ecuador and Venezuela were managed for semi-intensive production with feed inputs of 1,000–4,000 kg ha^{-1} per year and fertilizer applications of 2–4 kg $\cdot$ N and P_2O_5 ha^{-1} each week. Shrimp ponds in Thailand and the Philippines were used for intensive production with feed inputs of 10,000–25,000 kg ha^{-1} per year.

It was impossible to consider differences in concentrations of soil variables based on type of aquaculture, intensity of production, original site soils, quality of water supply, and pond age, because data for one or more of these variables were not available for many ponds. Comparison of data by region failed to show differences among most variables. Therefore, the decision to compare only freshwater samples with brackish-water samples seems reasonable.

3.12.1 Nitrogen

Average nitrogen concentrations (Table 3.8) were 0.28% and 0.30% for freshwater and brackish-water ponds, respectively. Percentage frequency plots of soil nitrogen concentration were remarkably similar for freshwater and brackish water (Fig. 3.12). Percentage nitrogen was below 0.5% in 90% of the samples.

C:N ratios were 6.4 and 6 for freshwater and brackish-water soils, respectively, and few samples had C:N ratios above 10. Nitrogen content of surface layers of mineral terrestrial soils averages 0.15% with a range of 0.02–0.5%.[10] Pond soils had the same nitrogen range as surface terrestrial soils, but average nitrogen concentration tended to be higher. The C:N ratio in terrestrial soils is usually 10–12. Plants in aquaculture ponds are primarily planktonic algae, which contain 5–10% of dry weight as nitrogen.[8] Feed applied to aquaculture ponds normally contains 5–6% nitrogen. Land plants usually contain no more than 2–4% nitrogen. Input of higher-nitrogen-content organic matter to aquaculture ponds probably accounts for pond soils having a lower C:N ratio than surface soils of terrestrial ecosystems.

3.12.2 Phosphorus

Phosphorus Concentrations

Brackish-water ponds had greater concentrations of phosphorus than for freshwater ponds (Table 3.8). Phosphorus concentrations were less than 40 ppm in

Table 3.8   Average Concentrations ±95% Confidence Intervals for Different Chemical Variables in 358 Freshwater Fish Ponds and 346 Brackish-Water Shrimp Ponds (values reported on an air-dry basis)

	Type of Pond	
Variable	Freshwater	Brackish-Water
Total nitrogen (%)	0.28 ± 0.01	0.30 ± 0.01
Total sulfur (%)	0.06 ± 0.01	0.49 ± 0.06
Acid-extractable phosphorus (ppm)	40 ± 10	164 ± 36
EPC[a] (ppm)	0.86 ± 0.13	0.82 ± 0.16
PBC[b]	3.82 ± 0.58	20.33 ± 10.41
Silicon (ppm)	60 ± 6	316 ± 46
Calcium (ppm)	3,539 ± 348	3,450 ± 363
Magnesium (ppm)	157 ± 18	2,258 ± 142
Potassium (ppm)	86 ± 11	822 ± 61
Sodium (ppm)	73 ± 10	15,153 ± 2,980
Iron (ppm)	140 ± 19	626 ± 113
Manganese (ppm)	48 ± 8	157 ± 27
Zinc (ppm)	3.10 ± 0.41	7.98 ± 1.05
Copper (ppm)	8.66 ± 4.50	7.60 ± 7.16
Boron (ppm)	0.78 ± 0.06	31 ± 21
Cobalt (ppm)	0.47 ± 0.08	2.35 ± 0.42
Molybdenum (ppm)	0.23 ± 0.02	0.94 ± 0.33

Source: Boyd et al.[19]

[a]EPC = equilibrium phosphorus concentration.

[b]PBC = phosphorus buffering capacity.

90% of the freshwater pond soils, and only 30% of the brackish-water pond soils had below 40 ppm phosphorus (Fig. 3.12). The average phosphorus concentration was 40 ppm in freshwater pond soils, because a small percentage of samples had very high phosphorus concentrations.

Phosphorus was extracted from samples with dilute acid (0.05 N HCl + 0.025 N H$_2$SO$_4$). Dilute acid solutions are efficient at extracting calcium phosphates from soils. Higher concentrations of acid-extractable phosphorus in soils from brackish-water ponds than in those of freshwater ponds suggests that brackish-water pond soils generally contain more calcium phosphates than freshwater ponds. However, there were not large differences in calcium concentrations between freshwater and brackish-water ponds (Fig. 3.12), and other factors may have been involved.

Acid-extractable phosphorus concentrations in surface horizons of terrestrial soils have a normal range of 50–2,000 ppm with an average of about 1,000 ppm.[10] This is based on soils from all climates. In humid, semitropical, and tropical climates, soils are highly leached and phosphorus concentrations are lower. Highly leached soils also tend to be acidic. Acidic soils contain primarily iron and aluminum phosphates, which are less soluble in dilute acid than calcium

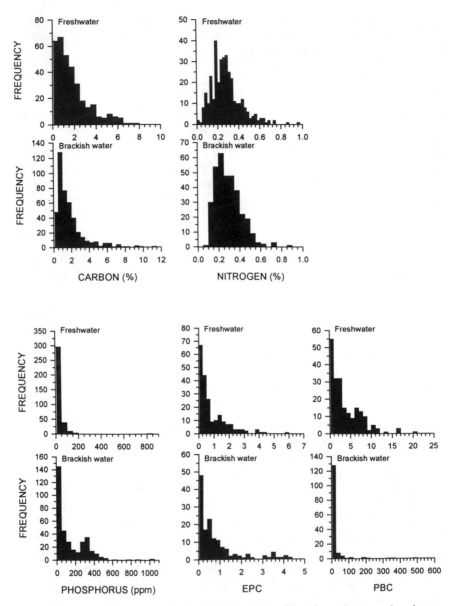

FIGURE 3.12. Frequency distribution histograms for pH, carbon, nitrogen, phosphorus, equilibrium phosphorus concentration (EPC), and phosphorus buffering capacity (PBC) in soils from freshwater and brackish-water aquaculture ponds (*Source: Boyd et al.*[19])

phosphates of neutral and calcareous soils. Pond soils were from humid, tropical, and subtropical sites. This probably is the reason that phosphorus concentrations for pond soils were lower than those reported for terrestrial soils in general. This conclusion is supported by data from Cope and Kirkland, which show that dilute acid-extractable phosphorus concentrations in agricultural soils from Alabama ranged from 0 to 150 ppm.[20] Alabama has a humid, subtropical climate and highly leached soils.

Equilibrium Phosphorus Concentration

EPC had similar distributions in freshwater and brackish-water soils, and EPC of more than one-half of the samples was below 1 ppm. Similarities in EPC between brackish-water and freshwater soils suggest that phosphorus is more water soluble in freshwater soils, because soil phosphorus concentrations were greater in brackish water than in freshwater. In agricultural soils EPC should be above 0.2 ppm for good crop growth.[10] Crop plants are rooted in the soil, and plant roots can absorb phosphorus directly from the soil solution. In ponds, phosphorus equilibrium occurs between soil particles and interstitial water, but phosphorus must diffuse into the pond water and be mixed within the pond volume by water currents. Transport of phosphorus from bottom soil to sites of absorption by phytoplankton is more complex logistically and not as well understood as phosphorus uptake from soil by crop plants. Freshwater ponds in this study that had soils with EPC values of 3–4 ppm were known to require phosphorus fertilization to stimulate phytoplankton growth. Nevertheless, a soil with a high EPC release phosphorus into aqueous solution easier than a soil with a low EPC.

Regression analyses between EPC (Y-variable) and other soil properties revealed weak correlations with soil phosphorus concentration ($r^2 = 0.29; P < 0.01$) and pH ($r^2 = 0.29; P < 0.01$). No other measured properties were significantly correlated with EPC.

Phosphorus Buffering Capacity

Average PBC values were 3.82 and 20.33 for freshwater and brackish-water soils, respectively. This suggests that brackish-water soils adsorb phosphorus more strongly than freshwater soils.

3.12.3 Sulfur

Sulfur concentrations in freshwater pond soils were seldom above 0.1%, and only eight samples contained more than 0.5% sulfur (Fig. 3.13). Average sulfur concentration was 0.06% (Table 3.8). Sulfur concentrations were much higher in brackish-water ponds. Almost one-half of the samples contained more than 0.5% sulfur, and more than 10% of the ponds contained in excess of 1% sulfur (Fig. 3.12). The average sulfur concentration in brackish-water pond soils was 0.49% (Table 3.8).

There are three reasons for higher concentrations of sulfur in brackish-water pond soils than in soils of freshwater ponds. Brackish water contains much more sulfate than freshwater, and more sulfate was present in the pore water of brackish-water pond samples than in pore water of freshwater pond soils. Sulfate in the pore water remained in samples after drying. Potential acid-sulfate soils are more common in coastal areas than at inland locations, and a greater percentage of brackish-water ponds were constructed on potential acid-sulfate soils that contained iron pyrite (FeS_2). Potential acid-sulfate soils often contain 1–5% total sulfur.[21] In anaerobic zones of brackish-water pond soils, conditions exist for the formation of iron pyrite, so ponds not constructed on potential acid-sulfate soil may contain sulfidic deposits.

Surface layers of terrestrial soils usually contain 5–10 times as much nitrogen as sulfur,[10] so sulfur concentrations usually are 0.015–0.03%. Non-acid-sulfate pond soils often contained higher concentrations of sulfur than normally found in terrestrial soils. Most of the sulfur in non-acid-sulfate soils is contained in organic matter. Pond soils tend to have greater accumulations of organic matter than do terrestrial soils, and hence more sulfur.

3.12.4 Calcium

Calcium concentration had similar distributions in freshwater and brackish-water soils (Fig. 3.13). Average calcium concentrations were 3,450 ppm and 3,539 ppm in brackish-water and freshwater soils, respectively (Table 3.8). There were two types of samples with regard to calcium concentration. One group came from ponds with acidic soils that contained little or no calcium carbonate, and the other group came from ponds located on calcareous soils. Calcium carbonate concentrations in soils from some ponds in the Black Belt Prairie of Albama were as high as 40%. In acidic soils, calcium is present as exchangeable cations, but in samples from calcareous soils much of the calcium is bound in calcium carbonate.

There was no obvious difference in pond soil calcium concentrations and those found in surface horizons of terrestrial soils. Terrestrial soils contain from <100 ppm calcium to several percent calcium.[10]

3.12.5 Magnesium

Average magnesium concentrations were greater in soils from brackish-water ponds than in soils of freshwater ponds (Table 3.8). Around 90% of brackish-

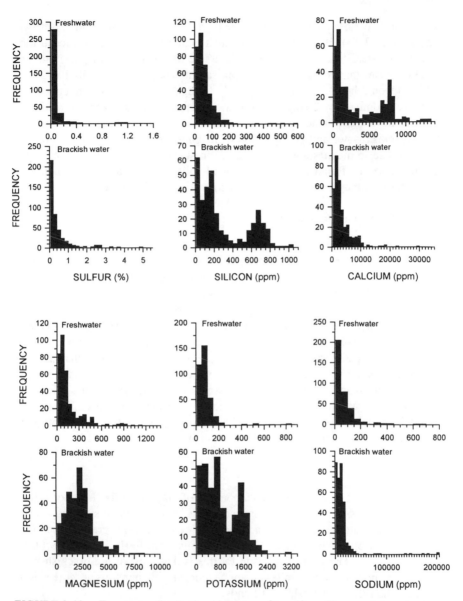

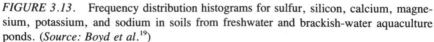

FIGURE 3.13. Frequency distribution histograms for sulfur, silicon, calcium, magnesium, potassium, and sodium in soils from freshwater and brackish-water aquaculture ponds. (*Source: Boyd et al.*[19])

water soils had more magnesium than the highest concentration found in freshwater soil (Fig. 3.13). Brackish water is much higher in magnesium concentration than freshwater. Seawater contains about 1,350 mg L^{-1} of magnesium, and brackish water that contains only 3.5 ppt salinity (seawater normally has 35 ppt salinity) would have around 135 mg L^{-1} of magnesium. Magnesium concentrations are seldom above 20 mg L^{-1} in freshwater ponds.[8] High magnesium concentrations in brackish-water pond soils resulted from adsorption of magnesium from brackish water and precipitation of magnesium during drying of samples.

The range in magnesium concentration in surface layers of terrestrial soils[10] is similar to that found in freshwater pond soils. Terrestrial soils normally do not contain as much magnesium as brackish-water soil samples.

3.12.6 Potassium and Sodium

Potassium and sodium concentrations were much higher in the soils of brackish-water ponds than in those of freshwater ponds (Table 3.8; Fig. 3.13). Average potassium and sodium concentrations were about 8 and 200 times higher, respectively, in soils of brackish-water ponds. Average ocean water contains 10,500 mg L^{-1} sodium and 380 mg L^{-1} potassium. Freshwaters used in aquaculture ponds seldom contain more than 10 mg L^{-1} potassium and 100 mg L^{-1} sodium.[8] Brackish-water soils adsorbed large amounts of potassium and sodium on cation-exchange sites and considerable amounts of sodium and potassium salts deposited in brackish-water soil samples when pore water was evaporated during drying.

Surface horizons of terrestrial soils also have a wide range of sodium and potassium concentrations.[10] In humid regions, concentrations are similar to those found in soils of freshwater ponds. Saline soils have sodium and potassium concentrations similar to those in brackish-water soils.

3.12.7 Silicon

Silicon is present in all soils as a component of clay minerals. It is not usually considered an essential nutrient for land plants, but in aquatic systems silica is essential for diatoms. Silicate concentrations were generally greater in extracts of soils from brackish-water ponds than in soils from freshwater ponds (Table 3.8; Fig. 3.13). The probable reason for this is that most of the brackish-water ponds were constructed on soils containing more silicate minerals or more soluble silicates than those of freshwater ponds.

3.12.8 Micronutrients and Other Elements

Concentrations of boron are much greater in brackish water and seawater than in freshwater of humid and semiarid regions,[22] and higher concentrations of boron in soils of brackish-water ponds (Table 3.8; Fig. 3.14) were likely related to this fact.

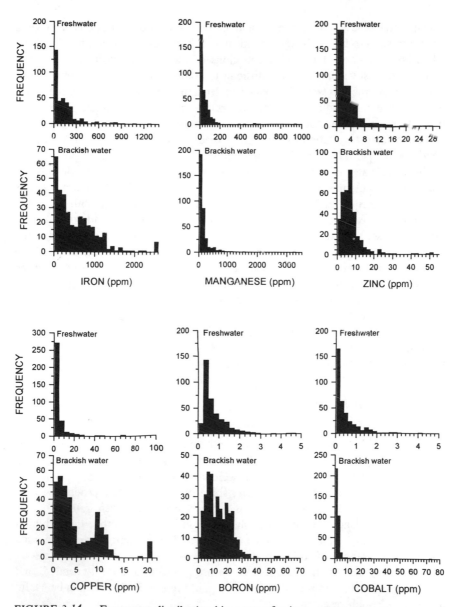

FIGURE 3.14. Frequency distribution histograms for iron, manganese, zinc, copper, boron, and cobalt in soils from freshwater and brackish-water aquaculture ponds. (*Source: Boyd et al.*[19])

Iron, manganese, zinc, cobalt, and molybdenum concentrations also were greater in soils of brackish-water ponds than in those of freshwater ponds (Table 3.8; Figs. 3.14 and 3.15). Concentrations of these micronutrients are highly variable in freshwaters, and it is difficult to compare freshwater concentrations with those of brackish water and seawater. High pH and alkalinity favor precipitation of these micronutrients. Pond waters that are naturally acidic usually are limed to raise pH and total alkalinity, and conditions in aquaculture ponds are favorable for precipitation of most trace metals. Water exchange of 5–20% of pond volume per day normally is used in brackish-water ponds to improve water quality, but water exchange rarely is applied in freshwater ponds. Because of the greater input of water, there is better opportunity in brackish-water ponds for precipitation of micronutrients and their accumulation in the soil.

The higher average concentration of copper in freshwater pond soils than in brackish-water soils resulted from copper sulfate application to a number of the channel catfish ponds. Farmers often apply copper sulfate to catfish ponds in attempts to combat off-flavor in fish caused by certain blue–green algae and the copper precipitates and accumulates in the soil.[23] If catfish ponds were omitted from the data set, copper concentrations would be higher in brackish-water pond soils than in freshwater ones.

Greater average concentrations of chromium and lead in brackish-water soils than in freshwater soils (Table 3.8; Fig. 3.15) also is probably attributable to use of water exchange in brackish-water ponds.

3.12.9 Application

It is not possible to make judgments on optimal concentrations of soil properties from data presented in this study, because data on aquacultural production in ponds could not be obtained. All ponds were in use for aquaculture or aquacultural research. According to pond managers, some ponds were more productive than others, but it could not be determined if these differences were related to soil characteristics. Data presented here show that levels of soil properties exhibit a tremendous range, and it is possible to produce fish or shrimp across the wide concentration ranges of soil properties. Research to define relationships among bottom soil characteristics and fish or shrimp production is badly needed, and the background data presented in Table 3.8 and Figures 3.12–3.15 can be useful in designing such research.

Relative abundance categories were computed for the soil samples (Tables 3.9 and 3.10) to facilitate quick evaluation of any future soil analysis with the data base provided by Boyd et al.[19] For example, if a soil sample from a freshwater pond contains 8 ppm dilute acid extractable phosphorus and 1,375 ppm calcium, Table 3.9 indicates that, compared to the data set, the soil is low in phosphorus and medium in calcium.

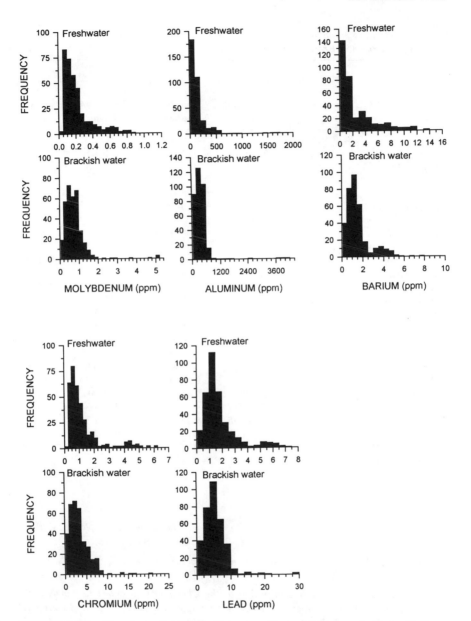

FIGURE 3.15. Frequency distribution histograms for molybdenum, aluminum, barium, chromium, and lead in soils from freshwater and brackish-water aquaculture ponds. (*Source: Boyd et al.*[19])

Table 3.9   Relative Abundance Categories of Soil Chemical Variables in a Series of Soil Samples from 358 Freshwater Aquaculture Ponds

	Decile				
Variable	1st (very low)	2nd and 3rd (low)	4th–7th (medium)	8th and 9th (high)	10th (very high)
pH	<5	5–6	6–7	7–8	>8
Carbon (%)	<0.5	0.5–1	1–2	2–3.5	>3.5
Nitrogen (%)	<0.2	0.2–0.3	0.3–0.4	0.4–0.5	>0.5
Sulfur (%)	<0.01	0.01–0.025	0.025–0.05	0.05–0.125	>0.125
Phosphorus (ppm)	<5	5–10	10–20	20–40	>40
Calcium (ppm)	<600	601–1,200	1,200–3,400	3,400–7,600	>7,600
Magnesium (ppm)	<45	45–80	80–120	120–230	>230
Potassium (ppm)	<30	30–60	60–80	80–110	>110
Sodium (ppm)	<15	15–35	35–60	60–100	>100
Iron (ppm)	<10	10–50	50–130	130–210	>210
Manganese (ppm)	<5	5–20	20–40	40–75	>75
Zinc (ppm)	<0.2	0.2–1.5	1.5–2.5	2.5–5	>5
Copper (ppm)	<0.3	0.3–1.25	1.25–2.5	2.5–6	>6
Silicon (ppm)	<20	20–40	40–60	60–100	>100
Boron (ppm)	<0.3	0.3–0.5	0.5–0.75	0.75–1.25	>1.25
Colbalt (ppm)	<0.10	0.1–0.2	0.2–0.35	0.35–0.8	>0.8
Molybdenum (ppm)	<0.1	0.11–0.15	0.15–0.2	0.21–0.35	>0.35
Aluminum (ppm)	<3.5	3.5–75	75–120	120–200	>200
Barium (ppm)	<0.5	0.5–1	1–1.5	1.5–4	>4
Chromium (ppm)	<0.5	0.5–0.75	0.75–1	1–1.75	>1.75
Lead (ppm)	<1	1–1.25	1.25–1.5	1.5–2.5	>2.5

Source: Boyd et al.[19]

References

1. Adams, F. 1974. Soil solution. In *The Plant Root and Its Environment,* W. E. Carson, ed., pp. 441–481. University Press of Virginia, Charlottesville, Va.

2. Garrels, R. M., and C. L. Christ. *Solutions, Minerals, and Equilibria.* 1965. Harper & Row, New York.

3. Hem, J. D. *Study and Interpretation of the Chemical Characteristics of Natural Water.* 1970. U.S. Geological Survey Water-Supply Paper 1473, U.S. Government Printing Office, Washington, D.C.

4. Adams, F. 1971. Ionic concentrations and activities in soil solutions. *Soil Sci. Soc. Amer. Proc.* **35**:420–426.

5. Boyd, C. E. 1981. Effects of ion-pairing on calculations of ionic activities of major ions in freshwater. *Hydrobiologia* **80**:91–93.

6. Schindler, P. W. 1967. Heterogenous equilibria involving oxides, hydroxides, carbon-

Table 3.10   Relative Abundance Categories of Soil Chemical Variables in a Series of Soil Samples from 346 Brackish-Water Aquaculture Ponds

Variable	1st (very low)	2nd and 3rd (low)	4th–7th (medium)	8th and 9th (high)	10th (very high)
			Decile		
pH	<4	4–6	6–8	8–9	>9
Carbon (%)	<0.5	0.5–1	1–2.5	2.5–4	>4
Nitrogen (%)	<0.15	0.15–0.25	0.25–0.4	0.4–0.5	>0.5
Sulfur (%)	<0.05	0.05–0.1	0.1–0.5	0.5–1.5	>1.5
Phosphorus (ppm)	<20	20–40	40–250	250–400	>400
Calcium (ppm)	<1,000	1,000–2,000	2,000–4,000	4,000–8,000	>8,000
Magnesium (ppm)	<700	700–1,500	1,500–3,000	3,000–4,000	>4,000
Potassium (ppm)	<100	100–400	400–1,200	1,200–1,700	>1,700
Sodium (ppm)	<2,500	2,500–7,000	7,000–15,000	15,000–25,000	>25,000
Iron (ppm)	<60	60–200	200–750	750–1,200	>1,200
Manganese (ppm)	<10	10–50	50–150	150–350	>350
Zinc (ppm)	<2	2–5	5–8	8–14	>14
Copper (ppm)	<1	1–2	2–8	8–11	>11
Silicon (ppm)	<30	30–100	100–500	500–750	>750
Boron (ppm)	<4	4–8	8–18	18–24	>24
Colbalt (ppm)	<0.5	0.5–1	1–2.5	2.5–3.5	>3.5
Molybdenum (ppm)	<0.3	0.3–0.5	0.5–0.9	0.9–1.2	>1.2
Aluminum (ppm)	<100	100–200	200–500	500–600	>600
Barium (ppm)	<0.5	0.5–1	1–1.5	1.5–3.5	>3.5
Chromium (ppm)	<1	1–2	2–4	4–7	>7
Lead (ppm)	<2	2–4	4–7	7–9	>9

Source: Boyd et al.[19]

ates, and hydroxide carbonates. In *Equilibrium Concepts in Natural Water Systems*, R. F. Gould, ed., pp. 196–221. Advances in Chemistry Series 67, American Chemical Society, Washington, D.C.

7. Sillen, L. G., and A. E. Martell. *Stability Constants of Metal-Ion Complexes*. 1971. Supplement No. 1, Special Publication No. 25, Chemical Society of London, England.

8. Boyd, C. E. *Water Quality in Ponds for Aquaculture*. 1990. Alabama Agricultural Experiment Station, Auburn University, Ala.

9. Alexander, M. *Introduction to Soil Microbiology*. 1977. John Wiley & Sons, New York.

10. Brady, N. C. *The Nature and Properties of Soils*. 1990. Macmillan, New York.

11. Olsen, S. R., and L. E. Sommers. 1982. Phosphorus. In *Methods of Soil Analysis, Part 2, Chemical and Microbiological Properties*, A. L. Page, R. H. Miller, and D. R. Kenney, eds., pp. 539–594. American Society of Agronomy and Soil Science Society of America, Madison, Wisc.

12. Hieltjes, A. H. M., and L. Liklema. 1982. Fractionation of inorganic phosphate in calcareous sediments. *J. Environ. Qual.* **9**:405–407.

13. Jackson, M. L. *Soil Chemical Analysis.* 1958. Prentice-Hall, Englewood Cliffs, N.J.

14. Hue, N. B., and C. E. Evans. *Procedures Used for Soil and Plant Analyses by the Auburn University Soil Testing Laboratory.* 1986. Dept. Agronomy and Soils, Departmental Series No. 106, Alabama Agricultural Experiment Station, Auburn University, Ala.

15. Chiou, C., and C. E. Boyd. 1974. The utilization of phosphorus from muds by the phytoplankter, *Scenedesmus dimorphus,* and the significance of these findings to the practice of pond fertilization. *Hydrobiologia* **45**:345–355.

16. Redshaw, C. J., C. F. Mason, C. R. Hayes, and R. D. Roberts. 1990. Factors influencing phosphate exchange across the sediment-water interface of eutrophic reservoirs. *Hydrobiologia* **192**:233–245.

17. Bortleson, G. C., and F. L. Lee. 1974. Phosphorus, iron manganese distribution in sediment cores of six Wisconsin lakes. *Limnol. Oceanogr.* **19**:794–801.

18. Pagenkopf, G. K. *Introduction to Natural Water Chemistry.* 1978. Marcel Dekker, New York.

19. Boyd, C. E., M. E. Tanner, M. Madkour, and K. Masuda. 1994. Chemical characteristics of bottom soils from freshwater and brackishwater aquaculture ponds. *J. World Aquaculture Soc.* **25**:517–534.

20. Cope, J. T., and D. L. Kirkland. *Fertilizer Recommendations and Computer Program Key Used by the Soil Testing Laboratory.* 1975. Circular 176, Alabama Agricultural Experiment Station, Auburn University, Ala.

21. Fleming, J. F., and L. T. Alexander. 1961. Sulfur acidity in South Carolina tidal marsh soils. *Soil Sci. Soc. Amer. Proc.* **25**:94–95.

22. Boyd, C. E., and W. W. Walley. 1972. Studies of the biogeochemistry of boron. I. Concentrations in surface waters, rainfall, and aquatic plants. *Amer. Midland Naturalist* **88**:1–14.

23. Masuda, K., and C. E. Boyd. 1993. Comparative evaluation of the solubility and algal toxicity of copper sulfate and chelated copper. *Aquaculture* **177**:287–302.

4

Exchange of Dissolved Substances between Soil and Water

4.1 Introduction

Pond soil can be a sink or a source of dissolved substances for pond water. An equilibrium exists between the concentration of a substance in the soil and its concentration in the water. If the concentration in the water increases, the soil will adsorb the substance until equilibrium is reestablished. Conversely, if the concentration in water decreases, the soil will desorb the substance until the aqueous concentration is again at equilibrium. In its simplest form, exchange of dissolved substances between soil and water can be expressed as

$$C_{(W)} = C_{(S)} \tag{4.1}$$

and

$$\frac{C_{(S)}}{C_{(W)}} = K_{SD} \tag{4.2}$$

where $C_{(S)}$ = amount of substance adsorbed per unit weight of dry soil (g g^{-1})
$\quad C_{(W)}$ = concentration of substance dissolved in water (g m^{-3})
$\quad K_{SD}$ = soil distribution coefficient (m^3 g^{-1})

Exchange processes are usually more complex than indicated by the soil distribution coefficient equation (4.2). In ponds, they involve inputs and loses of substances, movements of substances within pond water and soil, transfer of substances across the soil–water interface, and uptake or release of substances by the soil (Fig. 4.1).

In this chapter a general discussion of processes influencing the exchange of

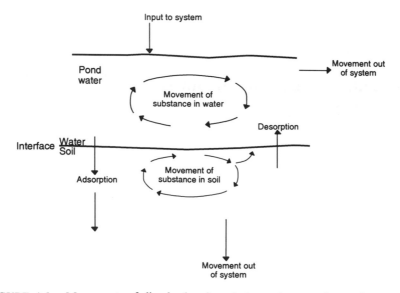

FIGURE 4.1. Movements of dissolved and particulate substances in pond water and soil.

dissolved substances between soil and water is followed by some specific examples of exchange phenomena important in pond management.

4.2 Exchange Between Soil Particles and Water

4.2.1 Adsorption–Desorption

According to Tchobanoglous and Schroeder, adsorption is a three-step process whereby the dissolved substance (1) moves by mass movement of the water (advection) to the soil surface, (2) diffuses to the layer of water next to soil particles, and (3) attaches to or is adsorbed by the soil surface.[1] Several different mechanisms of attraction between the soil surface and solute are possible. All molecules exert mutual attraction through *van der Waals forces,* in which physical adsorption results from electrical attraction between molecules of solid particles and dissolved or dispersed molecules. Negatively charged electrons of one molecule and positively charged nuclei of another molecule attract each other. This attraction is partially neutralized by repulsion of electrons by electrons and nuclei by nuclei, and van der Waals forces are weak. *Hydrogen bonds* occur between polar substances. They are usually illustrated by the attraction between the negatively charged side of one water molecule with the positively charged side of another water molecule (Fig. 4.2). Oxygen atoms, hydroxyl groups, and amino groups on solid particles also can form hydrogen bonds with polar substances and ions in the surrounding solution. In physical adsorption, attraction between

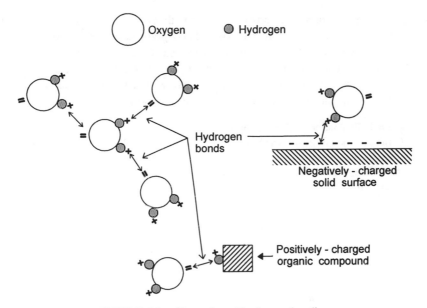

FIGURE 4.2. Examples of hydrogen bonding.

opposite electrical charges extends over a distance, and the adsorbed substance tends to form one or more layers over the adsorbing surface. The adsorbed substance is readily desorbed when its concentration in the surrounding solution decreases. In chemical adsorption, solutes (*adsorbates*) react chemically with solid surfaces of particles that are the *adsorbents*. Chemical forces do not act over a distance, and molecules of dissolved substances and adsorbents must come in contact to form chemical bonds. Substances are adsorbed more strongly by chemical adsorption than by physical adsorption.

A popular way to analyze adsorption data is to plot the mass of substance adsorbed per unit mass of solid versus the equilibrium concentration of the substance in solution on log–log graph paper. The equation for this plot is called the *Freundlich adsorption isotherm:*

$$\frac{X}{m} = K_f C^{1/n} \tag{4.3}$$

or

$$\log \frac{X}{m} = \log K_f + \frac{1}{n} \log C \tag{4.4}$$

where X = amount of a substance adsorbed
m = mass of adsorbent
K_f = Freundlich adsorption coefficient
C = concentration of substance in solution at equilibrium

Values of K_f and n for a given adsorbent vary with temperature and the substance being adsorbed, and they may be extrapolated graphically from a plot of the experimental data. In a log–log plot of X/m versus C, a straight line is obtained if adsorption follows the Freundlich isotherm. The slope of the line is $1/n$, and the ratio of X/m to C at $X/m = 1$ is K_f. Where $n = 1.0$ for the isotherm, the value of K_f is the same as the value of K_{SD} in Equation 4.2.

Example 4.1: Evaluation of Adsorption Data
  Different amounts of soil were placed in 1-L volumes of 50 mg L^{-1} phosphorus solution, and phosphorus concentrations at equilibrium were determined. From these data, values for X and X/m are calculated, X/m plotted versus C, and values obtained for K_f and n.

Solution:

Soil (g)	C (mg L^{-1})	X (mg)	X/m (mg g^{-1})
1	30	20	20
2	22	28	14
4	15	35	8.75
8	9	41	5.12
20	4	46	2.3

  The plot of C versus X/m is provided in Figure 4.3. The slope of the line and the value of C where $X/m = 1.0$ were determined graphically as 2.0 and 1.97, respectively.

$$1/n = \text{slope} = 1.12$$
$$n = 0.893$$

$$K_f = \frac{\left(\dfrac{X}{M}\right)}{C} = \frac{1.0}{2.0} = 0.50$$

  The Freundlich adsorption isotherm reveals that adsorbents are most efficient when there is a high concentration of adsorbate relative to the amount of adsorbent. The efficiency of an adsorbent must be sacrificed to reduce the equilibrium concentration of an adsorbate to a very low concentration. In Example 4.2, when the adsorbent was used at 1 g L^{-1}, each gram of adsorbent removed 20 mg adsorbate, but the equilibrium concentration of adsorbate was 30 mg L^{-1}. To diminish the adsorbate concentration to 4 mg L^{-1} required 20 g adsorbent L^{-1}, and each gram of adsorbent removed only 2.3 mg adsorbate. The ratio of adsorbent to adsorbate is a critical factor determining the extent of adsorption and the equilibrium concentration. When adsorption follows the Freundlich isotherm, the solid surface will not saturate over a wide range of solute concentration, but the efficiency of the adsorbent will differ greatly with solute concentration.

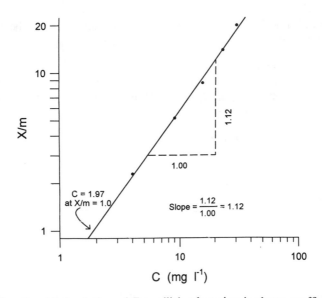

FIGURE 4.3. Graphical solution of Freundlich adsorption isotherm coefficients (see Example 4.1).

The *Langmuir adsorption isotherm* also can be used to evaluate adsorption data. It is similar to the Freundlich adsorption isotherm and has the form

$$X = \frac{X_m bC}{1 + bC} \tag{4.5}$$

or

$$\frac{1}{X} = \frac{1}{X_m} + \frac{1}{X_m bC} \tag{4.6}$$

where X and C are the same as for the Freundlich equation
X_m = maximum amount of a substance that can be adsorbed in a monolayer on the solid surface
b = an empirical coefficient

To evaluate adsorption data by the Langmuir isotherm, we plot $1/X$ versus $1/C$ on arithmetic graph paper. If a straight line is obtained, adsorption is according to the Langmuir isotherm. Adsorption data presented in Example 4.1 do not form a straight line when $1/X$ is plotted versus $1/C$.

Substances held to surfaces by physical attraction desorb to solution when the concentration of the adsorbate drops below the equilibrium concentration. In soils desorption may actually involve the dissolution of specific compounds in substances chemically bonded to the absorbent. Nevertheless, the process can still be dealt with in terms of adsorption isotherms.

Example 4.2: Desorption
A soil contains 0.005 g of substance X per gram. The value of n for the Freundlich isotherm is 1 and $K_f = 0.62$. What will be the equilibrium concentration of substance "X" in soil solution?

Solution: Because $n = 1$, $K_f = K_{SD} = 0.62$.

$$\frac{C_{(s)}}{C_{(w)}} = 0.62 \text{ m}^3 \text{ g}^{-1}$$

$$\frac{0.005 \text{ g g}^{-1}}{C_{(w)}} = 0.62 \text{ m}^3 \text{ g}^{-1}$$

$$C_{(w)} = 0.008 \text{ g m}^{-3}$$

4.2.2 Ion Exchange

Cation and anion exchanges occur in soil, but cation exchange dominates (see Chap. 2). In cation exchange the uptake of one ion species by a soil must be accompanied by the release of a chemically equivalent amount of a different ion species of the same charge. Ion exchange does not directly alter the ionic strength of the soil solution, but it can change the ionic composition. Ions displaced into the soil solution by ion exchange may be leached, whereas those retained on the soil particles are protected from leaching.

Example 4.3: Ion Exchange
A pond was fertilized heavily with a mixed fertilizer that contained potassium. After a time it was found that the exchangeable K^+ concentration in the soil had increased from 10 to 25 ppm. Estimate the exchange of K^+ for other cations in the soil. Assume that exchange has occurred in the upper 20-cm soil layer and that soil bulk density is 1.1 g cm^{-3} (1,100 kg m^{-3}).

Solution:

$$(25 - 10) \text{ ppm K}^+ = 15 \text{ ppm K}^+ = 15 \text{ mg K}^+ \text{ exchanged} \cdot \text{kg soil}^{-1}$$
$$(15 \text{ mg K}^+ \text{ kg}^{-1})(1,100 \text{ kg m}^{-3})(0.2 \text{ m}) = 3,300 \text{ mg K}^+ \text{ m}^{-2}$$
$$3,300 \text{ mg K}^+ \text{ m}^{-2} \div 39.1 \text{ mg K}^+ \text{ meq}^{-1} = 84.1 \text{ meq K}^+ \text{ m}^{-2}$$

Other cation species equivalent to 84.1 meq m^{-2} were displaced from the soil by potassium.

4.2.3 Precipitation–Dissolution

Soils contain many different minerals, and each of these minerals has a characteristic solubility (see Chap. 3). Many solutes in soil solution are in equilibrium with solid forms of minerals, and the minerals control solute concentrations.

Example 4.4: Mineral Control of Solute Concentration
A soil containing 3% gypsum is used to form a 10-cm-deep layer of soil in a tank, and the tank is filled with water containing only a trace of Ca and SO_4. How much Ca will the water contain when equilibrium with the gypsum ($CaSO_4 \cdot 2H_2O$) is attained?

Solution: Gypsum dissolves as follows:

$$CaSO_4 \cdot 2H_2O = Ca^{2+} + SO_4^{2-} + 2H_2O, \qquad K = 10^{-4.63}$$

$$(Ca^{2+})(SO_4^{2-}) = 10^{-4.63}$$

Letting $(Ca^{2+}) = (SO_4^{2-}) = x$ yields

$$(x)(x) = 10^{-4.63}$$

$$x = 10^{-2.315} M = 0.0048\ M\ Ca$$

$$0.0048\ M \times 40.08\ \text{g Ca mole}^{-1} = 0.192\ \text{g Ca L}^{-1} = 192\ \text{mg Ca L}^{-1}$$

Example 4.5: Mineral Control of Solute Concentration
The tank of water in Example 4.4 contains approximately 192 mg Ca L^{-1} and 461 mg SO_4 L^{-1}. If 100 mg L^{-1} of sodium carbonate (Na_2CO_3) are dissolved in the water, what is the approximate concentration of Ca at the new equilibrium?

Solution: Ca and Na_2CO_3 react as follows:

$$Ca^{2+} + SO_4^{2-} + Na_2CO_3 \rightarrow CaCO_3 + 2Na^+ + SO_4^{2-}$$

The amount of Ca that reacts is

$$\frac{X}{Ca} = \frac{100\ \text{mg L}^{-1}}{Na_2CO_3}$$
$$\frac{X}{40.08} = \frac{100\ \text{mg L}^{-1}}{106}$$

$$X = 37.8\ \text{mg L}^{-1}$$

The $CaCO_3$ formed will precipitate, and the new Ca concentration is

$$(192 - 37.8)\ \text{mg L}^{-1} = 154.2\ \text{mg L}^{-1}$$

Calcium carbonate dissolves in water, but at equilibrium with atmospheric carbon dioxide its concentration is about 24 mg L^{-1} of calcium.[2] The $CaCO_3$ precipitate would be practically insoluble in water containing more than 150 mg Ca L^{-1}.

As indicated in Example 4.5, precipitation reactions may occur in the water without involvement of the soil, but the precipitate will settle to the bottom and become part of the soil.

Redox potential and pH have a powerful influence on solubility of compounds, microbial transformations of substances, and the exchange of dissolved sub stances between soil and water. The hydrogen ion concentration is a master variable in soil chemical equilibria, because it is a reactant in many dissolution–precipitation reactions. The redox potential governs the direction that reactions will proceed, and pH affects the value of the redox potential.

4.2.4 Filtration

Soil comprises particles that filter particulate matter from water seeping downward. Fine precipitates that form in pond water or soil solution may remain suspended in the aqueous phase. As the water seeps through the soil, these precipitates are filtered out. A good illustration of this principle involves sand filtration. In fish hatcheries supplied by ground water, water is aerated to increase the dissolved oxygen concentration and to oxidize reduced substances. Water with a high concentration of ferrous iron contains ferric hydroxide particles after aeration. The water is passed through a sand filter to remove the ferric hydroxide before it is used in the hatchery.

4.3 Transport of Dissolved Substances

Pond bottom soil and water do not form a sharp, well-defined interface between soil and water phases. The soil surface is uneven, water enters pore spaces of the soil, and the soil surface may be covered by a flocculent layer up to several centimeters thick.[3] The flocculent layer comprises primarily dead plankton cells, uneaten, decomposing feed, fine soil particles, bacteria, and other microorganisms. It is slightly more viscous than overlaying water, and it is often brown in color. The surface layer of bottom soils has a high water content and behaves as a liquid with a high concentration of particles.[4] In shallow water where light penetrates to the bottom, there may be a well-developed layer of benthic algae. This mat of benthic algae produces oxygen in photosynthesis, so its surface is aerobic, but beneath the surface is normally a layer of decaying algae and anaerobic conditions may exist.

In a pond the surface water is mixed by wind. Surface water is driven against the leeward bank, and a return current flows along the bottom. Water circulation increases as a function of increasing wind velocity, and is enhanced by aerators, water circulators, and water exchange (flushing). Heating and cooling of the water surface causes thermally induced circulation. In the open water, dissolved and particulate substances are transported rapidly within the circulating water by advection. Frictional forces opposing flow are greatest near the soil–water interface, and water velocities decline exponentially in the boundary layer above the bottom. A molecule or ion released from the soil surface must move from the interface into the open water of the pond to be available to phytoplankton. The comparatively stagnant boundary layer between the soil–water interface and the freely circulating water limits the rate of exchange of substances between the soil and water above it. Substances move through the boundary layer by advection, diffusion, and turbation.

Pores exist among soil particles and provide a pathway for water flow. Movement of water across the soil–water interface and downward seepage of pore water is faster in coarse-grained soil than in fine-grained soil, but the processes are comparatively slow in soils of all textures. Considerable time is required for a molecule of oxygen or other substance to move from the soil–water interface to a point 1 cm below the interface. Diffusion plays the major role in the upward transport of substances in pore water. Downward movement of substances is caused by diffusion and seepage.

Almost all pond soils are devoid of oxygen below a few millimeters or centimeters because of impeded water circulation and microbial activity within the soil. Soils with large amounts of organic matter tend to be more highly anaerobic than soils with lower concentrations of organic matter. Maintenance of aerobic conditions at the soil surface is important in ponds. Aerobic conditions at the soil surface provide oxygen for respiration of fish, shrimp, and food organisms; promote microbial degradation of organic matter; and prevent release of toxic, reduced substances such as hydrogen sulfide and nitrite.

4.3.1 Diffusion

Howeler and Bouldin made several models to describe oxygen diffusion into aquatic soils.[5] They found that oxygen consumption could best be described by models that included oxygen consumption by microbial respiration in the aerobic zone and oxidation of Fe^{2+} and other reduced substances. In some cases, up to 50% of the oxygen diffusing into soils was used to oxidize reduced substances. The depth of the aerobic zone was usually 0.04–0.2 cm, and oxygen consumption rates of $(3–32) \times 10^{-6}$ mg O_2 cm^{-3} soil sec^{-1} were recorded. Diffusion of oxygen into the soils from overlaying water was assumed to equal oxygen consumption rate.

Example 4.6: Oxygen Diffusion Rate
  The aerobic zone in a soil is 0.1 cm deep, and the oxygen consumption rate
  is 32×10^{-6} mg O_2 cm^{-3} sec^{-1}. Estimate the oxygen diffusion rate per meter
  square per day.

Solution:

$$(32 \times 10^{-6} \text{ mg O}_2 \text{ cm}^{-3} \text{ sec}^{-1}) (0.1 \text{ cm}) (60 \text{ sec min}^{-1}) (1{,}440 \text{ min day}^{-1})$$
$$(10{,}000 \text{ cm}^2 \text{ m}^{-2}) (10^{-3} \text{ g mg}^{-1}) = 2.76 \text{ g O}_2 \text{ m}^{-2} \text{ day}^{-1}$$

In aquatic soils with high-organic-matter concentrations, the oxygen consumption rate can exceed the diffusion rate and cause the aerobic zone at the soil surface to be lost.

Dissolved substances are exchanged between soil and water in response to concentration gradients. For example, nitrate can be denitrified by soil microbes, and, according to Howeler and Bouldin, the diffusion of nitrate into bottom soil may be described by the equation.[5]

$$\frac{dQ}{dt} = -D\frac{C}{x} \tag{4.7}$$

where dQ/dt = rate of loss of nitrate from overlaying water
        per unit of soil surface area
    D = diffusion coefficient of nitrate in the oxidized soil zone
    C = concentration of nitrate in water at soil–water interface
    x = thickness of the oxidized soil layer

If nitrate concentration is uniform—a common situation in shallow aquaculture ponds—Equation 4.7 may be integrated to give

$$\frac{C}{C_0} = \ln\frac{D}{Bx}t \tag{4.8}$$

where C = concentration of nitrate at the soil–water interface at time t
    C_0 = concentration of nitrate at soil–water interface at time $t = 0$
    B = depth of overlaying water

Any substance consumed by microorganisms or removed from pore water by reactions within pond soil can diffuse into pore water in response to the resulting concentration gradient. Also, if substances are released by microorganisms or soil particles, they can diffuse out of the pond soil and into the pond water in response to the concentration gradient. It is difficult to use expressions such as Equation 4.8 to estimate pond soil influxes and outfluxes because reliable values of the diffusion coefficient (D) are difficult to obtain.

The physical system and the processes described are depicted in Figure 4.4. The rates of movement of substances from one point to another decrease as open water > boundary layer > pore water. Consider a soil particle containing phosphorus resting at the soil surface and at equilibrium with dissolved phosphate in the water volume above. If phytoplankton remove some phosphorus from water near the pond surface, the concentration of phosphate in the water volume will fall below equilibrium. This can result in the release of more phosphate from the soil to reestablish equilibrium (Fig. 4.5). The distance from the point of phosphate release by the soil to the point of phosphate uptake by the plants may be 1 m or more, and movement of phosphate by diffusion only would require a long time. The rate of mixing by processes such as turbulence and advection of water masses regulates the overall exchange mechanism for phosphate and other substances between pond soil and water.

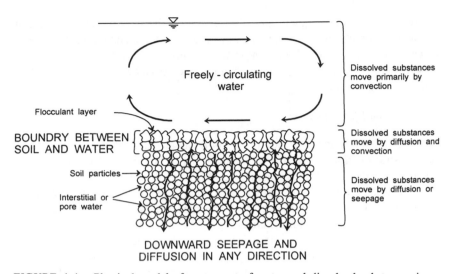

FIGURE 4.4. Physical model of movement of water and dissolved substances in an aquaculture pond.

The exchange of substances is even more difficult when pond soil at some depth below the soil–water interface is considered. Thorough mixing of the pond water volume will not affect water movement in soil below the soil–water interface. Strong water circulation in ponds disturbs the surface layer of bottom soil and suspended small soil particles and organic detritus in the water. Bioturba-

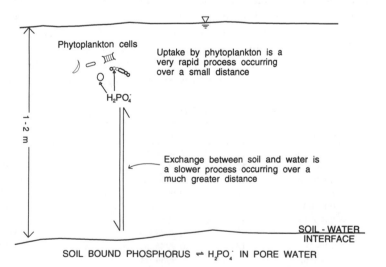

FIGURE 4.5. Illustration of rapid uptake of phosphate by phytoplankton cells and slower exchange of phosphate between pond soil and water.

tion by fish and other aquatic animals feeding on the bottom soil surface and burrowing animals living in the soil can influence the exchange of substances between soil and water. Bioturbation is obviously a greater factor in ponds stocked with bottom-feeding species than in ponds stocked with species feeding in the water column.

4.3.2 Thermal Stratification

Exchange of substances between pond soil and water is further complicated by thermal stratification. Water has a large capacity to hold heat. The specific heat of water is $1 \operatorname{cal} g^{-1} {}^{\circ}C^{-1}$, meaning that 1 calorie is required to raise the temperature of 1 g of water by 1°C. Absorption of solar energy as light passes through pond water heats the water. Light energy is absorbed exponentially with depth, so most heat is absorbed within the upper layer of water. This is particularly true in fish ponds because high concentrations of dissolved organic matter and particulate matter increase the absorption of energy, compared to less turbid water. Transfer of heat from upper to lower layers of water depends largely on mixing by wind.

Density of water depends on water temperature (Table 4.1). Ponds and lakes stratify thermally, because heat is absorbed rapidly in surface waters, and surface waters become warmer than bottom waters. The warm, surface stratum is less dense than the cooler, deeper bottom stratum. Stratification occurs when difference in density between upper and lower strata becomes so great that the two strata cannot be mixed by wind energy. The classical pattern of thermal stratification of lakes in temperate zones is described by Hutchinson[6] and Wetzel.[7] At the spring thaw, or at the end of winter in a lake or pond without ice cover, the water column has a relatively uniform temperature. Heat is absorbed at the surface on sunny days, but there is little resistance to mixing by wind and the entire volume of water circulates and warms. As spring progresses, surface waters heat more rapidly, and heat is transferred from surface waters to deeper waters by mixing. Surface waters finally become so much warmer and lighter than deeper waters

Table 4.1. Density of Water ($g \ cm^{-3}$) at Different Temperatures

°C	$g \ cm^{-3}$	°C	$g \ cm^{-3}$	°C	$g \ cm^{-3}$
0	0.9998679	11	0.9996328	22	0.9977993
1	0.9999267	12	0.9995247	23	0.9975674
2	0.9999679	13	0.9994040	24	0.9973256
3	0.9999922	14	0.9992712	25	0.9970739
4	1.0000000	15	0.9991265	26	0.9968128
5	0.9999919	16	0.9989701	27	0.9965421
6	0.9999681	17	0.9988022	28	0.9962623
7	0.9999295	18	0.9986232	29	0.9959735
8	0.9998762	19	0.9984331	30	0.9956756
9	0.9998088	20	0.9982323		
10	0.9997277	21	0.9980210		

that the wind no longer is powerful enough to mix the entire water body. The upper, warmer, stratum is called the *epilimnion*, and the lower, cooler, stratum is known as the *hypolimnion*. The stratum between the epilimnion and the hypolimnion has a marked temperature differential and is called the *metalimnion*,[7] but the term *thermocline* often is used to describe the middle stratum. A thermocline in lakes is defined as a layer across which temperature drops at a rate of $1°C \ m^{-1}$ or more. The depth of the thermocline below the surface depends on lake size, morphometry, wind exposure, and weather conditions. The thermocline may occur at depths of 20 m or more in large lakes or at 2 m or less in small lakes and ponds. Most large lakes do not destratify until autumn when air temperatures decline and heat is lost from surface water to the air. The difference in density between upper and lower strata decreases as the surface water cools until wind mixing causes the entire volume of water in the lake to circulate and destratify.

Ponds used for aquaculture are shallow, turbid, protected from wind, and have a smaller surface area than large reservoirs and natural lakes. The ordinary warmwater fish pond seldom has an average depth more than 2 m or a surface area more than a few hectares. Thermal stratification can develop even in very shallow ponds because turbid conditions result in rapid heating of surface waters on calm, sunny days. The classical definition of a thermocline is not applicable to ponds, because, even in winter, temperature gradients often exceed $1°C \ m^{-1}$ of depth.[8] During periods of thermal stratification in ponds, the thermocline is easily recognizable as the stratum where temperature changes most rapidly with depth (Fig. 4.6).

Stability of stratification is determined by the amount of energy required to mix the entire volume of a water body to a uniform temperature. The more energy required, the more stable is stratification. Aquaculture ponds are relatively

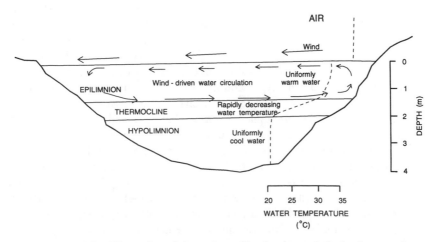

FIGURE 4.6. Illustration of thermal stratification in a relatively deep pond.

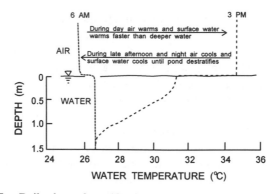

FIGURE 4.7.   Daily thermal stratification and destratification in a shallow pond.

small and shallow, and stratification is not as stable as in larger water bodies. For example, 0.04-ha ponds with average depths of about 1 m and maximum depths of 1.2–1.5 m on the Auburn University Fisheries Research Unit, Auburn, Alabama, stratify thermally during warm days only to destratify at night when upper layers of water cool by conduction (Fig. 4.7). Aquaculture ponds of 1–2 m maximum depth and areas up to 20 ha or more stratify and destratify daily in the same manner. Deeper aquaculture ponds (maximum depths of 3–5 m or more) with average depths of 1.5–2 m and of any surface area may remain stratified during warm months, but certain events may cause sudden destratification. Strong winds may supply enough energy to cause complete circulation, or cold, dense rain falling on the surface may sink through the warm epilimnion and cause upwelling and destratification. Disappearance of a heavy plankton bloom may allow heating to a greater depth and lead to complete mixing.

When ponds stratify for long periods, oxygen depletion occurs in hypolimnia. This results because the hypolimnion does not receive adequate light to stimulate oxygen production by photosynthesis, there is little diffusion and convection of oxygenated water from the epilimnion into the hypolimnion, and organic matter entering the pond or produced in the epilimnion continually settles into the hypolimnion to provide microbial substrate. In the absence of mechanical devices for enhancing the dissolved oxygen concentration, microbial activity in the hypolimnion usually causes oxygen depletion. In shallow ponds where destratification occurs daily, oxygen depletion may not occur, but dissolved oxygen concentrations in the bottom water are lower than in the surface water.

Exchange of bottom water with surface water does not occur during stratification, and dissolved substances in the hypolimnion are prevented from entering epilimnetic water. Some pond managers apply fertilizers by broadcasting them over pond surfaces. Fertilizer granules settle below the thermocline, and fertilizer nutrients stay in the hypolimnion and are unavailable to phytoplankton in the epilimnion. Soils located below the thermocline have no influence on processes occurring in the epilimnion.

4.3.3 Mixing and Destratification

Wind is the primary natural phenomenon causing mixing of pond water. Natural destratification is caused by the interaction of wind and ambient air temperature. Pond water temperatures closely follow air temperatures (Fig. 4.8), so when the air temperature declines, surface water temperature in a stratified pond falls until wind mixing can cause destratification.

Pond morphometry and orientation can be used to enhance wind mixing and reduce the tendency of ponds to stratify. If ponds are not more than 2 m deep, they generally exhibit daytime stratification only. Elimination of deep-water areas in watershed ponds during construction by filling these areas with soil is helpful in reducing depth and preventing stratification.[9] Ponds can be constructed of rectangular shape with long axes oriented in the direction of the prevailing wind to take maximum advantage of wind action.

Mechanically induced mixing of aquaculture ponds is a common practice. Mechanical aerators create water currents in ponds. A popular method of creating circular water movement in square ponds is shown in Figure 4.9. Mechanical devices that provide little aeration but create water currents also are used to induce circulation in ponds.[2] Round ponds have been advocated to enhance water circulation,[10, 11] but as can be seen from superimposing a circle over a square of equal area (Fig. 4.10), a square pond should not resist circular flow much more than a round one. The square pond actually is more similar than the plan view indicates, because the bottom is sloped in the corners of the square pond. A round pond is more expensive and difficult to build than a square one.

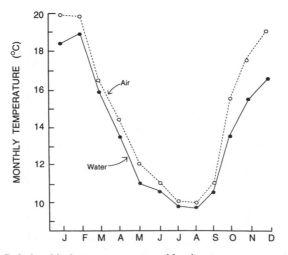

FIGURE 4.8. Relationship between mean monthly air temperatures and surface-water temperatures in ponds in Western Australia. (*Source: data from Morrissy.*[43])

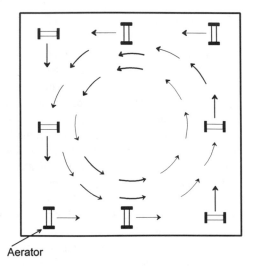

FIGURE 4.9.   Circular water currents created by paddlewheel aerators in a square pond.

Ponds in which aerators, destratifiers, or water circulators are operated have more uniform depth profiles of temperature and dissolved oxygen concentrations than ponds without induced water circulation. Temperature and dissolved oxygen profiles in ponds with and without mechanically induced water circulation are provided in Figure 4.11. In addition to reducing the tendency of stratification, mechanically induced mixing improves water circulation at the pond bottom

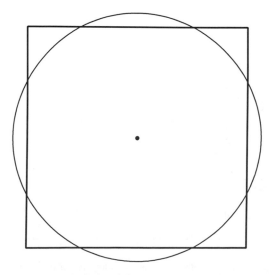

FIGURE 4.10.   Square and circle of the same area superimposed on each other.

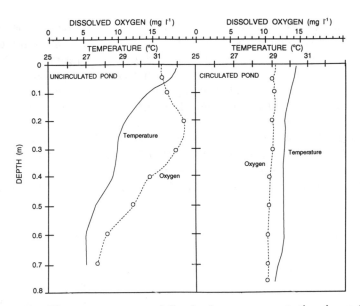

FIGURE 4.11. Water temperatures and dissolved oxygen concentrations in ponds with and without mechanically induced water circulation. (*From Fast et al.*[44])

and enhances exchange of dissolved substances between soil and water. Water circulation eliminates or reduces the thickness of the flocculent layer, but it will not drive water into the soil for the pore spaces are very small. Excessive water circulation can cause erosion of pond soils and create turbidity in the water column.

4.4 Examples of Exchange Phenomena in Ponds

4.4.1 Neutralization of Soil Acidity

Total alkalinity and total hardness of pond water are derived primarily from dissolution of calcium and magnesium carbonates in soil (Eqs. 3.11 and 3.12). Where watershed and pond soils contain appreciable concentrations of carbonates, pond waters have moderate or high concentrations of total alkalinity and total hardness. Acidic soils do not contain carbonates, colloids in acidic soils are base unsaturated, and pond waters in areas with acidic soils usually have low total alkalinity and total hardness.

Agricultural limestone is applied to ponds to neutralize soil acidity and increase total alkalinity and total hardness of the waters. Pond liming provides a good example of the exchange of substances between soil and water. For good fish production, pond waters should have total alkalinity and total hardness values of at least 20 mg L^{-1} equivalent calcium carbonate.[12] If a pond of 1 ha (10,000 m²) area and 1 m deep has a total alkalinity of 10 mg L^{-1}, the application of 10 mg L^{-1} of agricultural limestone ($CaCO_3$) should raise total alkalinity of pond water to 20 mg L^{-1}. This would require 10,000 m³ × 10 g m⁻³ = 100,000 g or 100 kg of $CaCO_3$. Application of this small quantity of limestone to the pond will temporarily increase total alkalinity and total hardness because it dissolves and provides Ca^{2+} and HCO_3^- ions. The HCO_3^- neutralizes H^+ in the acidic soil, and Ca^{2+} displaces Al^{3+} from soil colloids. The Al^{3+} precipitates as aluminum hydroxide. There will be no lasting increase in total alkalinity or total hardness because HCO_3^- is converted to CO_2 and H_2O through reaction with H^+, and the Ca^{2+} is adsorbed by soil colloids. Base unsaturation of the soil will decrease slightly, but 100 kg ha⁻¹ of agricultural limestone is not enough to appreciably lower base unsaturation. The reactions described are illustrated in Figure 4.12.

Total alkalinity and total hardness of water can be increased for several years if enough agricultural limestone is added to the pond, because reactions depicted in Figure 4.12 cause base unsaturation of the soil to decline substantially. The relationship between base unsaturation and total alkalinity is shown in Figure 4.13. Where pond soils are acidic because of exchange acidity, concentrations of total alkalinity and total hardness at equilibrium depend on the degree of base unsaturation of soil colloids. When a base unsaturation of about 0.2 is attained, total alkalinity and total hardness are 20 mg L^{-1} or above. For most pond soils an agricultural limestone application rate of 2,000–6,000 kg ha⁻¹ is necessary to obtain the target total alkalinity and total hardness.

Where soils contain calcium carbonate, they are nearly or completely base saturated, and equilibrium concentrations of total alkalinity and total hardness should be 50–60 mg L^{-1}, but because microbial decomposition of organic matter in pond soils and waters is a source of carbon dioxide, greater concentrations of hardness and alkalinity often occur.[2]

If an acidic pond soil is limed to increase its pH, mineralization of CO_2 and

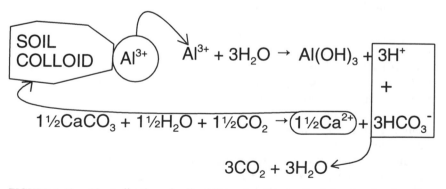

FIGURE 4.12. Neutralization of soil acidity and exchange of calcium ion for aluminum ion on soil colloids after agricultural limestone application.

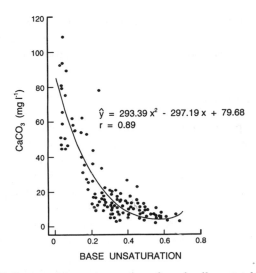

FIGURE 4.13. Influence of base unsaturation of pond soils on total hardness and total alkalinity (mg L^{-1} as $CaCO_3$) in water. (*Source: Boyd.*[12])

nitrification will cause the soil to become acidic again. The two processes cause acidity as follows:

Mineralization of CO_2

$$\text{Organic matter} + O_2 = CO_2 + H_2O + NH_4^+ \qquad (4.9)$$

$$CO_2 + H_2O = H^+ + HCO_3^- \qquad (4.10)$$

Nitrification

$$NH_4^+ + 2O_2 = NO_3^- + 2H^+ + H_2O \qquad (4.11)$$

Hydrogen ion reacts with $Al(OH)_3$ in soil to release Al^{3+}, Al^{3+} replaces Ca^{2+} adsorbed on soil colloids, base unsaturation increases, and soil pH decreases. In the pond water, hydrogen ion neutralizes HCO_3^- and causes alkalinity to decrease.

4.4.2 Phosphorus Dynamics

Many studies have been conducted on the exchange of phosphorus between lake sediment and water, and some studies of phosphorus exchange between pond soil and water have been made. Findings have been remarkably consistent in showing that lake or pond soils adsorb phosphorus strongly, and that the equilibrium concentration of phosphate between soil and water is low. The concentration of soil phosphorus is many times greater than that of dissolved phosphorus. Soils of lakes and ponds are almost always phosphorus sinks, but it is possible to saturate a soil with phosphorus so that its ability to adsorb phosphorus is greatly diminished.[13]

Phosphorus Adsorption

The ability of a pond soil to adsorb phosphorus is illustrated in Figure 4.14. When fertilizer is applied, there is a rapid increase in phosphorus concentration, but it quickly decreases to pretreatment levels. This results from uptake of phosphorus by aquatic plants and adsorption of phosphorus by soil. Adsorption by soil is usually the major pathway of phosphorus loss from pond water.

Phosphorus concentrations were measured in the bottom soil of a 0.04-ha pond on the Auburn University Fisheries Research Unit.[14] Phosphorus concentrations in the surface 5 cm of soil increased from shallow to deep water. Where water was less than 60 cm deep, total soil phosphorus was 100–600 ppm. Soil phosphorus concentrations increased to 1,600 ppm in the deepest part of the pond (120 cm). The correlation coefficient between water depth and total soil phosphorus was

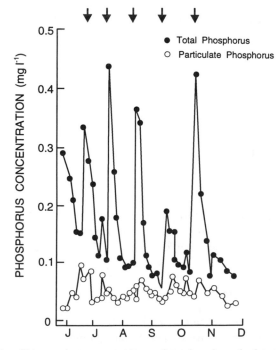

FIGURE 4.14. Changes in concentrations of total and particulate phosphorus after fertilizer applications (arrows) to a fish pond. (*Source: Maskey and Boyd.*[45])

0.73 ($P < 0.01$). It is generally accepted that phosphorus adsorption capacity in soils is associated with the fine particle-size fraction.[15] Sedimentation and resuspension of fine particles is a continual process in a pond bottom, and net movement of particles is in a horizontal direction from shallow to deep water. Draining of ponds and rain falling on bottoms of empty ponds intensifies accumulation of fine, phosphorus-rich particles in deeper-water areas.

The pond referred to had been used in fish culture experiments for 22 years without renovation.[14] Water supply for the pond was essentially free of turbidity, and the small amount of runoff that entered the pond was not turbid. Soft sediment in the pond resulted mainly from continued resuspension, erosion, and sedimentation of pond soil by internal processes. The depth at which sediment interfaced with original, uneroded bottom soil could be visually identified in the cores by a change in color, texture, and density of soil.

Phosphorus concentrations were measured in 5-cm layers from the soil surface to at least 5 cm below the original, undisturbed bottom soil. The highest concentration of phosphorus occurred in the 5–10-cm layer and not at the soil surface (Fig. 4.15). The vertical profile of phosphorus concentration resulted from the balance of downward and upward fluxes of phosphorus. Assuming that pond soil was initially of homogeneous phosphorus concentration, the shape of the vertical profile of sediment phosphorus concentrations should indicate the direction of the phosphorus flux. A net downward flux would produce a profile with a gradual decrease in phosphorus with sediment depth. A net upward flux with considerable release of phosphorus into the water would result in increasing phosphorus concentrations with sediment depth. The depth of the highest phosphorus concentration in a profile indicates the front of phosphorus accumulation. Phosphorus will diffuse into the water above the front and accumulate below the front. If accumulation was simply a result of sedimentation of phosphorus-bearing particles or adsorption, the front should occur at the sediment surface. The profile found in fish pond sediment (Fig. 4.15) shows that the net flux is

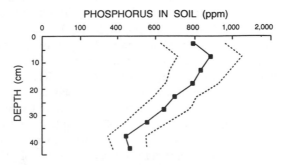

FIGURE 4.15. Changes in phosphorus concentration with depth in a pond soil. Dotted lines show 95% confidence limits. (*Source: Masuda and Boyd.*[14])

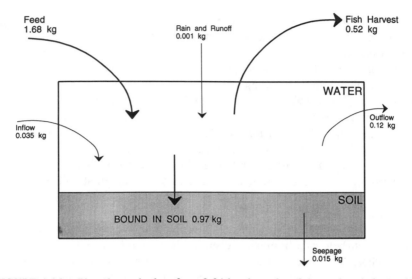

FIGURE 4.16. Phosphorus budget for a 0.04-ha channel catfish pond at Auburn, Alabama. (*Source: Boyd.*[16])

downward and that the sediment is a phosphorus sink. The presence of an accumulation front at 5–10-cm depth suggests that there is diffusion of phosphorus into the water but that the net flux is downward.

An annual phosphorus budget (Fig. 4.16) had been made for the pond mentioned above.[16] The annual input of phosphorus to the sediment was 0.97 kg. Over a 22-year period, annual inputs of phosphorus to the pond in feeds and fertilizers have been roughly equal, so the total input of phosphorus to the bottom soil was about 21.3 kg (0.97 kg $yr^{-1} \times$ 22 yr). The total amount of phosphorus that accumulated in the pond soil (sediment phosphorus concentration minus phosphorus concentration in original soil) was 24.4 kg and agreed well with the estimated input of 21.3 kg. The ability of pond soil to adsorb further phosphorus was estimated from phosphorus adsorption capacity (PAC) and net phosphorus sorption (NPS) tests. The PAC test employs a 50-mg L^{-1} phosphorus solution, and because of this high concentration it probably overestimates the true ability of a soil to adsorb phosphorus under natural conditions. Values of PAC for samples from the pond ranged from 480 to 500 ppm. The NPS, which may underestimate the true ability of a soil to adsorb phosphorus, was 140–200 ppm.

The amount of phosphorus in original bottom soil material, the amount accumulated in the soft sediment during 22 years of fish culture, and the remaining capacity of soft sediment to adsorb phosphorus is provided in Figure 4.17. Based upon NPS, the bottom could adsorb an additional 54 g m^{-2} of phosphorus, but PAC suggests that 200 g m^{-2} more phosphorus could be adsorbed. It is not known which amount is nearest to the true capacity of the pond sediment to remove phosphorus, but it still has a great capacity to adsorb phosphorus.

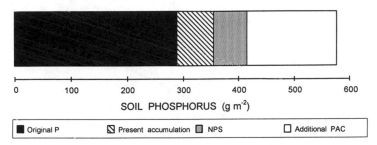

FIGURE 4.17. Original phosphorus, phosphorus accumulation over a 22-year period, and present capacity to adsorb phosphorus in a fish pond soil. Present capacity to adsorb phosphorus was measured by net phosphorus sorption (NPS) and phosphorus adsorption capacity (PAC) techniques. (*Source: Masuda and Boyd.*[14])

Phosphorus uptakes in laboratory soil–water systems containing soils from five physiographic regions in Alabama were studied. Soils differed greatly in composition as follows: pH 4.4–7.3; clay content, 1.7–44.7%; organic matter, 0.9–15%; dilute acid-extractable phosphorus, 1–52 ppm; calcium, 80–5,920 ppm. Nevertheless, all samples removed phosphorus at roughly equal rates (Fig. 4.18). Similar rates of phosphorus loss were observed in ponds treated with phosphate fertilizer (Fig. 4.19). These findings agree with the common experience that pond soils are sinks for added phosphorus.

In pond waters with high concentrations of Ca^{2+}, phosphorus may precipitate from water without direct involvement of soil.[18] Assuming hydroxyapatite as the controlling phosphate mineral,[19] we write

$$Ca_5(PO_4)_3OH \text{ (s)} + 7H^+ = 5Ca^{2+} + 3H_2PO_4^- + H_2O \qquad (4.12)$$

Concentrations of dissolved phosphate were computed for $Ca^{2+} = 20$ mg L^{-1} and different pH values. Concentrations of dissolved phosphate declined from near 0.05 mg L^{-1} at pH 7.5 to less than 0.00001 mg L^{-1} at pH 10. These calculated phosphate concentrations are similar to soluble reactive phosphate concentrations observed in fish ponds. Concentrations of soluble reactive phosphate may be as high as 0.25–0.5 mg L^{-1} after fertilizer applications, but they rapidly decline. The majority of phosphorus in pond water is adsorbed onto suspended particles or contained in plankton.

Avnimelech used the reaction between $Ca(OH)_2$ and H_3PO_4 to illustrate the importance of Ca^{2+} and pH on equilibrium concentrations of dissolved phosphate in ponds:[20]

$$nCa(OH)_2 + mH_3PO_4 = Ca_n(PO_4)m \qquad (4.13)$$

He regressed measured concentrations of $-\log[Ca(OH)_2]$ (*y* variable) versus $-\log(H_3PO_4)$ and obtained the following equation:

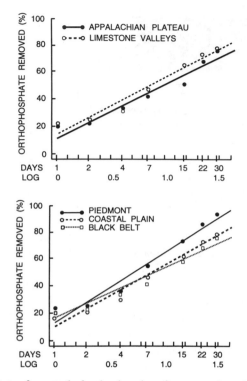

FIGURE 4.18. Rate of removal of orthophosphate from water by soils from five physiographic regions in Alabama in laboratory soil–water systems. (*Source: Boyd and Musig.*[17])

$$-\log[Ca(OH)_2] = 4.76 - 0.09687 \log (H_3PO_4) \qquad (4.14)$$

Avnimelech concluded that a state of equilibrium exists between a solid calcium phosphate phase and its soluble components, but this equilibrium is not always obvious because of daily changes in pH and frequent applications of phosphate in fertilizers and feeds.

Phosphorus Release

High concentrations of phosphorus occur in pore waters of bottom soils. Pore water in fish pond soils at Auburn, Alabama, contained 0.08–0.62 mg L^{-1} total dissolved phosphate, and almost all the dissolved phosphorus was soluble reactive phosphorus.[21] Concentrations of phosphorus in pore water of pond soils are not especially high when compared with concentrations of 0.1–10 mg L^{-1} reported for pore water from eutrophic lake sediment.[22] However, concentrations of total dissolved phosphorus in pore waters of pond soils were 10–50 times greater than those measured in overlaying pond waters. This suggests a large concentration

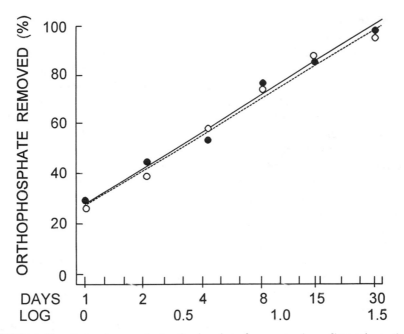

FIGURE 4.19. Rate of removal of orthophosphate from water by sediment in earthen ponds at Auburn, Alabama.

gradient that should encourage rapid diffusion of phosphorus from soil to pond water. Pore water in aquaculture ponds was highly reduced (average redox potential = −120 mV), and the high concentrations of dissolved phosphorus resulted from conversion of insoluble iron(III) phosphates to soluble iron(II) phosphates under anaerobic conditions.[21] Ferrous iron concentrations in pore water were usually around 20 mg L^{-1}. When phosphate and ferrous ions diffused into the aerobic zone at the mud–water interface, they quickly precipitated as iron(III) phosphate compounds and did not diffuse into the pond water.

Equilibrium phosphorus concentrations (EPC) for samples of soil from 385 ponds in several countries were determined. Values ranged from less than 0.01 mg L^{-1} to more than 4 mg L^{-1}, but most samples had EPC values below 1.0 mg L^{-1}. The average EPC was 0.76 mg L^{-1}. Soluble reactive phosphorus concentrations in ponds are seldom above 0.1 mg L^{-1}. In determining EPC, phosphorus equilibrium was forced by shaking, and the ratio of soil mass to water volume was much greater in the EPC test than the ratio of the mass of the thin, aerobic layer of surface soil participating in phosphorus exchange with overlaying water to the pond volume. The EPC provides a good index of relative capacities of different soils to release phosphorus. The EPC was positively correlated with increasing dilute acid-extractable phosphorus concentration and increasing soil pH. This suggests that soils containing more phosphorus tend to release more phosphorus to the water and that increasing soil pH favors release of phosphorus from soil to water.

Algal growth in cultures where soil was the only source of phosphorus was positively correlated with the concentration of phosphorus in soil added to the cultures.[23] When algae cultures were dependent on acidic soils for phosphorus, growth could be increased by liming soils to increase their pH.[24] Also, phytoplankton growth in ponds with acidic soils was increased by applying agricultural limestone to ponds.[24]

Aeration of fish ponds improves water circulation across the pond bottom and causes resuspension of sediment.[25] Aerated pond waters often have higher concentrations of total phosphorus than unaerated pond waters, so aeration is often thought to increase phosphorus release from the soil. To test this idea, the bottom of a 0.1-ha pond was manually disturbed.[26] Resuspension of sediment increased turbidity from 8 to 138 NTU (Fig. 4.20). There was almost no change in soluble reactive phosphorus concentration after resuspending sediment, but the total phosphorus concentration clearly increased (Fig. 4.20). Total phosphorus concentrations declined as soil particles precipitated and turbidity values fell. Soluble reactive phosphorus concentrations did not change appreciably as the solids settled. This suggests that resuspension of sediment by wind action, aeration, water circulation, and bioturbation does not increase availability of soil phosphorus. In fact, sediment resuspension may hasten removal of phosphorus from water by increasing contact between suspended soil particles and soluble reactive phosphorus.

Laboratory soil–water systems incubated in the dark were treated with 10 mg L^{-1} sodium acetate, 5 mg L^{-1} sodium nitrate, or sodium acetate plus sodium nitrate at three- to four-day intervals.[26] Nitrate provided a source of oxygen to microorganisms when dissolved oxygen concentrations are low[27] and poised the

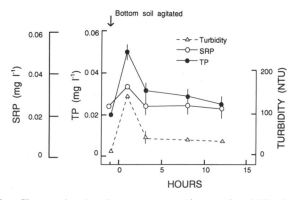

FIGURE 4.20. Changes in phosphorus concentrations and turbidity in a pond after agitating bottom mud manually for 1 h.

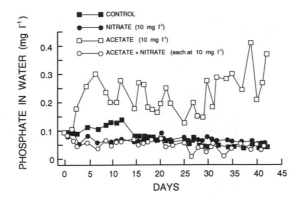

FIGURE 4.21. Phosphorus concentrations in water of laboratory soil–water systems treated with sodium acetate, sodium nitrate, sodium acetate plus sodium nitrate, and in controls.

redox potential so that it did not fall low enough to reduce iron(III) to iron(II). Less phosphorus was released from nitrate-treated systems than from the control (Fig. 4.21). Acetate treatment caused anaerobic conditions at the soil–water interface, and large amounts of phosphorus were released. Application of nitrate to acetate-treated systems essentially stopped phosphate release from soil (Fig. 4.21). Application of acetate to ponds also caused an increase in phosphorus release from soil. Phosphorus is released from soil when the oxidized layer at the soil–water interface is destroyed by rapid microbial activity. This permits pore water phosphorus to diffuse into the overlaying water. The development of anaerobic conditions on the pond bottom and in the water is harmful to fish and other aquatic organisms, so this phenomenon is of no value in recovering soil phosphorus. This phosphorus release mechanism possibly stimulates phytoplankton activity after episodes of oxygen depletion in ponds and favors rapid return of normal dissolved oxygen concentrations.

When the concentration of phosphorus in pond water falls below the equilibrium concentration established by the solubility of soil phosphorus compounds, phosphorus is released into the water from the aerobic layer of surface soil. In acidic soils, iron and aluminum phosphates control phosphorus solubility, and in neutral to alkaline soil calcium phosphates govern phosphorus solubility. Hepher demonstrated that the daily adsorption of phosphorus from pond water by phytoplankton in aquaculture ponds was 4.5–5.5 times greater than the daily influx of phosphorus from soil.[28] To maintain good phosphorus levels of phytoplankton productivity in fertilized ponds, one must add fertilizers at frequent intervals. In ponds with feeding, sorption of phosphorus by the soil is beneficial because removal of soluble reactive phosphorus from the water helps prevent excessive phytoplankton growth.

Table 4.2   Distribution of Soil and Water Phosphorus within Different Pools and Fractions for a Fish Pond at Auburn, Alabama

Phosphorus Pool	Phosphorus Fraction	Amount (g m^{-2})	(%)
Pond water[a]	Total phosphorus	0.252	0.19
	Soluble reactive phosphorus	0.019	0.01
	Soluble nonreactive phosphorus	0.026	0.02
	Particulate phosphorus	0.207	0.16
Soil[b,c]	Total phosphorus	132.35	99.81
	Loosely bound phosphorus	1.28	0.96
	Calcium-bound phosphorus	0.26	0.20
	Iron- and aluminum-bound phosphorus	17.30	13.05
	Residual phosphorus	113.51	85.60
Pond	Total phosphorus	132.60	100.0

Source: Masuda and Boyd[14]

[a]Average pond depth = 1.0 m.

[b]Soil Depth = 0.2 m.

[c]Soil bulk density = 0.797 g cm^{-3}.

Phosphorus Fractions

Sizes of soil and water phosphorus pools and amounts of phosphorus contained in different fractions of each pool are illustrated in Table 4.2. The soft soil layer in the pond averaged 36.8 cm deep, but only the upper 20-cm layer was used in these calculations. Soil in the upper 20-cm layer had an average bulk density of 797 kg m^{-3} and a total phosphorus concentration of 832 ppm. Residual phosphorus was computed by subtracting extractable phosphorus from total phosphorus. Residual phosphorus is not soluble in sequential extraction, and it is the most unavailable form of soil phosphorus.

Examination of Table 4.2 reveals that 99.81% of the phosphorus in the pond soil–water system was bound in soil. Extractable phosphorus in the soil accounted for only 14.18% of the pond phosphorus, and loosely bound soil phosphorus was only 0.96% of pond phosphorus. Most of the phosphorus in the water was particulate phosphorus; only 0.01% of the pond phosphorus was soluble reactive phosphorus.

The phosphorus pool in fish is not included in Table 4.2. Boyd reported that 520 g phosphorus was removed in fish when 410-m^2 channel catfish ponds at Auburn University Fisheries Research Unit were harvested.[16] Phosphorus in channel catfish harvests contained five times as much phosphorus as the pond water but only 0.01 times as much phosphorus as the soil. Most of the phosphorus added to ponds obviously ends up in the soil in highly insoluble form.

4.4.3 Inorganic Nitrogen

Ammonia N mineralized from soil organic matter dissolves in pore water. Klapwijk and Snodgrass found 10–50 mg L^{-1} ammonia N in pore water and 0.5–

5.0 mg L^{-1} ammonia N in the water column of Lake Ontario.[29] Pore water in a Canadian lake contained two to three times more ammonia N than water above the soil.[30] Schroeder reported 0.5 mg L^{-1} and 10 mg L^{-1} ammonia N, respectively, in the water column and pore water of aquaculture ponds in Israel.[31] Pore water in channel catfish ponds in Alabama had 25–75 times more ammonia N in pore water than in surface water.[21] Concentrations of ammonia N in pore water increased over time as fish grew and feeding rates increased.

Diab and Shilo found that the concentration of ammonia N adsorbed by pond soils was greater than the pore water concentration.[22] The amount of ammonia N in pond soils increased with time as fish grew and feed inputs increased. Ammonia N concentration also increased with soil depth. Reddy et al. also found ammonia N stratification in aquatic soil; concentration in pore water were 3 mg L^{-1} at 5-cm depth and 30–35 mg L^{-1} at 35-cm depth.[33] The lower ammonia N concentrations in the upper soil layer results from diffusion of ammonia into the water column and nitrification.[32]

Reddy et al. reported that 25% of ammonia N in soil of microcosms consisting of lake water, soil, and floating aquatic vegetation diffused into the water over a 183-day period and was absorbed by the plants.[33] The daily ammonia N flux averaged 0.048 g m^{-2}. The flux of ammonia N from soil to water in intensive polyculture ponds in Israel was 0.03 g N m^{-2} day^{-1}.[31] This was about 10% of the rate at which organic N was delivered to the pond bottom by sedimentation of dead algal cells. Blackburn et al. found that the nitrogen mineralization rate in soil of a marine fish pond was 0.25 g m^{-2} day^{-1}, and the ammonia N flux rate was 0.12 g m^{-2} day^{-1}.[34] They estimated that 70% of the flux was caused by diffusion and the rest resulted from bioturbation.

Nitrification in the surface layer of aquatic soils provides a concentration gradient for ammonia N to diffuse upward into the aerobic zone.[35] Nitrate produced by nitrification in the aerobic layer seeps downward into the anaerobic zone, where it is denitrified to N_2 and N_2O. Bouldin et al. reported that 7–15% of nitrate in soils of small ponds was lost daily through denitrification.[36] From 20% to 25% of the fertilizer nitrogen applied to rice fields was not recovered in soil, water, or plants, and it was probably denitrified.[37] The unaccounted nitrogen represented a loss of 0.02–0.025 g N m^{-2} day^{-1}. Boyd found that 43% of the nitrogen added to channel catfish ponds in feed could be recovered in water, soil, and fish.[6] The remainder of added nitrogen apparently was denitrified, at an average daily rate of 0.08 g N m^{-2} day^{-1}. Denitrification rates up to 0.1 g N m^{-2} day^{-1} were reported for highly eutrophic salt marshes.[38]

4.4.4 Copper Adsorption

The concentration of Cu^{2+} ion in water is governed by solubility of $Cu_2(OH)_2CO_3$ at pH < 7 and by CuO at pH > 7. Although copper ions form complexes with inorganic and organic ligands and dissolved copper (Cu_D) concentrations are much greater than those of Cu^{2+}, Cu_D is only a few micrograms per liter in most waters. High concentrations of Cu^{2+} and Cu_D following application of copper sulfate or chelated copper compounds to pond waters for phytoplankton control or fish disease treatment quickly decline.

The uptake of copper from copper sulfate and a chelated copper algicide by three soils was studied in laboratory soil–water systems.[39] The three soils were quite different in composition; they will be referred to as high-organic-matter clay, low-organic-matter clay, and sand. The CEC was greatest for the high-organic-matter clay and lowest for the sand.

Data on copper loss from water in laboratory soil–water systems over a 28-day period are provided in Figure 4.22. The fraction lost from the water represents copper that precipitated to bottoms of containers or was removed by soil. It will be termed Cu_S. Changes in Cu_D and Cu_T (dissolved copper plus particulate copper) concentrations also are provided. The shaded areas in Figure 4.22 represent the particulate copper (Cu_P) concentration. Particulate copper was solid copper compounds that formed in solution but did not precipitate; Cu_D consisted of Cu^{2+} and soluble copper complexes. Concentrations of Cu_D reported in Figure 4.22 exceeded concentrations expected for true equilibrium conditions. Centrifugation did not completely separate Cu_P and Cu_D, and it is doubtful that equilibrium was achieved between Cu^{2+} and stable mineral forms.

Copper was lost from water in all treatments through precipitation of copper compounds and by direct adsorption of copper by soil. Soil–water systems were incubated in the dark to prevent copper removal by algae. Physical adsorption occurs through van der Waals attraction at any surface between two phases, but van der Waals attractions are weak. Soil has a capacity to exchange adsorbed cations with dissolved cations, and metals can be tightly held to soil particles by cation exchange. Organic matter in soil can attract metal ions by van der Waals attractions and cation exchange, but it also can react chemically with Cu^{2+} and other metals to bind them tightly. Bacteria in soil can absorb and accumulate metals, and the abundance of bacteria in soil normally increases as the organic matter concentration increases.

Because there are several possible mechanisms of copper uptake by soil and it is not clear which mechanism is most important, copper uptake by soil will be referred to as copper adsorption. Data on concentration of copper adsorbed from water (Y axis) were regressed against time (X axis), and regression equations were expressed as the integrated form of Langumir's adsorption isotherm:[40]

$$Cu_S = V_e[1 - \exp(-aT)] \tag{4.15}$$

where Cu_S = concentration of copper lost from water (mg L^{-1})
V_e = equilibrium concentration (mg L^{-1})
a = coefficient
T = time (days)

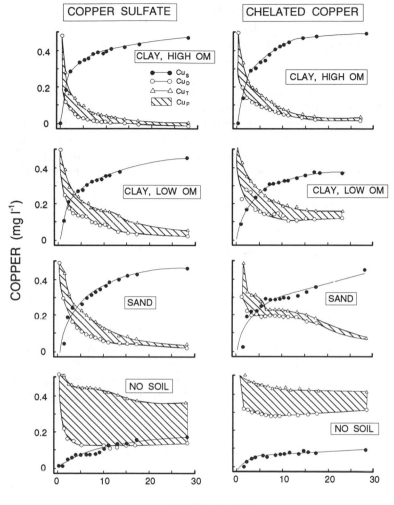

FIGURE 4.22. Concentrations of copper removed from the water (Cu$_S$), dissolved copper (Cu$_D$), and total copper remaining in the water (Cu$_T$) for laboratory soil–water systems and controls (no soil). The shaded areas between Cu$_T$ and Cu$_D$ lines represent particulate copper (Cu$_P$). Copper was applied at 0.5 mg Cu L^{-1} with chelated copper algicide and copper sulfate. (*Source: Masuda and Boyd.*[39])

In the adsorption isotherm equation V_e is the equilibrium concentration of Cu_S reached in each test. The larger the value of the coefficient a, the faster soil adsorbed copper. Values of V_e and a for the plots of Cu_S versus time (Fig. 4.22) are provided in Table 4.3). Coefficients of correlation (r^2) for all regressions were 0.95 or larger.

No adsorption of copper actually occurred in the no-soil treatment. Loss of copper from water resulted from precipitation of solid copper compounds on the bottoms of the containers. The amount of copper lost from the water in all treatments with soil was much greater than observed in the controls. If systems had been maintained longer, more of the Cu_P fraction would have precipitated. Smaller amounts of Cu_P in systems containing soil suggests that soil facilitates precipitation processes through unexplained factors.

The high-organic-matter clay had the greatest power to adsorb copper (Fig. 4.22; Table 4.3). This soil contained much clay and provided a large surface area for adsorption; it also had a higher cation-exchange capacity and larger organic-matter concentration than the other soils. All three of these factors favor copper adsorption. The low-organic-matter clay adsorbed copper faster than the sand, but equilibrium concentrations (V_e) were about the same. The low-organic-matter clay had a greater CEC than the sand, but the sand contained more organic matter. Because of its great copper-adsorbing power, V_e values did not differ between copper sulfate and chelated copper treatments of the high-organic-matter clay. The lower value of coefficient a for the chelated copper treatment revealed that copper from copper sulfate treatment was adsorbed more rapidly. In the other two soils and in the controls, Cu_S concentration was greater in the copper sulfate treatments (V_e values were larger) than in the chelated copper treatments.

Table 4.3   Values of V_e and a for the Equation $Cu_S = V_e [(1 - exp(-aT)]$ for the Regression of Copper Removed from Solution (Cu_s) versus Time (T)

Treatment	V_e [a]	a [a]	r^2
High-organic-matter-content clay			
Copper sulfate	0.456	0.506*	0.99
Chelated copper	0.458	0.326*	0.99
Low-organic-matter-content clay			
Copper sulfate	0.404*	0.274	0.99
Chelated copper	0.341*	0.278	0.99
Sand			
Copper sulfate	0.441*	0.208	0.99
Chelated copper	0.339*	0.240	0.99
No soil			
Copper sulfate	0.159*	0.089	0.96
Chelated copper	0.080*	0.019	0.95

Source: Masuda and Boyd.[39]

[a]Where a pair of V_e or a values has a asterisk, the members of the pair are different ($P < 0.05$).

However, the magnitude of a for copper sulfate and chelated copper did not differ for the low-organic-matter clay, sand, or controls (Table 4.3).

Results of algal toxicity tests suggested that copper sulfate or chelated copper applied at 0.1 mg Cu L^{-1} greatly inhibited algae. The regression equation may be used to estimate Cu_T concentration at different times after algicide treatment.

Example 4.7: Estimation of Copper Concentration

How long would it take the Cu_T concentration to fall to 0.1 mg Cu L^{-1} after applications of either copper sulfate or chelated copper at 5 mg Cu L^{-1}?

Solution: Equation 4.13 must be rearranged as follows:

$$Cu_S = V_e [1 - \exp(-aT)]$$

$$= V_e - V_e \exp(-aT)$$

$$\exp(-aT) = 1 - \frac{Cu_s}{V_e}$$

$$-aT = \ln\left(1 - \frac{Cu_s}{V_e}\right)$$

$$T = \left[\ln\left(1 - \frac{Cu_s}{V_e}\right)\right] - A$$

A concentration of 0.1 mg Cu L^{-1} in water corresponds to Cu_S of 0.4 mg L^{-1}. Using values of V_e and a for the high-organic-matter clay (Table 4.3), we have

$$T_{\text{copper sulfate}} = \frac{\ln(1 - 0.4/0.456)}{-0.506}$$

$$= 4.31 \text{ days}$$

and

$$T_{\text{chelated copper}} = \frac{\ln(1 - 0.4/0.458)}{-0.326}$$

$$= 6.33 \text{ days}$$

Therefore, it would take 1.47 times as long for a 0.5-mg Cu L^{-1} application from chelated copper to fall below 0.1 mg Cu L^{-1} as an equal application of copper from copper sulfate.

The study was conducted in the laboratory, and copper concentrations in ponds decline faster than in laboratory studies. In ponds, plants and bacteria in the water remove copper,[41] bacteria degrade the organic ligand of chelated algicides, and conditions for attaining equilibrium between Cu^{2+} and mineral forms are probably better. Button et al. found that 95% of the copper sulfate distributed over a lake surface dissolved in the surface 1.75 m, but Cu_T concentrations fell to pretreatment level within 24 h.[42]

Even though copper is lost from the water faster in ponds than in the laboratory studies described here, the same factors govern copper solubility in laboratory soil–water systems and in ponds. There appear to be two advantages of chelated copper algicide over copper sulfate. At the same copper dose, chelated copper algicide provides a higher concentration of copper in the water, and the loss of copper from the water to the pond soil is slightly slower with chelated copper algicide. It seems doubtful that these two advantages of chelated copper algicides compensate for their greater cost.

References

1. Tchobanoglous, G., and E. Schroeder. *Water Quality*. 1987. Addison-Wesley, Reading, Mass.

2. Boyd, C. E. *Water Quality in Ponds for Aquaculture*. 1990. Alabama Agricultural Experiment Station, Auburn University, Ala.

3. Delincé, G. *The Ecology of the Fish Pond Ecosystem*. 1992. Kluwer Academic Publishers, Dordrecht, The Netherlands.

4. Davison, W. 1982. Transport of iron and manganese in relation to the shape of their concentration depth profiles. *Hydrobiologia* **91**:463–471.

5. Howeler, R. H., and D. R. Bouldin. 1971. The diffusion and consumption of oxygen in submerged soils. *Soil Sci. Soc. Amer. Proc.* **35**:202–208.

6. Hutchinson, G. E. *A Treatise on Limnology,* Volume I., *Geography, Physics, and Chemistry*. 1957. John Wiley & Sons, New York.

7. Wetzel, R. G. *Limnology*. 1975. W. B. Saunders, Philadelphia.

8. Parks, R. W., E. Scarsbrook, and C. E. Boyd. *Phytoplankton and Water Quality in a Fertilized Fish Pond*. 1975. Circular 224, Alabama Agricultural Experiment Station, Auburn University, Ala.

9. Boyd, C. E. 1985. Hydrology and pond construction. In *Channel Catfish Culture*, C. S. Tucker, ed., pp. 107–133. Elsevier, The Netherlands.

10. Wyban, J. A., and J. N. Sweeny. 1989. Intensive shrimp growout trials in a round pond. *Aquaculture* **76**:215–225.

11. Wyban, J. A., J. N. Sweeny, and R. A. Kanna. 1988. Shrimp yields and economic potential of intensive round pond systems. *J. World Aquacultural Soc.* **19**:210–217.

12. Boyd, C. E. *Lime Requirements of Alabama Fish Ponds*. 1974. Bulletin 459, Alabama Agricultural Experiment Station, Auburn University, Ala.

13. Eren, Y., T. Tsur, and Y. Avnimelech. 1977. Phosphorus fertilization of fishponds in the Upper Galilee. *Bamidgeh* **29**:87–92.

14. Masuda, K., and C. E. Boyd. 1994. Phosphorus fractions in soil and water of aquaculture ponds built on clayey Ultisols at Auburn, Alabama. *J. World Aquaculture Soc.* **25**:379–395.

15. Holtan, H., L. Kamp-Nielsen, and A. O. Stuanes. 1988. Phosphorus in soils, water and sediment: An overview. *Hydrobiologia* **170**:19–34.

16. Boyd, C. E. 1985. Chemical budgets for channel catfish ponds. *Trans. Amer. Fish. Soc.* **114**:291–298.

17. Boyd, C. E., and Y. Musig. 1981. Orthophosphate uptake by phytoplankton and sediment. *Aquaculture* **22**:165–173.

18. Hepher, B. 1958. On the dynamics of phosphorus added to fishponds in Israel. *Oceanography* **3**:84–100.

19. Lindsay, W. L. *Chemical Equilibria in Soils.* 1979. John Wiley & Sons, New York.

20. Avnimelech, Y. 1975. Phosphate equilibrium in fish ponds. *Verh. Internat. Verein. Limnol.* **19**:2305–2308.

21. Masuda, K., and C. E. Boyd. 1994. Chemistry of sediment pore water in aquaculture ponds built on clayey, Ultisols at Auburn, Alabama. *J. World Aquaculture Soc.* **25**:396–404.

22. Enell, M., and S. Löfgren. 1988. Phosphorus in interstitial water: Methods and dynamics. *Hydrobiologia* **170**:103–132.

23. Chiou, C., and C. E. Boyd. 1974. The utilization of phosphorus from muds by the phytoplankter, *Scenedesmus dimorphus,* and the significance of these findings to the practice of pond fertilization. *Hydrobiologia* **45**:345–355.

24. Boyd, C. E., and E. Scarsbrook. 1974. Effects of agricultural limestone on phytoplankton communities of fish ponds. *Arch. Hydrobiol.* **74**:336–349.

25. Thomforde, H., and C. E. Boyd. 1991. Effects of aeration on water quality and channel catfish production. *Bamidgeh* **43**:3–26.

26. Masuda, K., and C E. Boyd. 1994. Effects of aeration, alum treatment, liming, and organic matter application on phosphorus exchange between soil and water in aquaculture ponds at Auburn, Alabama. *J. World Aquaculture Soc.* **25**:405–416.

27. Avnimelech, Y., and G. Zohar. 1986. The effect of local anaerobic conditions on growth retardation in aquaculture systems. *Aquaculture* **58**:167–174.

28. Hepher, B. 1966. Some aspects of the phosphorus cycle in fishponds. *Verh. Internat. Verein. Limnol.* **16**:1293–1297.

29. Klapwijk, A., and W. J. Snodgrass. 1982. Experimental measurement of sediment nitrification and denitrification in Hamilton Harbour, Canada, *Hydrobiologia* **91**:207–216.

30. Schindler, D. W., R. H. Hesslein, and M. A. Turner. 1987. Exchange of nutrients between sediments and water after 15 years of experimental eutrophication. *Can. J. Fish. Aquatic Sci.* **44**:26–33.

31. Schroeder, G. L. 1987. Carbon and nitrogen budgets in manured fish ponds on Israel's coastal plain. *Aquaculture* **62**:559–579.

32. Diab, S., and M. Shilo. 1986. Transformation of nitrogen in sediments of fish ponds in Israel. *Bamidgeh* **38**:67–88.

33. Reddy, K. R., R. E. Jessup, and P. S. C. Rao. 1988. Nitrogen dynamics in a eutrophic lake sediment. *Hydrobiologia* **159**:177–188.

34. Blackburn, T. H., B. Lund, and M. Krom. 1988. C- and N-mineralization in the sediments of earthen marine fishponds. *Marine Ecol. Prog. Ser.* **44**:221–227.

35. Patrick, Jr., W., and K. Reddy. 1976. Nitrification-denitrification reactions in flooded soils and water bottoms: Dependence on oxygen supply and ammonium diffusion. *J. Environ. Qual.* **5**:469–472.

36. Bouldin, D., R. Johnson, C. Burda, and C. Kao. 1974. Losses of inorganic nitrogen from aquatic systems. *J. Environ. Qual.* **3**:107–114.

37. Patrick, Jr., W., and K. Reddy. 1976. Fate of fertilizer nitrogen in a flooded rice soil. *J. Soil Sci. Soc. Amer.* **40**:678–681.

38. Valiela, I., and J. Teal. 1979. The nitrogen budget of a salt marsh ecosystem. *Nature* **280**:652–656.

39. Masuda, K., and C. E. Boyd. 1993. Comparative evaluation of the solubility and algal toxicity of copper sulfate and chelated copper. *Aquaculture* **117**:287–302.

40. Pagenkopf, G. K. *Introduction to Natural Water Chemistry*. 1978. Marcel Dekker, New York.

41. Toth, S. J., and D. N. Riemer. 1968. Precise chemical control of algae in ponds. *J. Amer. Water Works Assoc.* **60**:367–371.

42. Button, K. S., H. P. Hostetter, and D. M. Mair. 1977. Copper dispersal in a water supply reservoir. *Water Res.* **11**:539–544.

43. Morrissy, N. M. 1976. Aquaculture of marron, *Cherax tenuimanus* (Smith) Part 1. Site selection and the potential of marron for aquaculture. *Fish Res. Bull. Western Australia* **17**:1–27.

44. Fast, A. W., D. K. Barclay, and G. Akiyama. *Artificial Circulation of Hawaiian Prawn Ponds*. 1983. UNIH-SEAGRANT-CR-84-01, University of Hawaii, Honolulu.

45. Maskey, S., and C. E. Boyd. 1986. Seasonal changes in phosphorus concentrations in sunfish and channel catfish ponds. *J. Aquaculture Tropics* **1**:35–42.

5

Soil Organic Matter and Aerobic Respiration

5.1 Introduction

The primary source of organic matter in nature is photosynthesis by green plants. In *photosynthesis,* energy of sunlight is captured by plant pigments and used as an energy source to reduce inorganic carbon of CO_2 to organic carbon of simple sugars. Solar energy is transformed by photochemical reactions in plant cells to chemical energy of chemical compounds. Oxygen is released as a by-product. The summary reaction is

$$6CO_2 + 6H_2O \xrightarrow[\text{green plants}]{\text{light}} C_6H_{12}O_6 + 6O_2 \tag{5.1}$$

Chemoautotrophic bacteria also reduce carbon dioxide to organic matter. For example, nitrifying bacteria use energy released in the oxidation of ammonia to nitrate and reduce CO_2 to organic carbon of carbohydrate. On a global basis, the amount of organic matter synthesized by chemoautotrophy is very small in comparison with the quantity produced by photosynthesis.

The wide variety of organic compounds that constitute plants are elaborated from simple sugars made in photosynthesis. Carbohydrates from photosynthesis also are oxidized in plant cells to provide energy needed for plant growth and maintenance. Plants represent the primary production and the base of the food web in natural ecosystems. Animals cannot synthesize organic matter from carbon dioxide; they must obtain organic matter from their food. Some animals feed only on plants (herbivores), some animals eat only other animals (carnivores), and other animals eat both plants and animals (omnivores). Animals represent the secondary production of ecosystems. The organic remains of primary and secondary producers become organic matter, and serve as the food for decomposer organisms such as fungi and bacteria.

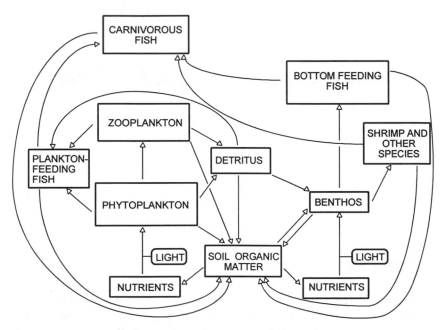

FIGURE 5.1. Natural food webs in aquaculture ponds.

Herbivorous, carnivorous, and omnivorous species are used in fish and shrimp culture (Fig. 5.1). Natural productivity within ponds is limited, and manufactured feeds often are applied to ponds to enhance production of fish and shrimp. Feeding does not alter the dependence of secondary production upon primary production because feeds are manufactured from plant and animal materials produced outside of the pond ecosystem (Fig. 5.2). The fraction of the natural productivity and feed that is not converted to harvestable aquatic animals becomes organic matter and is the food source for heterotrophic microorganisms.

All plants, animals, and microorganisms obtain energy for sustaining life processes through oxidation of organic compounds in *respiration*. The summary equation for aerobic respiration is the reverse of the summary equation for photosynthesis:

$$6_6H_{12}O_6 + 6O_2 \rightarrow 6CO_2 + 6H_2O + \text{energy} \tag{5.2}$$

Oxygen and organic matter, the products of photosynthesis, are consumed in respiration, and carbon dioxide and water, the reactants of photosynthesis, are released in respiration. Solar energy trapped in organic compounds by photosynthesis is released in respiration. The liberated energy is trapped in high-energy phosphate bonds and used to drive metabolic processes or is lost to the environment as heat.

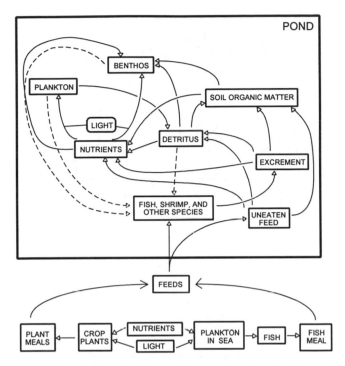

FIGURE 5.2. Food webs in aquaculture ponds where feed is applied.

There are cycles of carbon and oxygen in ecosystems (Fig. 5.3), and an equilibrium exists among carbon and oxygen in the atmosphere, living organisms, and organic matter. On a global basis the amount of carbon fixed and the quantity of oxygen released in photosynthesis are equal to the amount of carbon released and the quantity of oxygen consumed in respiration. Therefore, from year to year, oxygen and carbon dioxide concentrations in the atmosphere remain relatively constant. Aquaculture ponds contribute to the global carbon and oxygen cycles, but they also have internal cycles of these two gases. During daylight, carbon dioxide is removed from pond water for use in photosynthesis faster than it is released by respiratory processes. This causes CO_2 concentrations to decline during the day. At night photosynthesis stops, and the water is replenished with carbon dioxide from respiration. The behavior of dissolved oxygen concentrations is opposite that of CO_2 concentrations. In the daytime, photosynthesis produces oxygen faster than it is used in respiration and dissolved oxygen concentrations increase. At night no oxygen is produced to replace oxygen used in respiration, and dissolved oxygen concentrations decline. Organic compounds not consumed in respiration constitute plant and animal biomass and eventually become organic matter.

Organic matter accumulates in the bottoms of ponds. Organic matter is benefi-

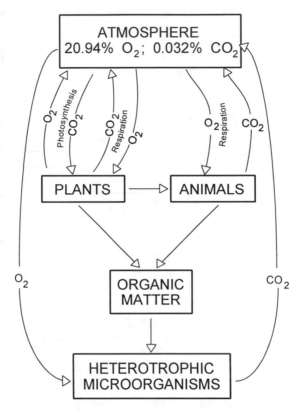

FIGURE 5.3. Carbon and oxygen cycles.

cial as a source of nutrition for benthic fish food organisms, but it also represents an oxygen demand that can have detrimental effects on water quality. This chapter provides information on soil organic matter dynamics in the aerobic layers of pond soil.

5.2 Microbial Physiology

The major decomposers in soil are bacteria, actinomycetes, fungi, and protozoa; of these four groups, bacteria are the most numerous and active. Decomposers require a source of nutrients for growth. Nutrients have three separate functions in growth and metabolism. They are raw materials for elaboration of biochemical compounds of which organisms are made. Carbon and nitrogen contained in organic nutrients are used in making protein, carbohydrate, and other components of microbial cells. Nutrients supply energy for growth and chemical reactions. Carbohydrates from organic nutrients are oxidized in respiration, and the energy released is used to drive chemical reactions to synthesize biochemical compounds necessary for growth and maintenance. Nutrients also can serve as electron and hydrogen acceptors in respiration.

Heterotrophic microorganisms obtain inorganic and organic nutrients from organic matter. The terminal electron and hydrogen acceptor for aerobic respiration is molecular oxygen. An organic metabolic product or inorganic substance replaces oxygen as electron or hydrogen acceptor in anaerobic respiration.

5.2.1 Growth

The soil contains many species of microorganisms capable of decomposing almost any biochemical substance. Some substances decompose faster than others, but few, if any, organic compounds completely resist decay by soil microorganisms. Microbial activity is slow in soil with a low concentration of organic matter, but actively growing microorganisms, resting spores, and other propagules are present. Application of organic matter to a soil of low-organic-matter concentration provides substrate for microbial growth, and the number of microorganisms increases. Most soil microorganisms reproduce by binary fission. One cell divides into two cells, and the new cells divide. The time between cell divisions is called the *generation time* or doubling time. The number of bacteria cells present after a given time (N_t) can be computed from the initial number of cells (N_0) and the number of generations (n):

$$N_t = N_0 \times 2^n \qquad (5.3)$$

Example 5.1: Increase in Bacterial Abundance
A soil that contains 10^3 bacterial cells per gram is treated with organic matter. Bacteria in the soil can double every 4 h. Calculate the number of bacterial cells after 36 h.

Solution: The number of generations is

$$n = \frac{36\,\text{h}}{4\,\text{h}} = 9$$

The number of cells after 36 h can be calculated with Equation 5.3:

$$N_t = 10^3 \times 2^9 = 512 \times 10^3 = 5.12 \times 10^5$$

When fresh organic substrate is added to a culture of microorganisms, it takes a short time for microorganisms to adjust to the new nutrient supply. The period during which there is little or no increase in cell number is called the *lag phase* (Fig. 5.4). Rapid growth follows as microorganisms utilize the new substrate; this period is known as the *logarithmic phase*. After a period of rapid growth, a *stationary phase* is attained during which cell number remains relatively con-

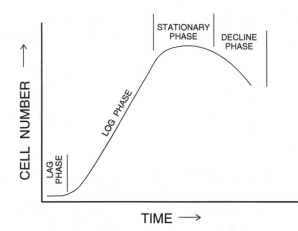

FIGURE 5.4. Characteristic growth phases of microbial cultures.

stant. As the substrate is used up or metabolic by-products accumulate, growth slows and a *decline phase* occurs in which the number of cells decreases.

The generation time for a bacterial population can be computed from data taken on cell numbers during the logarithmic phase. Equation 5.3 can be rewritten as

$$\log N_t = \log N_0 + n \log 2 \tag{5.4}$$

and

$$0.301n = \log N_t - \log N_0 \tag{5.5}$$

$$n = \frac{\log N_t - \log N_0}{0.301}$$

The generation time is computed from the time interval (t) over which a given number of generations (n) occurred:

$$g = \frac{t}{n} \tag{5.6}$$

and

$$g = \frac{0.301t}{\log N_t - \log N_0} \tag{5.7}$$

Example 5.2: Calculation of Generation Time
    A bacterial culture increases from 10^4 cells mL^{-1} to 10^8 cells mL^{-1} in 12 h. Calculate the generation time.

Solution:

$$g = \frac{0.301(12)}{8 - 4} = 0.903 \text{ h} = 54 \text{ min}$$

The generation time of microorganisms differs among species, and, for a given species, generation time varies with temperature, substrate availability, and other environmental factors.

When bacteria are growing rapidly, they also are rapidly decomposing substrate. In a soil where microorganisms are growing logarithmically, the evolution of carbon dioxide increases logarithmically.

5.2.2 Respiration

Respiration involves oxidation of carbohydrate and other organic compounds to CO_2 and water. The purpose of respiration is to release energy from organic compounds and store it in high-energy phosphate bonds of adenosine triphosphate (ATP):

$$\text{Adenosine diphosphate (ADP)} + \text{phosphate} + \text{energy} = \text{ATP} \qquad (5.8)$$

Energy stored in ATP can be used to drive chemical synthesis reactions in cells; when the energy is released from ATP, ADP is regenerated for making ATP again.

In aerobic carbohydrate metabolism, a glucose molecule passes through glycolysis and the citric acid cycle (Krebs cycle) before being completely oxidized. Glycolysis is a series of ordered reactions that transform one glucose molecule to two pyruvate molecules:

$$C_6H_{12}O_6 \rightarrow 2C_3H_4O_3 + 4H^+ \qquad (5.9)$$

A small amount of ATP is formed in glycolysis, but no carbon dioxide is formed. Glycolysis does not require oxygen.

In aerobic respiration each pyruvate molecule from glycolysis can react with coenzyme A (CoA) and nicotinamide adenine dinucleotide (NAD^+) to form one acetyl CoA molecule:

$$\text{Pyruvate} + \text{CoA} + NAD^+ \rightarrow \text{Acetyl CoA} + \text{NADH} + H^+ + CO_2 \quad (5.10)$$

The formation of acetyl CoA from pyruvate is an oxidation; two H^+ ions and one CO_2 molecule are released per molecule of pyruvate. Acetyl CoA formation links glycolysis to the Krebs cycle. Although glycolysis does not use oxygen, acetyl CoA production and the Krebs cycle require oxygen.

The first step in the Krebs cycle is the reaction of acetyl CoA and oxaloacetic acid to form citric acid. In the succeeding reactions of the Krebs cycle, a series of transformations of organic acids occur with release of eight H^+ ions and two CO_2 molecules for each molecule of citric acid entering the cycle. One molecule of oxaloacetic acid also is regenerated (recycled), and it can again react with acetyl CoA.

Hydrogen ions produced through oxidation of glucose in glycolysis and the Krebs cycle reduce the enzymes NAD, NADP (nicotinamide adenine dinucleotide phosphate), and FAD (flavin adenine dinucleotide). A representative reaction where H^+ is released in the Krebs cycle is

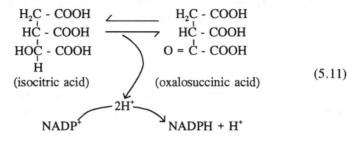

$$
\begin{array}{ll}
\text{H}_2\text{C - COOH} & \text{H}_2\text{C - COOH} \\
\text{HC - COOH} & \text{HC - COOH} \\
\text{HOC - COOH} & \text{O = C - COOH} \\
\text{H} & \\
\text{(isocitric acid)} & \text{(oxalosuccinic acid)}
\end{array}
$$

(5.11)

$$-2\text{H}^+$$

NADP$^+$ NADPH + H$^+$

Enzymes of the Krebs cycle are brought in contact with the electron transport system where they are reoxidized. Energy released in these oxidations is used for ATP synthesis. Regenerated NAD^+, NADP, and FAD are used again as hydrogen acceptors in the Krebs cycle. Reoxidation of enzymes from the Krebs cycle is accomplished by a series of cytochrome enzymes that pass electrons from one cytochrome compound to another. In the electron transport system, ADP combines with inorganic phosphate to form ATP and H^+ ions combine with molecular oxygen to form water.

A total of 38 ATP molecules can be formed from 1 mole of glucose in aerobic respiration (glycolysis and Krebs cycle). The ATP molecules contain approximately one-third of the theoretical energy released during glucose oxidation. Energy not used to form ATP is lost as heat. In aerobic respiration of glucose, 1 mole of glucose consumes 6 moles of oxygen and releases 6 moles of CO_2 and water. This stoichiometry is the same as that illustrated in the summary equation for respiration (Equation 5.2).

Example 5.3: Oxygen Requirements in Respiration
  Calculate the amount of oxygen needed to decompose glucose.

Solution: The overall equation for aerobic respiration is

$$C_6H_{12}O_6 + 6O_2 \rightarrow 6CO_2 + 6H_2O$$

The following ratio can be established for computing oxygen consumption:

$$\frac{1 \text{ g glucose}}{1 \text{ mole glucose } (180 \text{ g})} = \frac{X \text{ g } O_2}{6 \text{ moles } O_2 \, (192 \text{ g})}$$

$$X = 1.07 \text{ g } O_2$$

5.2.3 Respiratory Quotient

Oxidation of glucose was used as an example of aerobic respiration. Soil organic matter contains a wide array of organic compounds, and soil microorganisms oxidize all of these compounds. Substances more complex than glucose must be broken down by enzymatic action until carbon fragments suitable for entering the Krebs cycle. These fragments pass through the respiratory reactions in the same way as glucose fragments, but amounts of O_2 consumed and CO_2 released per gram differ among compounds.

The ratio of moles CO_2 produced to moles O_2 consumed is the *respiratory quotient* (*RQ*):

$$RQ = \frac{CO_2}{O_2} \tag{5.12}$$

The RQ for carbohydrates such as glucose is 1.0. A more highly oxidized compound will have $RQ > 1$, and a more highly reduced compound will have $RQ < 1$.

The degree of reduction of organic compounds increases as the percentage carbon increases. Values for the usual composition of carbohydrates, proteins, and fatty acids are

	%C	%H	%O
Carbohydrate	40	6.7	53.3
Protein	53	7	22
Fatty acids	77.2	11.4	11.4

The degree of reduction increases in the order carbohydrate < protein < fatty acids. The caloric content of foodstuffs is related to degree of reduction. Average calorific values are carbohydrate, 4.1 kcal g^{-1}; protein, 5.5 kcal g^{-1}; fat, 9.3 kcal g^{-1}. Plants typically have caloric contents of 4–4.5 kcal g^{-1}, reflecting the high percentage of carbohydrate in their structure. Animals consist primarily of protein, and they often contain more fat than plants. They tend to have higher calorific values than plants.

Example 5.4: Effect of Degree of Reduction of Organic Matter on Oxygen Consumption in Respiration

Calculate the amount of oxygen required to completely oxidize the carbon in 1 g each of carbohydrate, protein, and fat.

Solution: Assume that carbon concentrations are carbohydrate, 40%; protein, 53%, fat, 77.2%. One gram of mass represents 0.4 g C, 0.53 g C, and 0.772 g C for carbohydrate, protein, and fat, respectively. The stoichiometry for converting organic carbon to CO_2 is

$$C + O_2 \rightarrow CO_2$$

The appropriate ratio for computing oxygen consumption is

$$\frac{g\,C}{12} = \frac{O_2\,used}{32}$$

$$O_2\,used = \frac{32\,(g\,C)}{12}$$

$$O_2\,carbohydrate = \frac{(32)(0.4)}{12} = 1.07\,g$$

$$O_2\,protein = \frac{(32)(0.53)}{12} = 1.41\,g$$

$$O_2\,fat = \frac{(32)(0.772)}{12} = 2.06\,g$$

5.2.4 Factors Controlling Growth and Respiration

Microorganisms respire continuously, and respiration rate is a reliable index of growth rate. Factors that affect growth also influence respiration in the same degree and direction. Major factors are temperature, oxygen supply, moisture availability, pH, and nitrogen concentration.

Temperature

Temperature is a critical factor controlling microbial growth. Temperature effects of growth are illustrated in Figure 5.5. Growth is limited by low temperature, there is a narrow optimal temperature range, temperature can be too high for good growth, and temperature can reach the thermal death point. At suboptimal temperature, an increase of 10°C normally doubles respiration and growth. Most bacteria found in soils probably grow best at 30–35°C. The decomposition rate of organic matter in soil often doubles if the temperature is increased from 20°C to 30°C.

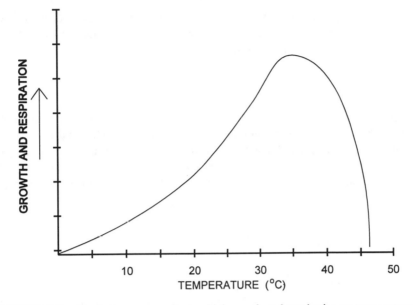

FIGURE 5.5. Typical response of microbial growth and respiration to temperature.

Oxygen

A continuous supply of oxygen is necessary for aerobic respiration. When soils are waterlogged, atmospheric oxygen cannot enter the soil readily, and the rate of aerobic respiration declines. Anaerobic respiration begins when oxygen is depleted in pore water. Anaerobic decomposition of organic substrates often is slower and less complete than aerobic decomposition. Pond soils usually are flooded, but the surface layer of the soil normally is aerobic. When ponds are drained for harvest, their bottoms dry and the soil is aerated.

Moisture

Flooded soils have low rates of respiration because water fills the pore spaces and atmospheric oxygen cannot enter them rapidly. Soils also can become too dry for rapid microbial activity. Smith et al. indicated that active soil biomass declines in dry soils, and soil respiration is greatly depressed.[1] Respiration of terrestrial soils is commonly greatest at 60–80% of the soil's moisture-holding capacity. At this moisture content the soil contains plenty of water for microbial activity, yet some pores are not full of water and air can enter to provide oxygen.

pH

Each species of microorganisms has its optimal pH for growth. However, soil respiration is normally greater in neutral or slightly alkaline soils than in low pH soils. Fungi grow best at low pH, but they are not efficient decomposers. Bacteria thrive at pH 7–8, and they are more efficient decomposers than fungi. Liming may be used to increase the pH of acidic soils.

Nutrients

Microorganisms must have adequate concentrations of essential mineral elements. Quantities of most mineral elements needed are quite low and are available from the substrate or soil, but a large amount of nitrogen is required. Microorganisms normally contain 5–10% nitrogen on a dry weight basis. Therefore, the nitrogen content of organic residues is an important factor controlling decomposition.

5.3 Biochemical Compounds

Organic matter consists of remains of plants, animals, animal excrements, and microorganisms. Any chemical compound synthesized by living things may be found in organic matter. Synthetic organic compounds may be applied to soils. In agricultural soils, it is not uncommon to find pesticide residues, and in industrial areas soils may be contaminated with toxic chemicals. Quantities of synthetic organic chemicals in soils are generally insignificant in comparison with amounts of natural organic matter, but exceedingly small concentrations of some synthetic substances have powerful toxic effects. Fortunately, soil microbes can gradually decompose toxic, synthetic chemicals. Synthetic chemical residues usually are not a problem in ponds because aquaculturists guard against accidental introduction of pollutants to ponds, and few synthetic chemicals are applied for management purposes.

5.3.1 Carbohydrate

As the name implies, carbohydrates have a C:H:O ratio of 1:2:1 (CH_2O) which suggests that they are hydrates of carbon (CH_2O). Common carbohydrates are sugars, starch, cellulose, and hemicellulose. Sugars are monosaccharides, the most important of which are six carbon sugars (hexoses) and five carbon sugars (pentoses). The four most common hexoses in plants are glucose, fructose, mannose, and galactose. Xylose, arabinose, ribose, and 2-deoxy-D-ribose are the most common pentoses of plants. Molecular structures of glucose, fructose, and xylose are provided in Figure 5.6. Other hexoses and pentoses differ in the location of the double-bonded oxygen or the position of hydroxyl group on a particular carbon atom (compare structures of glucose and fructose). The structures of sugars often are written as straight carbon chains, but most sugars have cyclic structures as shown for glucose:

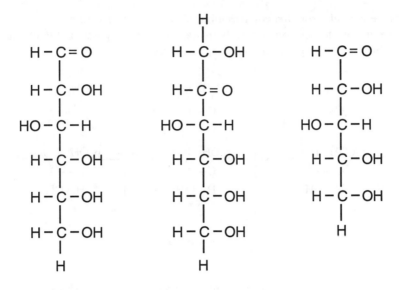

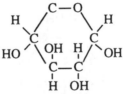

GLUCOSE

FIGURE 5.6.   Structure of glucose, fructose, and xylose.

Oligosaccharides are made from interconnected monosaccharide molecules. Oligosaccharides range in complexity from disaccharides made of two sugar molecules to polysaccharides made of a great many sugar molecules. The most common disaccharide is sucrose:

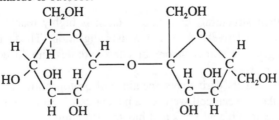

SUCROSE

Sucrose is made from the condensation of glucose and fructose.

The major storage form of carbohydrate in plants is starch. Starch can be thought of as a straight-chain polymer of glucose units:

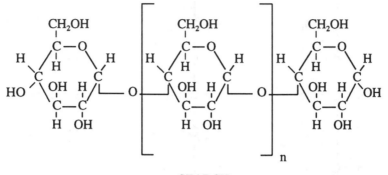

STARCH

The compound used by plants to form cell walls is cellulose. This compound is probably the most abundant organic substance in nature. The structure of cellulose is almost identical to that of starch except that the linkages between the ends of the individual glucose molecules are different:

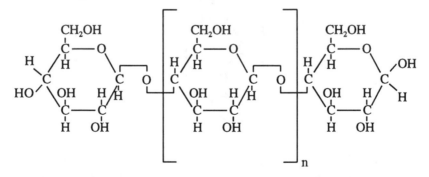

CELLULOSE

In spite of their structural similarities, starch is highly reactive and digestible and cellulose is relatively inert and resists digestion. Those interested in the carbon linkages and nomenclature of carbohydrates should consult a general biochemistry text.

Pectic acid and its derivatives, pectin and protopectin, are abundant in the middle lamella between cell walls of adjacent cells. Pectic acid is composed of linked D-galacturonic acid units and has the structure

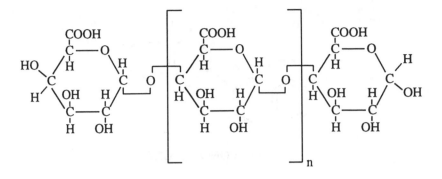

PECTIC ACID

Plants also have polymers of five-carbon sugars such as xylan and araban, which are composed of xylose or arabinose molecules. Xylan is the most common pentose polymer in plants and is a constituent of cell walls. Hemicellulose is a mixture of hexose and pentose sugars and also occurs in cell walls. Hydrolysis of hemicellulose yields glucose and a pentose sugar (usually xylose).

A number of sugars occur in nature in which a hydroxyl group has been replaced by an amino or acetylamino group. Two examples of amino sugars are glucosamine and acetyl glucosamine:

GLUCOSAMINE   ACETYL GLUCOSAMINE

Glucosamine molecules polymerize to form chitin, which is the skeletal material of insects and crustaceans.

5.3.2 Lignin

Lignin occurs in cell walls of plants. Lignin molecules contain only carbon, hydrogen, and oxygen, but lignin is not a carbohydrate. It has an aromatic structure made up of repeating units. It has been proposed that the repeating units in lignin are guaiacylpropane derivatives, dehydrodilisoeugenol, and polyflavonone (Fig. 5.7). Lignin is more resistant to decay than carbohydrates.

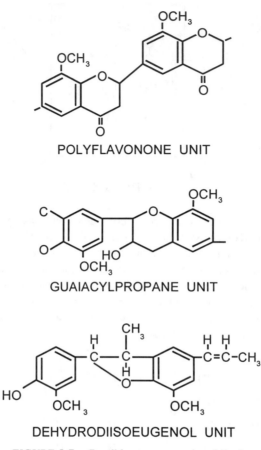

FIGURE 5.7. Possible structure units of lignin.

5.3.3 Lipids

Lipids include neutral fats, waxes and sterols, phospholipids, and lipoproteins. Numerous different lipid compounds are found in nature. Neutral fats are esters of glycerol and long-chain fatty acids. They are usually triglycerides. For example, esterification of glycerol and stearic acid gives the neutral fat, tristearin:

(5.13)

In a mixed triglyceride two or more fatty acids react with glycerol. Diglycerides and polyglycerides also occur in nature.

Triglycerides and other fats are readily hydrolyzed:

$$
\begin{array}{l}
\quad\quad\quad\quad O \\
\quad\quad\quad\quad \| \\
H_2C - O - C - C_{17}H_{35} \\
\;| \quad\quad\quad O \\
\quad\quad\quad\quad \| \\
HC - O - C - C_{17}H_{35} + 3H_2O \;\rightarrow\; 3C_{17}H_{35}COOH + C_3H_5(OH)_3 \quad (5.14) \\
\;| \quad\quad\quad O \\
\quad\quad\quad\quad \| \\
H_2C - O - C - C_{17}H_{35}
\end{array}
$$

The fatty acids found in nature usually have an even number of carbon atoms. Some common fatty acids are acetic, CH_3COOH; capric, $C_9H_{19}COOH$; palmitic, $C_{15}H_{31}COOH$; and stearic, $C_{17}H_{35}COOH$. These arc all saturated fatty acids. Saturated fatty acids have no double-bonded carbon atoms. Linseed oil is a common unsaturated fatty acid composed of linolenic acid, which has three double bonds. Waxes and sterols are esters of monatomic alcohols of high molecular weight in which the fatty acids are palmitic, stearic, or oleic. Waxes and sterols are found in both plant and animal tissues. The most important sterols are sitosterol and stigmasterol (Fig. 5.8). As their names imply, a phosphorus atom has been esterified in phospholipids, and a protein is associated with lipoproteins.

5.3.4 Proteins

Proteins are made up of amino acids. All amino acids contain the amino group NH_2. There are many amino acids, and 16–20 of them commonly occur in most organisms. Plants synthesize their amino acids. Animals and some bacteria cannot produce certain amino acids, but they have nutritional requirements for them. Such amino acids are considered *essential amino acids* in the diets of animals or microorganisms. Structural formulas of some common amino acids are provided in Figure 5.9.

The formation of protein from amino acids involves amino acids combining by the carboxyl group of one amino acid joining with the amino group of another:

$$
\begin{array}{l}
\quad\quad\quad\quad\quad NH_2 \quad\quad\quad\quad\quad\quad\quad\quad\quad CH_2COOH \quad\quad\quad (5.15) \\
\quad\quad\quad\quad\quad | \quad\quad\quad\quad\quad\quad\quad\quad\quad\quad\;\; | \\
CH_2COOH + HOOCC - CH_2 - \bigcirc\!\!\!\!\bigcirc \;\rightarrow\; HN \quad NH_2 \\
\;| \quad\quad\quad\quad\quad | \quad\quad\quad\quad\quad\quad\quad\quad\quad\quad \backslash \quad\; | \\
NH_2 \quad\quad\quad\quad\;\; H \quad\quad\quad\quad\quad\quad\quad\quad\quad\;\; CO - C - CH_2 - \bigcirc\!\!\!\!\bigcirc + H_2O \\
\quad\quad\quad\quad\quad\quad\quad\quad\quad\quad\quad\quad\quad\quad\quad\quad\quad\quad\quad\;\; | \\
\quad\quad\quad\quad\quad\quad\quad\quad\quad\quad\quad\quad\quad\quad\quad\quad\quad\quad\quad\;\; H
\end{array}
$$

A protein molecule is made up of many individual amino acid units. The average nitrogen concentration of protein is 16%. The nitrogen concentration of a material

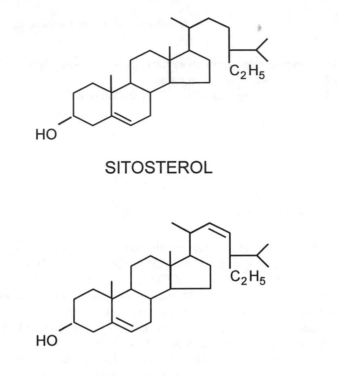

SITOSTEROL

STIGMASTEROL

FIGURE 5.8. Common plant sterols.

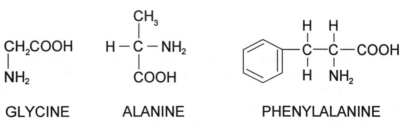

GLYCINE   ALANINE   PHENYLALANINE

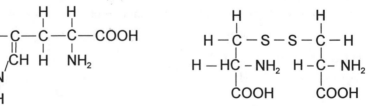

TRYPTOPHAN   CYSTINE

FIGURE 5.9. Common amino acids.

multiplied by 6.25 (1/16) is a rough estimate of its protein concentration and is called *crude protein*.

Proteins combine with many other compounds to give nucleoproteins, nucleic acids, glycoproteins, phosphoprotein, chromoproteins, and lipoproteins. Enzymes comprise proteins.

5.3.5 Other Substances

In addition to the major compounds discussed, organic matter contains vitamins, pigments, tannic acid, cutin, and many other biochemical compounds. Essential mineral elements are also contained in organic matter. For example, sulfur is contained in protein, and phosphorus is a component of phytin, phospholipids, nucleic acids, nucleoproteins, and related compounds.

5.4 Soil Organic Matter

5.4.1 Residues

The major input of organic matter to soil is plant material that consists primarily of cellulose, hemicellulose, lignin, protein, starch, fats, waxes, and oils. Structural (cell wall) material is the dominant component of vascular plants (Table 5.1). Algae contains less structural material and more protein than vascular plants; planktonic algae often contain 50% protein or more (Table 5.1).

Animal production in ecosystems normally is 10% or less than plant production.[2] Structural material of vertebrate animals is bone, which is composed mainly of calcium phosphate. Many invertebrate animals have exoskeletons made of the carbohydrate chitin. When compared to vascular plants, animals have much less carbohydrate and more fat and protein. Animals frequently contain 60–70% protein and 5–10% fat.

Table 5.1   Composition of Fresh Organic Matter from Several Sources (all components expressed on dry-weight basis)

Constituent	Cornstalks	Alfalfa Plants	Cow Manure	Vascular Aquatic Plants	Macrophytic Algae	Plankton Algae
Simple compounds (%)	20.13	27.65	13.11	—	—	—
Hemicellulose (%)	17.63	8.52	18.57	—	—	—
Cellulose (%)	29.67	26.71	25.23	26.7	22.6	15.7
Lignin (%)	11.28	10.78	20.21	—	—	—
Crude protein (%)	1.98	8.13	14.87	13.7	19.3	50.19
Ash (%)	7.53	10.33	12.95	13.0	19.3	6.20

Sources: Compiled from Waksman,[48] Boyd and Lawrence,[49] and Boyd[50, 51]

Table 5.2   Composition of Fish and Shrimp Feed

Component	Catfish Feed	Shrimp Feed
Crude protein (%)	32.0	35.0
Crude fiber (%)	7.0	4.0
Lipid (%)	6.0	10.0
Phosphorus (%)	0.7	1.5
Ash (%)	7.0	12.0

Composition of manures and other agricultural by-products used as fertilizers in aquaculture ponds resembles the composition of plants from which they were derived (Table 5.1). Fish and shrimp feeds are high-quality mixtures of plant meals, fish meal, and other expensive ingredients. Compositions of representative catfish and shrimp feeds are listed in Table 5.2. These feeds have a relatively large amount of protein, a small amount of fiber, and a large amount of starch that animals use as an energy source. A list of ingredients for a common channel catfish feed is 37% corn, 37% soybean meal; 15% cotton seed meal, 3% fish meal, 4% wheat middlings, 3% blood and bone meal, and 1% calcium phosphate. Fish and shrimp eat most of the feed applied to ponds; uneaten feed usually is less than 5–10% of the ration. About 80–90% of ingested feed is absorbed across the gut; the unabsorbed fraction becomes feces. Organic matter contains mineral elements required by plants and animals. For feedstuffs it is common to indicate the total mineral matter in the feed as ash content (Tables 5.1 and 5.2). *Ash* is the weight of the residue remaining after a sample of organic matter has been ignited in a furnace at 500–550°C.

Soil microorganisms become soil organic matter when they die. Bacteria contain mostly protein and a little cell-wall material. Fungi and actinomycetes contain more cell-wall material and less protein than bacteria. Animals living in the soil have a high percentage of protein, and many have chitinous exoskeletons.

The major source of organic residues in aquaculture ponds are planktonic algae and invertebrates. These organisms tend to have low concentrations of structural carbohydrates and high concentrations of protein. Uneaten feed also is low in structural carbohydrate (fiber) and high in protein and starch. Manures and feces of aquatic animals are of lower nutritional quality than the other residues normally found in aquaculture ponds.

Organic residues are substrate (food) for decomposer organisms. There is continuous input of organic residues to soil, and residues are decomposed to inorganic substances such as carbon dioxide, ammonia, phosphate, and water, which are lost from the soil by volatilization, leaching, or erosion. Soluble organic matter also may be lost by leaching and erosion. Inputs of organic matter are roughly balanced by losses, and a residual concentration of organic matter exists in the soil. This concept may be expressed in terms of carbon as follows:

$$\text{Organic C input} = [CO_2 + \text{organic C loss}] \pm \Delta \text{ soil organic C} \quad (5.16)$$

According to Anderson, variables controlling decomposition of organic matter are types of decomposers, environmental quality, and characteristics of the residue.[3] Soils have a microbial flora capable of decomposing almost any natural product, but some substances decompose more readily than others. Those that decompose quickly are called *labile* organic matter, and those that decompose slowly are known as *refractory* organic matter. Complex organic residues contain both labile and refractory compounds. An experiment on decomposition rates of different components of plant residues by Minderman illustrates the preceding point.[4] Percentage decomposition of different components of a plant residue in one year were sugar 99%, hemicellulose 90%, cellulose 75%, lignin 50%, waxes 25%, and phenols 10%.

Fresh organic residue must be broken down into soluble, relatively simple compounds before soil microorganisms can decompose it. Swift et al. divided decomposition into three processes: comminution, catabolism, and leaching.[5] *Comminution* is a reduction in particle size of organic residues by animals feeding on them and by physical forces. Animals engulf or ingest particulate organic matter, but the microbial flora must absorb soluble organic substances. *Catabolism* is the enzymatic degradation of substrate by soil bacteria, fungi, actinomycetes, and animals. Some soluble compounds leach out of residues, but much of the organic matter in residues consists of large, insoluble molecules. Hydrolytic bacteria can convert cellulose, hemicellulose, and pectin to sugars, hydrolyze fat, split proteins, break phenolic rings, and cause many other alterations in organic compounds. Hydrolysis of cellulose appears simple:

$$(C_610_{10}O_5)_n + nH_2O = nC_6H_{12}O_6 \quad (5.17)$$

However, hydrolysis of cellulose requires the action of the enzyme cellulase, and only microorganisms that produce cellulase can readily hydrolyze cellulose. An *enzyme* is a protein produced by living cells that functions as a catalyst in chemical reactions. Many different enzymes are necessary to fragment organic residues into carbon fragments small enough for use in respiration. Initial fragmentation of substrate molecules occurs outside microbial cells and is catalyzed by extracellular enzymes excreted by microbial cells. Organic carbon fragments from extracellular enzyme activity are absorbed by microbes. Enzymes inside microbial cells break absorbed substrate into smaller pieces for use in glycolysis. No energy is obtained from the preparation of organic substrate for respiration. *Leaching* is the downward movement of soluble organic matter in the soil. Particulate organic matter can be lost from the soil by erosion. Some organic compounds produced by decomposers inhibit decomposition, and removal of these substances by leaching and erosion prevents toxins from accumulating and inhibiting microbial activity.

5.4.2 Soil Organic Matter Pools

Fresh and partially decomposed residues, compounds excreted by soil microorganisms, and substances formed in reactions between degradation products and excretory products form soil organic matter. Soil organic matter can be conceptualized as belonging to various pools that are linked.[6] The most widely recognized pools are litter, light fraction, soluble organic compounds, biomass, and humus.

Litter consists of macroorganic matter at the soil surface. It includes plant and animal remains, excrement, and other organic particles of 2 mm or more in diameter. It usually is possible to identify litter as the remains of a specific organism or material. Manure particles, aquatic plant residues, or vegetative remains entering ponds from their watersheds are litter. Litter usually is not important in aquaculture ponds, because microscopic, planktonic algae are the major source of organic matter.

The term *biomass* normally refers to the soil microbial community, but it can be extended to include living animals and plant roots. The composition of the biota in a fertile, terrestrial soil was given by Metting as follows: bacteria, 10^8–10^9 g^{-1}; actinomycetes, 10^7–10^8 g^{-1}; fungi, 10^5–10^6 g^{-1}; microalgae 10^3–10^6 g^{-1}; protozoa, 10^3–10^5 g^{-1}; nematodes, 10^1–10^2 g^{-1}; worms, 30–300 m^{-2}; other invertebrate microfauna, 1–200 m^{-2}.[7] The biomass of soil organisms probably ranges from 10 to 2,000 kg dry weight ha^{-1}. Pond soils also contain large numbers of bacteria, fungi, actinomycetes, and protozoa.

Soluble organic matter consists of sugars, fatty acids, fragments of proteins, amino acids, and other compounds leached from fresh residues, produced by activity of extracellular enzymes on complex compounds, or excreted by soil organisms. Soluble organic matter is contained in the soil solution and subject to leaching.

The most characteristic and unique pool of soil organic matter is humus. *Humus* is a product of synthesis and decomposition by soil microorganisms.[8] It exists as a series of acidic, yellow-to-black macromolecules of unknown but high molecular weight. Soil organic matter is 60–80% humus. Humus chemistry is not well understood. According to Stevenson and Elliot, humus consists of a heterogeneous mixture of molecules that form systems of polymers.[6] Molecular weights of polymers in humus range from several hundred to more than 300,000. One hypothesis holds that in humus formation, polyphenols derived from lignin and others synthesized by soil microorganisms polymerize alone or with amino compounds to form polymers of variable molecular weights with functional acidic groups. Humus often is considered to be composed of three classes of compounds: humic acids, fulvic acids, and humin. Fulvic acids have higher oxygen contents, lower carbon contents, lower molecular weights, and higher degrees of acidity than humic acids. Fulvic acids are yellowish, whereas humic acids are dark brown or black. The properties of humin are poorly defined. Humin differs from fulvic and humic acids by being insoluble in alkali.

Humus is slowly decomposable by soil microorganisms and represents a large proportion of the soil organic matter. It is an important food source for soil microorganisms, and it influences soil properties. Humus is closely associated with the mineral fraction. It affects soil structure because it is adsorbed by mineral particles and binds them. Nitrogen mineralized from humus by soil microorganisms is a plant nutrient. Humus contributes to soil acidity and cation-exchange capacity. Humus also can chelate trace metals and enhance their availability to plants.

The *light fraction* of soil organic matter comprises particles smaller than those of litter (<2 mm diameter), and it is not closely associated with mineral matter like true humus. Stevenson and Elliott defined the light fraction as organic matter associated with soil particles of 0.25–2 mm diameter.[6] The 0.25-mm lower limit for the light fraction particle size was suggested because organic matter associated with microparticles (<0.25 mm) decomposes slowly, whereas organic matter associated with macroparticles decomposes more rapidly.[9] Gregorich et al. obtained the following carbon mineralization rates during a 20-day incubation period for organic matter associated with different particle separates of a soil: sand, 49 mg C g^{-1}; silt, 15 mg C g^{-1}; clay, 7 mg C g^{-1}.[10] Organic matter associated with sand decomposed more readily than that associated with silt and clay. The separation between sand and smaller silt and clay particles is made at 0.05 mm in the USDA soil particle classification scheme. This separation requires a 270 mesh sieve. Based on findings of Gregorich et al.,[10] the separation of the light fraction at a lower particle size limit of 0.05 mm, as done by Cambardella and Elliott,[11] is a more convenient and logical practice than using a lower particle limit of 0.25 mm. Cambardella and Elliott showed that the light fraction organic matter in a grassland soil ranged from 39% of soil organic matter in plots covered by native sod to 18% of soil carbon in bare, cultivated soils.[11]

5.4.3 Carbon Flow Models

Many models have been proposed to describe the flow of organic carbon through terrestrial soils. Parton et al. presented a model (Fig. 5.10) with organic matter pools similar to those already described.[2] Plant residue was divided into structural (cell-wall) and nonstructural (non-cell-wall) material. In the model, structural material consists of cellulose, hemicellulose, and lignin, and the nonstructural material is made up of sugars, starch, protein, fatty acids, and other protoplasmic components. Nonstructural material decomposes faster than structural material. Soil organic matter was separated into three pools—active, slow, and passive, which indicate relative rates of decomposition. The active soil carbon pool is roughly comparable to the light fraction. The slow carbon pool corresponds to the more reactive humus, and the passive carbon pool is the most stable component of the humus. Although the model of Parton et al. was developed for agricultural or grassland soils, it is useful in conceptualizing carbon flow through pond soils.[12]

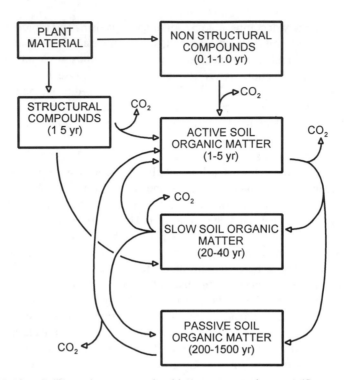

FIGURE 5.10.   Soil organic matter pools with turnover rates in years. (*Source: modified from Parton et al.*[12])

5.4.4 C:N Ratio

When soil organic matter decomposes, nitrogen in organic compounds is either assimilated by soil microorganisms and converted to microbial biomass or mineralized to ammonia N and excreted into the environment. The amount of ammonia N excreted depends largely on the nitrogen concentration of the organic substrate. Soil bacteria usually are 10% nitrogen and 50% carbon on a dry-matter basis. They have a *carbon assimilation efficiency* of 5–10%, which means that 5–10% of substrate carbon is converted to microbial biomass. The remaining substrate carbon is transformed to carbon dioxide or other metabolites. The amount of nitrogen required per unit of substrate carbon decomposed is estimated in the following example.

Example 5.5: Nitrogen Requirements of Soil Bacteria
    The amount of nitrogen needed by bacteria in decomposing 1 g of substrate carbon will be computed for carbon assimilation efficiencies of 5% and 10%.

Solution:
5% carbon assimilation efficiency

$$1 \text{ g substrate C} \times 0.05 \text{ g bacterial C (g substrate C)}^{-1}$$
$$= 0.05 \text{ g bacterial C}$$

10% carbon assimilation efficiency

$$1 \text{ g} \times 0.10 \text{ g} = 0.10 \text{ g bacterial C}$$

Bacteria are 50% carbon, so 0.1–0.2 g bacteria biomass will be produced per gram of substrate C.
Bacteria are 10% nitrogen; 0.1–0.2 g bacteria contains 0.01–0.02 g N.
The nitrogen requirements for bacteria are 0.01–0.02 g N (g substrate C)$^{-1}$.

Actinomycetes and fungi are less efficient than bacteria in decomposing organic matter. Carbon assimilation efficiencies are 15–30% for actinomycetes and 30–40% for fungi. Actinomycetes and fungi convert a larger percentage of substrate carbon to biomass than do bacteria. Actinomycetes and fungi are about 50% carbon and 5% nitrogen. Repetition of calculations made in Example 5.5 for actinomycetes and fungi give nitrogen requirements of 0.03–0.04 and 0.03–0.06 g N (g substrate C)$^{-1}$, respectively. Nitrogen in excess of microbial requirements will be released to the environment or *mineralized*. Nitrogen contained in microbial biomass will be mineralized when microbes die and become soil organic matter.

Example 5.6: Mineralization of Organic N
    A substrate containing 40% C and 2% N is decomposed by bacteria. Bacteria are 50% C, 10% N, and have a carbon assimilation efficiency of 7.5%. The amount of organic N mineralized per gram of residue decomposed will be computed.

Solution:

$$\text{Substrate C} = 1 \text{ g} \times 0.4 \text{ g C (g substrate)}^{-1} = 0.4 \text{ g}$$
$$\text{Substrate N} = 1 \text{ g} \times 0.02 \text{ g N (g substrate)}^{-1} = 0.02 \text{ g}$$
$$\text{Bacterial C} = 0.4 \text{ g C} \times 0.075 \text{ g bacterial C (g substrate C)}^{-1} = 0.03 \text{ g}$$
$$\text{Bacterial biomass} = 0.03 \text{ g bacterial C} \div 0.5 \text{ g C (g bacteria)}^{-1} = 0.06 \text{ g}$$
$$\text{Bacterial N} = 0.06 \text{ g bacteria} \times 0.1 \text{ g N (g bacteria)}^{-1} = 0.006 \text{ g}$$
$$\text{N mineralized} = \text{Substrate N} - \text{bacterial N} = 0.02 \text{ g} - 0.006 \text{ g} = 0.014 \text{ g}$$

Decomposition of an organic residue with a high concentration of nitrogen results in greater nitrogen mineralization than decomposition of a residue with a low nitrogen concentration.
    A residue of low nitrogen concentration cannot supply enough nitrogen to satisfy bacterial nitrogen requirements. If in Example 5.6 the residue had a nitrogen concentration of 0.5% instead of 2%, 1 g substrate would contain only 0.005 g nitrogen. Decomposition of 1 g substrate would require 0.006 g of nitrogen to satisfy bacterial needs, and there would be a nitrogen shortage of

0.001 g. The substrate would not decompose rapidly. Complete decomposition of the residue would require that some bacteria die and their nitrogen be mineralized for use by new generations of bacteria to complete decomposition of the residue. When a nitrogen deficient residue is added to soil, microbes remove nitrate or ammonia N from the soil solution and use it to supplement their nitrogen needs. Removal of inorganic N from soil by microorganisms is known as nitrogen *immobilization*.

A residue with a narrow C:N ratio decomposes quickly and completely with mineralization of organic N. A residue with a wide C:N ratio normally decomposes slowly and incompletely with immobilization of soluble inorganic N. In most soils an equilibrium is reached where inputs of organic matter and nitrogen are balanced by losses of these two substances, and soil organic matter concentrations remain stable. A residue with a narrow C:N ratio will completely decompose, and ammonia N released from the residue can be used by soil microbes to accelerate decomposition of organic matter present before the residue was added. Soil organic matter concentration will decline. A residue of low nitrogen concentration will not decompose quickly or completely, and the soil organic matter concentration will increase. Also, the soluble inorganic N concentration in soil will decline because of nitrogen immobilization induced by nitrogen deficiency.

Boyd used bioassays with the algae *Scenedesmus dimorphus* to determine availability of nitrogen for algal growth from different kinds of decomposing organic matter.[13] Growth of algae in cultures increased in proportion to percentage nitrogen in aquatic plant residues. Algal abundance in a culture containing 10 mg of an aquatic plant residue with 4.36% N was more than 40 times greater than in a culture with 10 mg of aquatic plant residue containing 1.09% N. Similar results were obtained when pond soils of different nitrogen concentrations were used as nitrogen sources in cultures. Growth was about four times greater in an algal culture where soil contained 0.30% N than in a culture where soil contained 0.01% N. These results suggest that ammonia N released from organic residues and soil organic matter can be used by plants in aquatic ecosystems.

Decomposition rates of aquatic plant residues increased in proportion to concentrations of nitrogen in the residues,[14] and residues with low nitrogen concentrations immobilized nitrogen from culture solutions (Fig. 5.11). One reason that plant residues with a high nitrogen concentration decompose faster is that more nitrogen is available for microorganisms. Another reason is that the proportion of cell-wall material to non-cell-wall material in plants decreases with increasing nitrogen content.[15]

5.5 Soil Respiration

Soil respiration is an index of the activity of soil microorganisms. Respiration of aquatic soils is particularly important because it represents an oxygen demand that can result in dissolved oxygen depletion in ponds and other water bodies.

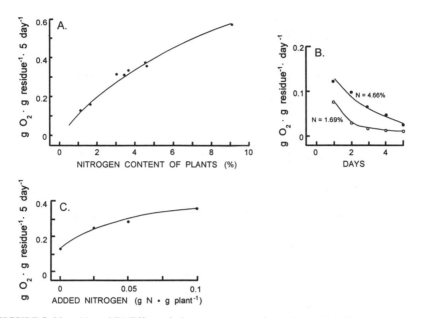

FIGURE 5.11.   (A and B) Effect of nitrogen content of aquatic plant residues on oxygen consumption rates during decomposition. (C) Effect of nitrogen supplementation on oxygen consumption during decomposition of low-nitrogen-content (1.48%) aquatic plant residue.

5.5.1 Terrestrial Soils

Many studies have been conducted on the respiration rate of terrestrial soils, and only a brief summary is provided here. The usual range of respiration is 10–100 kg CO_2 ha^{-1} day^{-1}, and most soils have values of 25–40 kg CO_2 ha^{-1} day^{-1}. Rates are greater in warm weather and vary with soil pH, moisture content of soil, oxygen supply, and nitrogen availability. Rates are highest near the soil surface, because organic matter concentrations are greater near the surface and oxygen is more available. Soil respiration often can be increased by drying flooded soils, increasing the pH of acidic soils by liming, cultivating to improve aeration, and using nitrogen fertilization. Cultivation causes soil organic matter concentrations to decline. In the southeastern United States, organic matter concentrations decrease from 3% to 5% to 1% to 2% after 20–30 years of cultivation. Cultivation causes an especially rapid decline in organic matter of tropical soils.

Example 5.7: Estimation of Organic Matter Loss
    The average CO_2 loss from a soil is 8 g CO_2 m^{-2} day^{-1}, and the annual organic input to the soil is estimated as 100 g m^{-2}. The initial soil organic matter concentration was 3.5%. What is the concentration of organic matter after one year. Assume that bulk density is 1.4 ton m^{-3}, all gains and losses of organic matter occur in the surface 20 cm, and organic matter = carbon × 1.9.

Solution:
Organic matter (OM) loss

$$8 \text{ g CO}_2 \text{ m}^{-2} \text{ day}^{-1} \times \frac{12 \text{ g C}}{44 \text{ g CO}_2} \times 1.9 \text{ g OM g C}^{-1} \times 365 \text{ day}^{-1}$$
$$= 1,513 \text{ g OM m}^{-2} \text{ yr}^{-1}$$

Initial OM

$$1,400 \text{ kg soil m}^{-3} \times 0.2 \text{ m} \times 0.035 \text{ kg OM kg soil}^{-1} \times 10^3 = 9,800 \text{ g OM m}^{-2}$$

Change in OM

$$\text{Input} - \text{loss} = (100 - 1,513) \text{ g m}^{-2} \text{ yr}^{-1} = -1,413 \text{ g OM m}^{-2}$$

New OM concentration

$$(9,800 - 1,413) \text{ g OM m}^{-2} = 8,387 \text{ g OM}^{-2}$$

$$\frac{8.387 \text{ kg OM m}^{-2}}{1,400 \text{ kg soil m}^{-3} \times 0.2 \text{ m}} \times 100 = 3.00\%$$

5.5.2 Natural Aquatic Soils

Respiration by aquatic soils usually is reported in terms of dissolved oxygen uptake rate.[16] Bottom soil respiration in natural bodies of water has been studied extensively. The influences of temperature and pH on respiration of agricultural soil mentioned earlier have also been noted for aquatic soils. Most studies have shown that respiration was independent of oxygen concentration between 3 and 8 mg L^{-1} of dissolved oxygen.[17,18] When other factors are optimal for respiration, its rate apparently depends on the amount of organic substrate and the rate at which oxygen can enter the soil. Oxygen delivery rate is critical. Where sludge covers the bottom, there is no relationship between sludge depth and oxygen uptake.[16] This observation suggests an oxygen limitation. Resuspension of sediment also increases oxygen uptake, apparently because of greater availability of oxygen in the water above the soil than in pore water. Aquatic soil respiration also increases as the velocity of water passing over the bottom increases.[19] Water flowing across the bottom replenishes the oxygen supply, drives oxygen into the bottom soil surface, and resuspends sediment. In shallow-water areas, photosynthetic activity of benthic algae supplies oxygen for use in soil respiration during daylight, but algal biomass contributes to respiration.

Not all oxygen consumed by aquatic soils is used in microbial respiration. Invertebrates living in the soil consume oxygen,[16] and some oxygen is used to oxidize reduced inorganic substances at the interface between aerobic and anaerobic soil layers.[20] Oxygen movement into aquatic soil can be facilitated by activities of borrowing invertebrates. Bottom-feeding fish can resuspend large amounts of sediment.[21]

In labratory systems where adequate dissolved oxygen concentrations were maintained in the water above aquatic soil samples, respiration rates up to 20–30 g O_2 m^{-2} day^{-1} have been measured.[16] High uptakes of oxygen by aquatic soils have been measured *in situ* with continuous-flow respiration chambers,[19] but *in situ* soil oxygen consumption rates for natural bodies of water were generally below 5 g O_2 m^{-2} day^{-1} and most were much lower.

5.5.3 Pond Soils

Rates of oxygen uptake by soils of Alabama ponds stocked with 3,000 channel catfish ha^{-1} ranged from 0.19 to 2.74 g m^{-2} day^{-1}; 1.46 mg O_2 m^{-2} day^{-1} was the median value.[22] Schroeder reported bottom soil respiration rates of 1–3 g O_2 m^{-2} day^{-1} in carp ponds in Israel stocked at 7,000–14,000 fish ha^{-1}.[23] In another study of Israeli carp ponds, soil respiration increased as the growing season progressed, but water-column respiration did not change greatly with time.[24] The greatest rate of soil respiration in a pond stocked with 20,000 carp ha^{-1} was 2.2 g O_2 m^{-2} day^{-1}, compared with 1.3 g O_2 m^{-2} day^{-1} for a pond stocked with 7,000 carp ha^{-1}. Soil respiration rates of 1.5–3.8 g O_2 m^{-2} day^{-1} were reported for brackish-water ponds in Alabama stocked at 5,000 ha^{-1} striped bass.[25] Soil respiration rates up to 1.2 g O_2 m^{-2} day^{-1} were measured in fertilized ponds stocked with 20,000 tilapia ha^{-1} in Honduras.[26] Schroeder studied the carbon budget of heavily measured, polyculture ponds (11,000–13,000 fish ha^{-1}) in Israel.[27] Soil respiration rates were 3–4 g O_2 m^{-2} day^{-1}. He concluded that the maximum rate of oxygen diffusion into pond soil was about 4 g m^{-2} day^{-1} and suggested that aerobic respiration in the bottoms of intensive aquaculture ponds is limited by oxygen supply. Schroeder observed that the thickness of the aerobic soil layer was less than 0.5 mm. This implies that all of the oxygen diffusing through the soil surface was used in aerobic respiration before it reached a depth of 0.5 mm. Howeler and Bouldin reported that the thickness of the oxidized soil in laboratory soil–water systems ranged from 0.5 to 2.4 mm.[20]

Data for pond soil respiration just summarized reveal that oxygen consumption in semi-intensive aquaculture ponds is about 1–2 g m^{-2} day^{-1}, whereas values up to 4 g m^{-2} day^{-1} occur in intensive ponds. To compare these data with typical terrestrial soil respiration data, we must convert them to a carbon dioxide basis.

Example 5.8: Conversion of Oxygen Consumption Rate to Carbon Evolution Rate
A factor for converting 1 g O_2 m^{-2} day^{-1} to g CO_2 m^{-2} day^{-1} will be calculated.

Solution: A RQ of 1 will be assumed, and for each mole of O_2 consumed a mole of CO_2 will be released.

$$1 \text{ mg } O_2 \text{ m}^{-2} \text{ day}^{-1} \times \frac{44 \text{ g } CO_2}{32 \text{ g } O_2} = 1.375 \text{ g } CO_2 \text{ m}^{-2} \text{ day}^{-1}$$

Typical CO_2 evolution rates for terrestrial soils were given as 10–100 kg CO_2 ha^{-1} day^{-1}. Pond soil values ranged from 0.2 to 4 g O_2 m^{-2} day^{-1}; this corresponds to 0.28–5.5 g CO_2 m^{-2} day^{-1}. This range of CO_2 evolution is equal to 2.8–55 kg CO_2 ha^{-1} day^{-1} and indicates that CO_2 evolution from pond soils may occur at a slower rate than for most terrestrial soils.

Aeration adds oxygen to pond water and improves the circulation of oxygenated water over the bottom. Although organic matter concentrations increased in pond soils during grow-out periods of six to eight months, the increase was less in aerated ponds than in unaerated ponds.[28,29] Ghosh and Mohanty showed that aeration of laboratory soil–water systems increased nitrogen mineralization rates by 20–41%.[30] This implies that soil respiration also increased.

Soil organic matter concentrations decline after harvest because fine soil particles containing organic matter are swept from the pond bottom by outflowing water.[29,31] Fish farmers often dry pond bottoms after draining to oxidize the soil and accelerate decomposition of organic matter,[32,33] but the effectiveness of this practice is not well documented. A slight decrease in soil organic matter concentrations were noted after a five-week fallow period in ponds in Honduras.[29] Soil respiration during the fallow period in ponds in Honduras was compared with respiration during the grow-out period (Fig. 5.12). Soil respiration averaged 4.7 g CO_2 m^{-2} day^{-1} during the fallow periods and only 0.75 g CO_2 m^{-2} day^{-1} during the grow-out period.[26] The total soil respiration during the first fallow period of 12 days was 68% of the total respiration during the 152-day grow-out period. Note that estimates are for aerobic respiration only.

Boyd and Pippopinyo studied the influence of drying, liming, tilling, bacterial augmentation, and nitrogen fertilization on respiration of Alabama pond soils exposed to air in laboratory respiration chambers.[34] The optimal soil moisture concentration for respiration was 12–20%, and further drying decreased soil respiration (Fig. 5.13). Optimal soil moisture content differs with the kind and amount of clay in pond soils. The optimal moisture content was 35–45% in fallow pond soils in Honduras.[26] The Alabama pond soils had approximately the same percentage clay as the pond soils in Honduras. In Alabama, clay consisted of kaolinite, mica, and hydrous oxides, and the soil had a specific surface area around 100 m^2 g clay^{-1}. The clay of pond soils in Honduras had a large proportion of smectite, and the surface area was 439 m^2 g clay^{-1}. Because of the greater surface area, soils from ponds in Honduras hold water more tightly and have a larger amount of biologically unavailable water than soils of ponds at Auburn. The optimal moisture concentration for soil respiration varies from site to site, depending on the type and amount of clay and other colloidal particles in the soil.

Soil respiration was greatest at pH 7.5–8.0 (Fig. 5.14). Both calcium hydroxide and calcium carbonate were effective in enhancing soil respiration. However, calcium hydroxide causes a high initial pH, which retards soil respiration for a few days until the pH declines to the optimal range.

Even when soils were held at the optimal pH and moisture concentration for

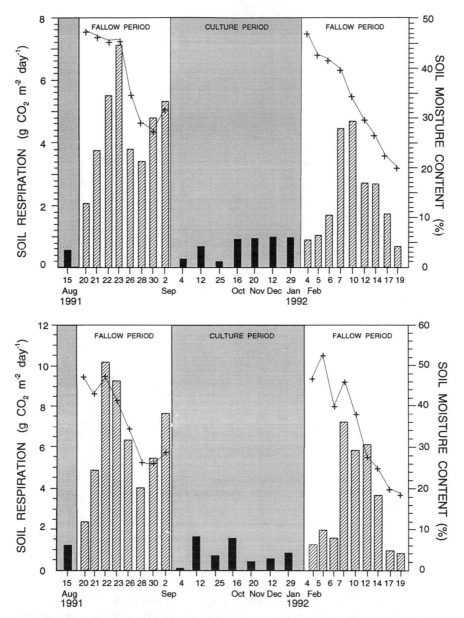

FIGURE 5.12. Soil respiration in ponds during grow-out and fallow periods. Soil moisture concentrations during the fallow period are provided. (*Source: Boyd and Teichert–Coddington.*[26])

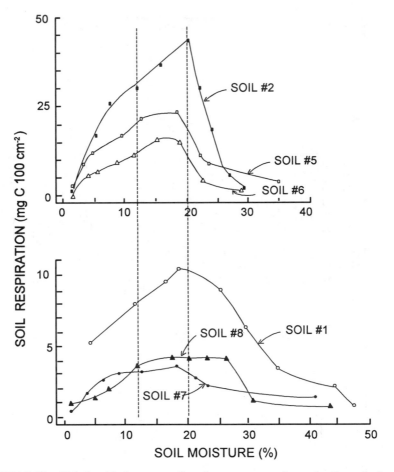

FIGURE 5.13. Relationship between soil moisture concentration and soil respiration in laboratory respiration chambers. (*Source: Boyd and Pippopinyo.*[34])

decomposition, the respiration rate declined with time. After 32 days the respiration rate was only 21.8% of the initial rate. About 70% of the respiration for a 32-day incubation period occurred during the first 16 days. This suggests a steady decline in the amount of readily oxidizable organic substrate as microbial activity progressed.

Pulverization of the soil surface and application of 600 mg kg^{-1} of ammonium nitrate both enhanced soil respiration. Application of a bacterial amendment had no influence on respiration rate.

5.6 Organic Matter Concentrations

Concentrations of organic matter in terrestrial soil vary with climate, vegetative cover, topography, and use. Organic matter in soil originates primarily from

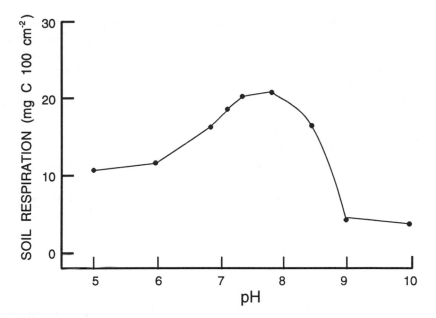

FIGURE 5.14. Relationship between soil pH and soil respiration in laboratory respiration chambers. (*Source: Boyd and Pippopinyo.*[34])

vegetation, so soils with scant vegetative cover are deficient in organic matter. At comparable inputs of vegetal remains, soils in warm climates usually contain less organic matter than soils of cooler climates. For example, soils of the midwestern United States have appreciably more organic matter than soils of the southeastern United States. Poorly drained soils in wetland areas accumulate organic matter, and organic soils often form in wetlands. Erosion of steep slopes removes organic matter and deposits it in level areas. River valley soils may contain more organic matter than upland soils. Cultivation of soils generally causes a reduction in amounts of organic matter. Concentrations of organic matter in terrestrial soils range from essentially 0% in some desert soils to nearly 100% in some organic soils. Most agricultural soils in humid, subtropical, and tropical areas contain less than 5% organic matter, whereas agricultural soils in colder regions may contain more organic matter. This can be illustrated by considering soil organic matter in the southern United States (Texas), midwestern United States (Iowa), and Canada (Saskatchewan). Loams in Texas often contain about 1–2% organic matter; loams in Iowa contain 2–5% organic matter; loams in Saskatchewan had an average organic matter content of 7.86% (range 2.68–11.44%).[35]

Surface layers of soil often are removed from pond bottom areas and used in constructing levees. Soil organic matter concentrations are greater in surface layers and decrease rapidly with soil depth, and removal of the surface soil from the bottom area during construction tends to give a bottom soil of low-organic-matter content. Organic soils are not ideal for aquaculture ponds, but they are sometimes used because sites with mineral soils are not available. The layer of organic soil may be from less than a meter to several meters deep. It usually is not possible to provide a low-organic-matter content bottom for new ponds built at a site with organic soils.

Soil organic matter in new ponds is the remains of terrestrial or wetland vegetation. In mineral soils the organic matter is composed primarily of humus. Vegetation and litter are removed from the surface, but plant roots may remain in the soil mass. Vegetation and litter also are removed from pond bottoms at sites with organic soils, but soil contains large amounts of woody and fibrous material.

Although the organic matter input to soils in aquaculture ponds consists mostly of low-fiber-content material that is readily decomposable by soil microorganisms, it does not decompose completely. Studies have shown that soil organic matter concentrations in aquaculture ponds constructed on mineral soils tend to increase over time.[13,29,36] Organic carbon input to the bottom of an intensive polyculture pond was 0.5 g m^{-2} day^{-1} greater than the release of carbon in respiration.[27] Zur showed that the increase in organic carbon in carp ponds was 2,540–8,900 kg ha^{-1} yr^{-1}.[37] Large amounts of organic matter mixed with mineral sediment can accumulate in bottoms of intensive shrimp ponds.[38,39]

The decomposition of organic matter in aquatic soils usually follows first-order kinetics.[40,41]:

$$\ln\frac{C}{C_i} = -Kt \quad \text{or} \quad C = C_i e^{-Kt} \tag{5.18}$$

where t = time

C_i = initial organic matter concentration at $t = 0$

C = final organic matter concentration at $t = 1$

K = rate constant

e = 2.203 (base of natural logarithms)

Rate constants (K) were 10^{-2}–10^{-3} yr^{-1} for organic carbon decomposition in lake and large reservoir soil.[41] Soil in a pond containing 7,000 kg ha^{-1} of carp and tilapia had $K = 0.2$ yr^{-1}; and $K = 0.15$ day^{-1} for sediment collected from the bottoms of intensive fish production units.[42] Sediment collected from intensive fish grow-out units was much more concentrated in organic matter than soil from a pond. This sediment consisted of uneaten feed and feces that had not been subjected to appreciable decay before decomposition rates were measured.

In a pond soil the rate of accumulation of organic matter is the difference between input rate and decomposition rate. According to Avnimelech, effects of both inputs and decomposition on concentrations of soil organic matter can be estimated as follows:

$$C = \frac{B - e^{-Kt}(B - KC_i)}{K} \tag{5.19}$$

where B is the organic matter influx to the soil.[42] This equation implies that the organic matter concentration in the sediment will eventually reach an equilibrium concentration because

$$\lim_{t \to \infty} C = \frac{B}{K} \tag{5.20}$$

The potential organic matter concentration increases linearly over time in response to influx B, but the potential organic matter concentration cannot be obtained because of decomposition. At a given influx rate the accumulation rate depends on the magnitude of K.

Schroeder found that the carbon influx to the soil in an intensive polyculture pond averaged 7.7 g C m^{-2} day^{-1}.[27] Disregarding seasonal effects, this represents an influx of 2.81 kg C m^{-2} yr^{-1}. This may be considered a very high input; most ponds would have influxes of 1.5 kg C m^{-2} yr^{-1} or less.

Example 5.9: Estimation of Organic Carbon Accumulation over Time

Assume that pond soil contains 1 kg C m^{-2} in the upper 10-cm layer, its bulk density is 950 kg m^{-3}, the carbon influx is 1 kg C m^{-2} yr^{-1}, and $K = 0.2$ yr^{-1}. Estimate the change in carbon content of the soil over time until equilibrium is attained.

Solution: Equation 5.19 must be solved for each year. The value of C computed for the end of the first year must be used as C_0 for calculating C at the end of the second year, and so on.

Year 1 $C = \dfrac{B - e^{-Kt}(B - KC_i)}{K}$

$$C = \frac{1 - 2.303^{-(0.2)(1)}(1 - (0.2)(1))}{0.2} = 1.61 \text{ g C m}^{-2}$$

Year 2 $$C = \frac{1 - 2.303^{-(0.2)(1)}(1 - (0.2)(1.61))}{0.2} = 2.13 \text{ kg C m}^{-2}$$

Year 3 $$C = \frac{1 - 2.303^{-(0.2)(1)}(1 - (0.2)(2.13))}{0.2} = 2.57 \text{ kg C m}^{-2}$$

.
.
.

Year 29 $$C = \frac{1 - 2.303^{-(0.2)(1)}(1 - (0.2)(4.96))}{0.2} = 4.97 \text{ kg C m}^{-2}$$

Year 30 $$C = \frac{1 - 2.303^{(-0.2)(1)}(1 - (0.2)(4.97))}{0.2} = 4.97 \text{ kg C m}^{-2}$$

The quantities of carbon can be converted to percentage carbon as follows:

Year 1 $\qquad \dfrac{1.61\ \text{kg C m}^{-2}}{(950\ \text{kg m}^{-3})(0.10\ \text{m})} \times 100 = 1.69\%$

Year 2 $\qquad \dfrac{2.13\ \text{kg C m}^{-2}}{(950\ \text{kg m}^{-3})(0.10\ \text{m})} \times 100 = 2.24\%$

Year 3 $\qquad \dfrac{2.57\ \text{kg C m}^{-2}}{(950\ \text{kg m}^{-3})(0.10\ \text{m})} \times 100 = 2.70\%$

.
.
.

Years 29, 30 $\qquad \dfrac{4.97\ \text{kg C m}^{-2}}{(950\ \text{kg m}^{-3})(0.10\ \text{m})} \times 100 = 5.23\%$

Data from Example 5.9 are plotted in Figure 5.15 to illustrate the rate of organic accumulation in a pond. The situation depicted in Figure 5.15 is for a pond with a constant influx or organic carbon and a constant organic carbon mineralization rate. These assumptions are reasonable, for most ponds are stocked and fertilized or fed at about the same rate each year. The organic carbon concentration increased rapidly during the first 10 years (85% of final concentration) and then increased at a much slower rate until equilibrium was obtained in 29 years.

For those wanting a nonmathematical explanation of the attainment of equilibrium organic carbon concentrations in soils, consider that there is an influx of "fresh" organic carbon each year and a pool of "old" organic carbon that has accumulated from previous years. Decomposition of organic carbon involves mineralization of fresh organic carbon and old organic carbon. The influx of fresh carbon is constant each year, there is a constant amount of fresh carbon decomposed each year, and a constant amount of fresh carbon is converted to old carbon at the end of each year. The pool of old carbon increases each year, and, as a result, the amount of old carbon decomposed each year increases. After a period of years, the annual decomposition of old organic matter equals the undecomposed fraction of the annual input of fresh organic carbon. When this situation occurs, there can be no further increase in the amount of organic carbon in the soil and equilibrium is attained.

Changes in the organic carbon influx or changes in K will alter the accumulation pattern of organic carbon in soil. The organic matter influx varies with the intensity of aquaculture; ponds with high production have larger organic matter influxes than ponds with low production. The size of the decay constant K varies with the kinds of organic matter introduced into ponds and environmental conditions affecting microbial activity. Few estimates of carbon influxes and K values have been made, but based on data of Avnimelech, Schroeder, and co-

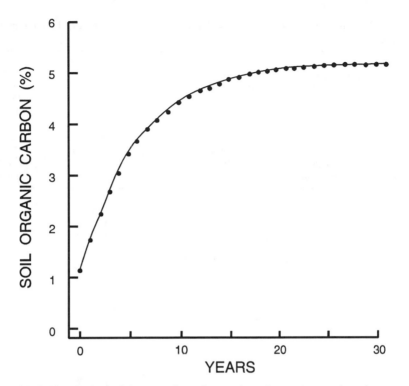

FIGURE 5.15. Estimated increase in soil organic carbon concentrations in a pond. Assumptions: Initial organic C = 1.05%; organic C influx = 1 kg C m^{-2} yr^{-1}; K = 0.2 yr^{-1}.

workers, some possible ranges for carbon influxes and K were estimated for different types of aquatic animal production systems (Table 5.3).

Example 5.10: Effect of K on Organic Carbon Accumulation in Pond Soils
 Assume the same conditions for C_i, soil depth, bulk density, and B as in Example 5.9, and compute changes in carbon content of soil over time until equilibrium is reached. Use K = 0.1, 0.2, 0.3, and 0.5 yr^{-1}.

Solution: Calculate C with Equation 5.19 as in Example 5.9. Use each value of K in a separate series of computations.
 K = 0.1   Equilibrium will be reached in 48 yrs. The equilibrium carbon concentration will be 10.5%.
 K = 0.2   From Example 5.7, an equilibrium carbon concentration of 5.23% is attained in 29 yr.
 K = 0.3   An equilibrium carbon concentration of 3.47% will be reached in 18 yr.
 K = 0.5   An equilibrium carbon concentration of 1.99% will be attained in 9 yr.

Table 5.3   Estimates of Organic Carbon Influx and Decay Constant (K) for Different Aquaculture Systems

Source of Organic Matter	Fish Production	Organic Carbon Influx	K
	$(ton\ yr^{-1})$	$(kg\ C\ m^{-2}\ yr^{-1})$	(yr^{-1})
Extensive ponds	<1	<1	0.02–0.1
Semi-intensive pond	1–5	1–2	0.1–0.3
Intensive ponds	5–20	2–3	0.3–0.5
	$(kg\ m^{-3})$	$(k\ C\ m^{-3}\ day^{-1})$	(day^{-1})
Intensive grow-out units	10–100	0.02–0.2	0.1–0.2
Resuspended organic sediment	—	—	0.4–0.6

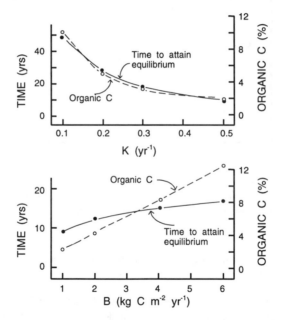

FIGURE 5.16.   (Upper) Influence of the decay constant (K) on equilibrium organic carbon concentration and time required to reach equilibrium in a fish pond. Assumptions: Initial organic C concentration = 1.05%; organic C influx = 1 kg C m^{-2} yr^{-1}. (Lower) Influence of the organic carbon influx (B) on equilibrium organic carbon concentration and time to reach equilibrium in a fish pond. Assumptions: Initial organic C = 1.05%; K = 0.5 yr^{-1}.

Calculations of Example 5.10 show that at the same influx rate, a higher K value will lower the length of time to attain equilibrium and result in a lower carbon concentration at equilibrium (Fig. 5.16).

Example 5.11: Effect of Carbon Flux on Organic Matter Accumulation in Pond Soils
    Assume the same conditions for C_i, soil depth, and bulk density as in Example

5.9, and compute the carbon content of soil over time until equilibrium is reached. Use $K = 0.5$ yr^{-1} and $B = 1$, 2, and 4 kg C m^{-2} yr^{-1}.

Solution: Calculate C with Equation 5.19 as in Examples 5.9 and 5.10. Use each value of B in a separate series of computations.

$B = 1$ Equilibrium will be reached after 9 yr with 2.09% organic carbon.

$B = 2$ An equilibrium carbon concentration of 4.20% will be attained in 12 yr.

$B = 4$ An equilibrium carbon concentration of 8.41% will be attained in 14 yr.

$B = 6$ An equilibrium carbon concentration of 12.62% will be attained in 16 yr.

Example 5.11 shows that a larger carbon influx results in a higher equilibrium concentration at the same value of K (Fig. 5.16). A slightly longer time is required to attain equilibrium at a higher carbon influx.

Results of computations shown in Figures 5.15 and 5.16 are in general agreement with observations on organic carbon concentrations in aquaculture ponds. Soil organic carbon tends to gradually increase in catfish ponds.[36] Old, unfertilized sportfish ponds contained up to 6% soil organic carbon,[13] and old, fertilized ponds did not contain more than 3.5% soil organic carbon.[43] Fertilized ponds had higher nutrient concentrations and larger organic carbon influxes than unfertilized ponds. Channel catfish ponds with feeding (3–6 ton production ha^{-1} yr^{-1}) in Mississippi and Alabama normally contain 1–4% soil organic carbon. Channel catfish ponds with feeding have larger nutrient and organic carbon influxes than do fertilized ponds. Ponds in Israel, which generally have higher production than channel catfish ponds in the United States, have soil organic concentrations as high as 6%.[42] Large amounts of undecomposed organic matter often accumulate in intensive shrimp ponds.[38] Intensive shrimp ponds have larger nutrient and organic carbon influxes than many intensive fish ponds. However, if bottoms of intensive shrimp ponds are dried between crops, most of the organic matter that accumulated during the previous crop decomposes.[44]

The average organic carbon concentrations in 358 soil samples from freshwater ponds and 346 samples from brackish-water ponds were 1.78% and 1.79%, respectively.[45] However, some samples from freshwater ponds contained up to 8% organic carbon, and some samples from intensive brackish-water shrimp ponds contained more than 10% organic carbon (Fig. 5.17). All of these values fall within the range of the computed values in Figures 5.15 and 5.16.

This discussion has focused on organic carbon accumulation in ponds built on mineral soils with low initial organic carbon concentrations. Sometimes ponds are constructed on organic soils. For example, a series of soil samples from shrimp ponds constructed in an area of mostly organic soils near Tumaco, Colombia, contained 10–40% organic C. Environmental conditions in organic soils are not good for rapid decomposition. Soils are usually acidic, low in nitrogen

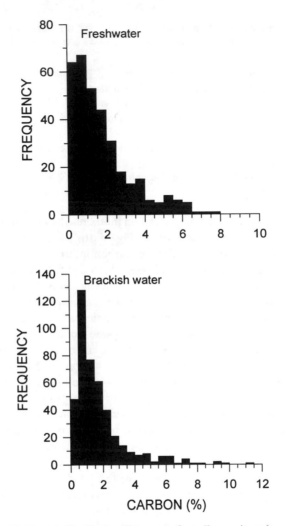

FIGURE 5.17.   Frequency distribution histograms for soil organic carbon concentrations in freshwater and brackish-water aquaculture ponds. (*Source: Boyd et al.*[44])

content, and anaerobic. When converted to shrimp ponds, soils are limed, nitrogen is added in feeds and fertilizers, and efforts are made to maintain aerobic conditions at the soil–water interface. This improves conditions for decomposition, and soil organic matter concentrations usually decline over time in ponds built on organic soils.

Example 5.12: Change in Organic Carbon Concentration in a Pond Bottom Constructed of Organic Soil
   A pond is constructed in a uniform organic soil with a bulk density of 0.5 ton

m^{-3} and an organic carbon concentration of 25%. Assume that the carbon influx will be 1 kg C m^{-2} yr^{-1} and K is 0.2. Changes in organic carbon concentrations in the surface 10-cm layer of soil will be computed.

Solution: The organic carbon content of the surface layer is:

$$(0.25 \text{ kg C kg soil}^{-1})(500 \text{ kg soil m}^{-3})(0.1 \text{ m soil depth}) = 12.5 \text{ kg C}$$

After one year,

$$C = \frac{1 - 2.303^{-(0.2)(1)}(1 - (12.5 \times 0.2))}{0.2}$$

$$= 11.3 \text{ kg C m}^{-2}$$

After two years,

$$C = \frac{1 - 2.303^{-(0.2)(1)}(1 - (11.3 \times 0.2))}{0.2}$$

$$= 10.3 \text{ kg C m}^{-2}$$

Continuing these computations: after 5 yr, $C = 8.26$; after 10 yr, $C = 6.41$; after 15 yr, $C = 5.61$; after 20 yr, $C = 5.26$; after 30 yr, $C = 5.05$ kg C m^{-2}.

Calculate the percentage organic carbon
After one year,

$$\text{Organic C} = \frac{11.3 \text{ kg C m}^{-2}}{(500 \text{ kg m}^{-3})(0.1 \text{ m})} \times 100 = 22.6\%$$

Continuing,
After two years, $C = 20.6\%$; 5 yr, $C = 16.5\%$; 10 yr, $C = 12.8\%$; 15 yr, $C = 11.2\%$; 20 yr, $C = 10.5\%$; 30 yr, $C = 10.1\%$.

Organic matter concentrations in pond soil vary from place to place in the bottom, and gradients in organic matter concentrations also exist. Boyd showed that organic matter concentrations in the 0–5-cm soil layer increased with increasing water depth (Fig. 5.18).[46,47] He concluded that fine organic particles tended to move downslope and accumulate in the deeper water. Decomposition of organic matter also was probably slower in the deeper water because of lower dissolved oxygen concentrations. The pond stratified during summer months, and the hypolimnia were depleted of dissolved oxygen. Organic matter concentrations are greater in the surface layers of soil and tend to decrease with depth.

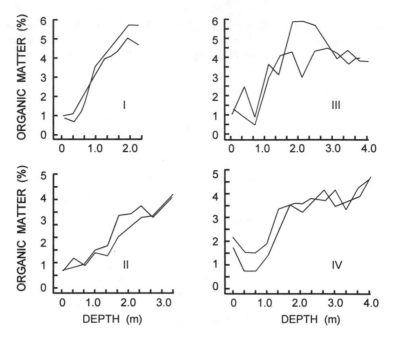

FIGURE 5.18.   Concentrations of organic matter for pond soil samples taken along two transects from shallow to deep water in each of four ponds. (*Source: Boyd, unpublished data.*)

References

1. Smith, J. L., R. I. Papendick, D. F. Bezdicek, and J. M. Lynch. 1993. Soil organic matter dynamics and crop residue management. In *Soil Microbial Ecology,* F. B. Metting, Jr., ed., pp. 65–94. Marcel Dekker, New York.

2. Odum, E. P. *Basic Ecology.* 1983. CBS College Pub., Philadelphia.

3. Anderson, J. M. 1987. Production and decomposition in aquatic ecosystems and implications for aquaculture. In *Detritus and Microbial Ecology in Aquaculture,* D. J. W. Moriarty and R. S. V. Pullin, eds., pp. 123–147. ICLARM Conference Proceedings 14, International Center for Living Aquatic Resources Management, Manila, Phillipines.

4. Minderman, G. 1968. Addition, decomposition, and accumulation of organic matter in forests. *J. Ecol.* **56**:355–362.

5. Swift, M. J., O. W. Neal, and J. M. Anderson. *Decomposition in Terrestrial Ecosystems.* 1979. Blackwell Scientific Publications, Oxford.

6. Stevenson, F., and E. T. Elliott. Methodologies for assessing the quantity and quality of soil organic matter. In *Dynamics of Soil Organic Matter in Tropical Ecosystems.* D. C. Coleman, J. M. Oades, and Goro Vehera, eds., pp. 173–199. University of Hawaii Press, Honolulu.

7. Metting, Jr., F. B. 1992. Structure and physiological ecology of soil microbial communities. In *Soil Microbial Ecology*, F. B. Metting, Jr., ed., pp. 3–25. Marcel Dekker, New York.

8. Alexander, M. *Introduction to Soil Microbiology*. 1977. John Wiley & Sons, New York.

9. Elliott, E. T. 1986. Aggregate structure and carbon, nitrogen, and phosphorus in native and cultivated soils. *Soil Sci. Soc. Amer. J.* **50**:627–633.

10. Gregorich, E. G., R. G. Kachanoski, and R. P. Voroney. 1989. Carbon mineralization in soil size fractions after various amounts of aggregate disruption. *J. Soil Sci.* **40**:649–659.

11. Cambardella, C. A., and E. T. Elliott. 1992. Particulate soil organic-matter changes across a grassland cultivation sequence. *Soil Sci. Soc. Amer. J.* **56**:777–783.

12. Parton, W. J., D. S. Schimel, C. V. Cole, and D. S. Ojima. 1987. Analysis of factors controlling soil organic matter levels in Great Plains grasslands. Division S-3—soil microbiology and biochemistry. *Soil Sci. Soc. Amer. J.* **51**:1173–1179.

13. Boyd, C. E. 1974. The utilization of nitrogen from the decomposition of organic matter in cultures of *Scenedesmus dimorphus*. *Arch. Hydrobiol.* **73**:361–368.

14. Almazon, G., and C. E. Boyd. 1978. Effects of nitrogen levels on rates of oxygen consumption during decay of aquatic plants. *Aquatic Bot.* **5**:119–126.

15. Polisini, J. M., and C. E. Boyd. 1972. Relationships between cell wall fractions, nitrogen, and standing crop in aquatic macrophytes. *Ecology* **53**:484–488.

16. Bowman, G. T., and J. J. Delfino. 1980. Sediment oxygen demand techniques: A review and comparison of laboratory and *in situ* systems. *Water Res.* **14**:491–499.

17. Hargraves, B. T. 1969. Similarity of oxygen uptake by benthic communities. *Limnol. Oceanogr.* **14**:801–805.

18. Patmatmat, M. M., R. S. Jones, H. Samborn, and A. M. Bhagwat. *Oxidation of Organic Matter in Sediments*. 1973. Ecological Research Series, USEPA, EPA-660/3-73-005, U.S. Government Printing Office, Washington, DC.

19. James, A. 1974. The measurement of benthal respiration. *Water Res.* **8**:955–959.

20. Howeler, R. H., and D. R. Bouldin. 1971. The diffusion and consumption of oxygen in submerged soils. *Soil Sci. Soc. Amer. Proc.* **35**:202–208.

21. Keen, W. H., and J. Gagliardi. 1981. Effect of brown bullheads on release of phosphorus in sediment and water systems. *Prog. Fish-Cult.* **43**:183–185.

22. Mezainis, V. E. 1977. Metabolic rates of pond ecosystems under intensive catfish cultivation. M.S. thesis, Auburn University, Ala.

23. Schroeder, G. L. 1975. Nighttime material balance for oxygen in fish ponds receiving organic wastes. *Bamidgeh* **27**:65–74.

24. Shapiro, J., and O. Zur. 1981. A simple *in situ* method for measuring benthic respiration. *Water Res.* **15**:283–285.

25. Daniels, H. V., and C. E. Boyd. 1989. Chemical budgets for polyethylene-lined brackishwater ponds. *J. World Aquaculture Soc.* **20**:53–60.

26. Boyd, C. E., and D. Teichert-Coddington. 1994. Pond bottom soil respiration during fallow and culture periods in heavily-fertilized tropical fish ponds. *J. World Aquaculture Soc.* **25**:417–423.

27. Schroeder, G. L. 1987. Carbon and nitrogen budgets in manured fish ponds on Israel's coastal plain. *Aquaculture* **62**:259–279.

28. Gately, R. 1990. Organic carbon concentrations in bottom soils of ponds: Variability, changes over time, and effects of aeration. M.S. thesis, Auburn University, Ala.

29. Ayub, M., C. E. Boyd, and D. Teichert-Coddington. 1993. Effects of urea application, aeration, and drying on total carbon concentrations in pond bottom soils. *Prog. Fish-Cult.* **55**:210–213.

30. Ghosh, S. R., and A. R. Mohanty. 1981. Observations on the effect of aeration on mineralization of organic nitrogen in fish pond soil. *Bamidgeh* **33**:50–56.

31. Schwartz, M. F., and C. E. Boyd. 1994. Effluent quality during harvest of channel catfish from watershed ponds. *Prog. Fish-Cult.* **56**:25–32.

32. Neess, J. C. 1946. Development and status of pond fertilization in Central Europe. *Trans. Amer. Fish. Soc.* **76**:335–358.

33. Wurtz, A. G. *Methods of Treating the Bottom of Fish Ponds and Their Effects of Productivity.* 1960. General Fisheries Council Mediterranean, Studies and Reviews, No. 11, FAO, Rome, Italy.

34. Boyd, C. E., and S. Pippopinyo. 1994. Factors affecting respiration in dry pond bottom soils. *Aquaculture* **120**:283–294.

35. Millar, C. E. *Soil Fertility.* 1955. John Wiley & Sons, New York.

36. Tucker, C. S. *Organic Matter, Nitrogen, and Phosphorus Content of Sediments from Channel Catfish, Ictalurus Punctatus, Ponds.* 1985. Research Report 10, Mississippi Agriculture and Forestry Experiment Station, Mississippi, State Univ., Mississippi State, Miss.

37. Zur. O. 1981. Primary production in intensive fish ponds and a complete organic carbon balance in the ponds. *Aquaculture* **23**:197–210.

38. Boyd, C. E. 1992. Shrimp pond bottom soil and sediment management. In *Proceedings of the Special Session on Shrimp Farming*, J. A. Wyban, ed., pp. 166–181. World Aquaculture Society, Baton Rouge, La.

39. Hopkins, J. S., P. A. Sandifer, and C. L. Browdy. 1994. Sludge management in intensive pond culture of shrimp: Effect of management regime on water quality, sludge characteristics, nitrogen extinction, and shrimp production. *Aquacultural Eng.* **13**:11–30.

40. Johnson, T. C., J. E. Evans, and S. J. Eisenreich. 1982. Total organic carbon in Lake Superior sediments: Comparisons with hemipelagic and pelagic marine environments. *Limnol. Oceanogr.* **27**:481–491.

41. Avnimelech, Y., J. R. McHenry, and J. D. Ross. 1984. Decomposition of organic matter in lake sediments. *Environ. Sci. Technol.* **18**:5–11.

42. Avnimelech, Y. *Reactions in Fish Pond Sediments as Inferred from Sediment Cores*

Data. 1984. Publication No. 341, Technion Israel Institute of Technology, Soils and Fertilizers Research Center, Haifa, Israel.

43. Boyd, C. E. 1970. Influence of organic matter on some characteristics of aquatic soils. *Hydrobiologia* **36**:17–21.

44. Boyd, C. E., P. Munsiri, and B. F. Hajek. 1994. Composition of sediment from intensive shrimp ponds in Thailand. *World Aquaculture* **25**:53–55.

45. Boyd, C. E., M. E. Tanner, M. Madkour, and K. Masuda. 1994. Chemical characteristics of bottom soils from freshwater and brackishwater aquaculture ponds. *J. World Aquaculture Soc.* **25**:517–534.

46. Boyd, C. E. 1976. Chemical and textural properties of muds from different depths in ponds. *Hydrobiologia* **48**:141–144.

47. Boyd, C. E. 1977. Organic matter concentrations and textural properties from different depths in four fish ponds. *Hydrobiologia* **53**:277–279.

48. Waksman, S. A. *Soil Microbiology*. 1952. John Wiley & Sons, New York.

49. Boyd, C. E., and J. M. Lawrence. 1966. The mineral composition of several freshwater algae. *Proc. Annual Conf. Southeastern Assoc. Game Fish Comm.* **20**:413–424.

50. Boyd, C. E. 1968. Fresh-water plants: A potential source of protein. *Econ. Bot.* **22**:359–368.

51. Boyd, C. E. 1973. Amino acid composition of freshwater algae. *Arch. Hydrobiol.* **72**:1–9.

6

Soil Organic Matter, Anaerobic Respiration, and Oxidation–Reduction

6.1 Introduction

Oxidation–reduction reactions are important in pond aquaculture because many biological processes that affect soil condition, water quality, and aquatic animal production are biologically mediated oxidations and reductions. Photosynthesis is a well-known reduction reaction; inorganic carbon in carbon dioxide is reduced to organic carbon in carbohydrate with the capture of energy. Aerobic respiration is an oxidation reaction in which carbon in organic matter is oxidized to carbon dioxide with the release of energy. Respiration by microorganisms decomposing organic matter in pond soil consumes oxygen faster than it can penetrate the soil mass, and only the surface layer is aerobic. Pond soil is vertically stratified according to electron acceptors used by microbes in respiration. Below the thin, aerobic surface layer where oxygen is the terminal electron acceptor in respiration are successive layers where nitrate, iron and manganese, sulfate, and carbon dioxide, respectively, are used as electron acceptors or oxidants in respiration. Fermentation also occurs in anaerobic soil.

By-products of anaerobic respiration are soluble organic compounds, carbon dioxide, ammonia, nitrite, nitrogen gases, ferrous iron, manganous manganese, hydrogen sulfide, hydrogen, and methane. These substances are transported within the soil profile and into water above the soil by diffusion, seepage, and sediment disturbances. Some reduced substances, such as nitrite and hydrogen sulfide, are highly toxic to aquatic animals.

The oxidation–reduction potential or *redox potential* is an index of the degree of oxidation or reduction in chemical systems. Redox potential of pond soils is sometimes measured by shrimp and fish farmers as an index of soil condition. The writer is convinced that most practical aquaculturists do not understand the principles of oxidation–reduction or know how to interpret redox potential measurements. Furthermore, aquaculture technicians and scientists seldom have more than a vague concept of oxidation–reduction principles. This chapter explains the redox potential as simply as possible, discusses anaerobic respiration, and comments on the role of redox potential in aquaculture.

6.2 Oxidation–Reduction Reactions

A simple oxidation reaction is

$$H_2^0 + Cl_2^0 \rightarrow 2H^+ + 2Cl^- \tag{6.1}$$

In this reaction, hydrogen is oxidized because it loses electrons and becomes more positive in valence. Chlorine is reduced because it gains the electrons lost by hydrogen and becomes more negative in valence. Two equations can be written to show the transfer of electrons:

$$H_2^0 = 2H^+ + 2e^- \tag{6.2}$$

$$Cl_2^0 + 2e^- = 2Cl^- \tag{6.3}$$

Hydrogen is the reducing agent and chlorine is the oxidizing agent. The oxidizing agent is always reduced, and the reducing agent is always oxidized. The number of electrons gained by the oxidizing agent must equal the number of electrons lost by the reducing agent.

A more complex oxidation–reduction reaction is

$$2KMnO_4 + 10FeSO_4 + 8H_2SO_4 \rightarrow 5Fe_2(SO_4)_3 + K_2SO_4 + 2MnSO_4 + 8H_2O \tag{6.4}$$

In this reaction, Fe^{2+} in $FeSO_4$ is oxidized to Fe^{3+} in $Fe_2(SO_4)_3$, and Mn^{7+} in $KMnO_4$ is reduced to Mn^{2+} in $MnSO_4$. The separate oxidation and reduction equations for illustrating electron transfer are

$$10Fe^{2+} \rightarrow 10Fe^{3+} + 10e^- \tag{6.5}$$

$$2Mn^{7+} + 10e^- \rightarrow 2Mn^{2+} \tag{6.6}$$

Ferrous(II) ion is the reducing agent and manganese(VII) ion is the oxidizing agent. Notice again that the reducing agent is oxidized and the oxidizing agent is reduced, and that electrons lost by the reducing agent are accepted by the oxidizing agent. In addition to the transfer of electrons, some oxidation–reduction reactions need hydrogen ion, hydroxyl ion, or water to proceed. Hydrogen ion was required in Equation 6.4, and water was produced.

Oxygen is a strong oxidizing agent, and hydrogen is a strong reducing agent.

These two elements are used as standards against which strengths of other oxidizing and reducing agents are compared. One can identify an oxidation–reduction reaction by transfers of electrons and changes of valence. Oxidations also can involve loss of hydrogen or gain of oxygen, and loss of oxygen and acquisition of hydrogen can occur in a reduction. However, all gains or loses of hydrogen and oxygen do not constitute oxidations or reductions.

6.3 Oxidation–Reduction Potential

Principles of the law of mass action, free-energy change of reaction, and the equilibrium constant can be applied to oxidation–reduction reactions. The driving force for oxidation–reduction reactions can be expressed in terms of a measurable electrical current. Consider a typical oxidation–reduction reaction such as

$$I_2 + H_2 = 2H^+ + 2I^- \tag{6.7}$$

Iodine is reduced to iodide, and hydrogen is oxidized to hydrogen ion. The reaction can be divided into two parts, one showing the loss of electrons (e^-) in oxidation and one showing the gain of electrons in reduction as follows:

$$I_2 + 2e^- = 2I^- \tag{6.8}$$

$$H_2 = 2H^+ + 2e^- \tag{6.9}$$

Equations 6.8 and 6.9, called *half-cells* can be added to give Equation 6.7. The electrons cancel during addition. Equation 6.7 is called a *cell* because it can be separated into two half-cell reactions.

The standard free energies of reaction for Equations 6.8 and 6.9 are

$$\Delta Fr^{\circ}_{(6.8)} = 2\Delta F_f^{\circ}I^- - \Delta Fr_f^{\circ}I_2 - 2\Delta F_f^{\circ}e^-$$

$$\Delta Fr^{\circ}_{(6.9)} = 2\Delta F_f^{\circ}H^+ + 2\Delta F_f^{\circ}e^- - \Delta F_f^{\circ}H_2$$

The $\Delta F_f^{\circ}e^-$ term cancels during addition of the right-hand sides of these expressions, and their sum

$$2\Delta F_f^{\circ}H^+ + 2\Delta_f^{\circ}I^- - \Delta F_f^{\circ}H_2 - \Delta F_f^{\circ}I_2 \tag{6.10}$$

is the same as ΔF_r° for Equation 6.7. Because the $\Delta F_f^{\circ}e^-$ term cancels when the two half-cells are added, it is permissible to assign a value of zero to $\Delta F_f^{\circ}e^-$. Also, from Table 3.2, $\Delta F_f^{\circ}H^+ = \Delta F_f^{\circ}H_2 = 0$. Thus, $\Delta F_r^{\circ} = 0$ for the reaction in Equation 6.9.

6.3.1 Standard Hydrogen Electrode

Equation 6.9 is called the *hydrogen half-cell* and is often written as

$$\tfrac{1}{2}H_2 = H^+ + e^- \tag{6.11}$$

The hydrogen half-cell is a very useful expression. Any oxidation–reduction reaction can be written as two half-cell reactions with the hydrogen half-cell functioning as either the electron-donating or electron-accepting half-cell:

$$Fe^{3+} + e^- = Fe^{2+}, \qquad \tfrac{1}{2}H_2 = H^+ + e^- \tag{6.12}$$

$$Fe(s) = Fe^{2+} + 2e^-, \qquad 2H^+ + 2e^- = H_2(g) \tag{6.13}$$

In Equation 6.12, hydrogen is the reductant because it donates electrons and reduces Fe^{3+} to Fe^{2+}. In Equation 6.13, H^+ is the oxidant because it accepts electrons and oxidizes Fe(s) to Fe^{2+}.

The flow of electrons between half-cells can be measured as an electric current. The chemical reaction involving reduction of I_2 by H_2 (Eq. 6.7) can be used to make a cell of two half-cells in which the flow of electrons can be measured (Fig. 6.1). This cell consists of a solution of $1M$ hydrogen ion and a solution of $1M$ iodine. A Platinum electrode coated with platinum black and bathed in

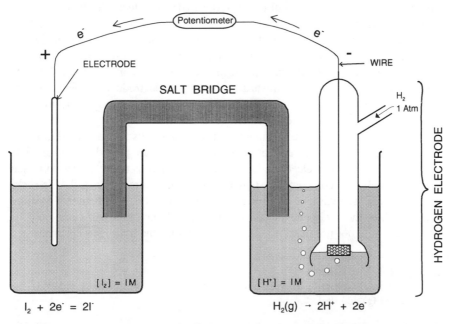

FIGURE 6.1. The hydrogen half-cell or electrode connected to an iodine–iodide half-cell.

hydrogen gas at 1 atm is placed in the solution of $1M$ hydrogen ion to form the hydrogen half-cell or *hydrogen electrode*. A shiny platinum electrode is placed in the iodine solution to form the other electrode. A platinum wire connected between the two electrodes allows free flow of electrons between the two half-cells. A salt bridge connected between the two solutions allows ions to migrate from one side to the other and maintain electrical neutrality. The flow of electrons is measured with a potentiometer. Electron flow in the cell shown in Figure 6.1 is from the hydrogen electrode to the iodine solution. The iodine-iodide half-cell is initially more oxidized than the hydrogen electrode, and electrons move to the iodine solution and reduce I_2 to I^-. The oxidation of H_2 to H^+ is the source of the electrons. Electrons continue to flow from the hydrogen electrode to the iodine–iodide half-cell until equilibrium is reached.

The voltage for the cell in Figure 6.1 is initially 0.62 V. This voltage is called the *standard electrode potential* (E°) for the cell. The value of E° declines as the reaction proceeds, and at equilibrium $E^\circ = 0$ V. The standard electrode potential refers to the voltage that develops between the standard hydrogen electrode (half-cell) and any other half-cell under standard conditions (unit activities and 25°C). The electrons transferred between the half-cells to drive the oxidation–reduction reaction to equilibrium do not always flow in the direction shown in Figure 6.1. In some cases the standard hydrogen electrode may be more oxidized than the other half-cell, and electrons will flow toward the hydrogen electrode and the reduction will occur at the hydrogen electrode. It is necessary to give a sign to E°. The positive sign usually is applied when electrons flow away from the hydrogen electrode and toward the other half-cell as in Figure 6.1. The positive sign means that the other half-cell is more oxidized than the hydrogen electrode. When the hydrogen electrode is more oxidized than the other half-cell, electrons flow toward the hydrogen electrode and a negative sign is applied to E°. The sign is sometimes applied to E° in the opposite manner, and the sign of E° can cause considerable confusion.

Values of E° have been determined for many half-cells and tabulated in reference works. Selected half-cell reactions and E° values are provided in Table 6.1. The E° of the hydrogen electrode is 0 V, and this value is the reference to which other half-cells are compared. A half-cell with an E° greater than 0 V is more oxidized than the hydrogen electrode, and one with an E° less than 0 V is more reduced than the hydrogen electrode. The E° of any two half-cells may be compared; it is not necessary to compare only with the hydrogen electrode.

Example 6.1: Evaluation of E° Values

Refer to E° values to determine if Mn^{4+} is more oxidized or more reduced than $O_2(g)$ and Fe^{3+}.

Solution: The half-cells and E° values from Table 6.1 are

Table 6.1   Standard Electrode Potentials at 25°C

Reaction	$E°$ (V)
$O_3(g) + 2H^+ + 2e^- \rightleftharpoons O_2(g) + H_2O$	+2.07
$Mn^{4+} + e^- \rightleftharpoons Mn^{3+}$	+1.65
$2HOCl + 2H^+ + 2e^- \rightleftharpoons Cl_2(aq) + 2H_2O$	+1.60
$MnO_4^- + 8H^+ + 5e^- \rightleftharpoons Mn^{2+} + 4H_2O$	+1.51
$Cl_2(aq) + 2e^- \rightleftharpoons 2Cl^-$	+1.39
$Cl_2(g) + 2e^- \rightleftharpoons 2Cl^-$	+1.36
$Cr_2O_7^{2-} + 14H^+ + 6e^- \rightleftharpoons 2Cr^{3+} + 7H_2O$	+1.33
$O_2(aq) + 4H^+ + 4e^- \rightleftharpoons 2H_2O$	+1.27
$2NO_3^- + 12H^+ + 10e^- \rightleftharpoons N_2(g) + 6H_2O$	+1.24
$MnO_2(s) + 4H^+ + 2e^- \rightleftharpoons Mn^{2+} + 2H_2O$	+1.23
$O_2(g) + 4H^+ + 4e^- \rightleftharpoons 2H_2O$	+1.23
$Fe(OH)_3(s) + 3H^+ + e^- \rightleftharpoons Fe^{2+} + 3H_2O$	+1.06
$NO_2^- + 8H^+ + 6e^- \rightleftharpoons NH_4^+ + 2H_2O$	+0.89
$NO_3^- + 10H^+ + 8e^- \rightleftharpoons NH_4^+ + 3H_2O$	+0.88
$NO_3^- + 2H^+ + 2e^- \rightleftharpoons NO_2^- + H_2O$	+0.84
$Fe^{3+} + e^- \rightleftharpoons Fe^{2+}$	+0.77
$I_2(aq) + 2e^- \rightleftharpoons 2I^-$	+0.62
$MnO_4 + 2H_2O + 3e^- \rightleftharpoons MnO_2(s) + 4OH^-$	+0.59
$SO_4^{2-} + 8H^+ + 6e^- \rightleftharpoons S(s) + 4H_2O$	+0.35
$SO_4^{2-} + 10H^+ + 8e^- \rightleftharpoons H_2S(g) + 4H_2O$	+0.34
$N_2(g) + 8H^+ + 6e^- \rightleftharpoons 2NH_4^+$	+0.28
$Hg_2Cl_2(s) + 2e^- \rightleftharpoons 2Hg(l) + 2Cl^-$	+0.27[a]
$SO_4^{2-} + 9H^+ + 8e^- \rightleftharpoons HS^- + 4H_2O$	+0.24
$S_4O_6^{2-} + 2e^- \rightleftharpoons 2S_2O_3^{2-}$	+0.18
$S(s) + 2H^+ + 2e^- \rightleftharpoons H_2S(g)$	+0.17
$CO_2(g) + 8H^+ + 8e^- \rightleftharpoons CH_4(g) + 2H_2O$	+0.17
$H^+ + e^- \rightleftharpoons \frac{1}{2}H_2(g)$	0.00
$6CO_2(g) + 24H^+ + 24e^- \rightleftharpoons C_6H_{12}O_6 \text{ (glucose)} + 6H_2O$	−0.01
$SO_4^{2-} + 2H^+ + 2e^- \rightleftharpoons SO_3^{2-} + H_2O$	−0.04
$Fe^{2+} + 2e^- \rightleftharpoons Fe(s)$	−0.44

Source: Snoeyink and Jenkins.[18]

[a] $E° = 0.242$ V for saturated KCl and 0.282 V for 1 M KCl in a calomel electrode.

$$Mn^{4+} + e^- = Mn^{3+} \qquad E° = 1.65 \text{ V}$$

$$O_2(g) + 4H^+ + e^- = H_2O \qquad E° = 1.23 V$$

$$Fe^{3+} + e^- = Fe^{2+} \qquad E° = 0.77 \text{ V}$$

Values of $E°$ indicate that Mn^{4+} is the most oxidized and Fe^{3+} is the least oxidized of the three substances.

Example 6.2: Evaluation of $E°$ Values

$E°$ values reveal if two substances can exist together. For example, can dissolved oxygen and hydrogen sulfide exist together?

Solution: The half-cells and $E°$ values from Table 6.1 are

$$O_2(aq) + 4H^+ + 4e^- = 2H_2O \qquad\qquad E° = 1.27 \text{ V}$$

$$SO_4^{2-} + 10H^+ + 8e^- = H_2S(g) + 4H_2O \qquad E° = 0.34 \text{ V}$$

The $E°$ for the oxygen half-cell is greater than that for the half-cell reaction that yields hydrogen sulfide. If dissolved oxygen is present, any hydrogen sulfide produced will be oxidized to sulfate. The two substrates should not occur in the same environment.

6.3.2 $E°$ and $\Delta Fr°$

The relationship between voltage in cell reactions and standard free-energy change is

$$\Delta Fr° = -nE°F \qquad\qquad (6.14)$$

where $E°$ = voltage of reaction in which all substances are at unit activity
n = number of electrons transferred in cell reaction
F = Faraday constant (23.06 kcal V-gram equivalent^{-1})

Example 6.3: Proof of Equation 6.14
Measured $E°$ for the reaction in Equation 6.8 is 0.62 V (Table 6.1). It will be shown that Equation 6.14 gives the same value of $\Delta F_r°$ as calculated by Equation 3.3 using tabular values of $\Delta F_f°$.

Solution: From Equation 6.14, $\Delta F_r°$ is

$$\Delta F_r° = 2 \times 0.62 \text{ V} \times 23.06 \text{ kcal g-V}^{-1} = 28.6 \text{ kcal}$$

The $\Delta F_r°$ may be calculated from $\Delta F_f°$ values with Equation 3.3:

$$\Delta F_r° = 2\Delta F_f°H^+ + 2\Delta F_f°I^- - \Delta F_f°I_2 - \Delta F_f°H_2$$

Using $\Delta F_f°$ values from Table 3.2 gives

$$\Delta F_r° = 0 + 2(13.35) - (-3.93) - 0 = 28.6 \text{ kcal}$$

Thus, $\Delta F_r°$ calculated from $E°$ equals $\Delta F_r°$ calculated with $\Delta F_f°$ values.

It was shown in Equation 3.4 that when a reaction is at equilibrium,

$$\Delta F_r° = -RT \ln K$$

Substituting Equation 6.14 into Equation 3.4 gives

$$E° = -\frac{RT}{nF} \ln K \tag{6.15}$$

Equation 6.15 is for standard conditions. A general expression for other conditions is

$$-nEF = -nE°F + RT \ln \frac{(C)^c(D)^d}{(A)^a(B)^b} \tag{6.16}$$

or

$$E_h = E° - \frac{RT}{nF} \ln \frac{(C)^c(D)^d}{(A)^a(B)^b} \tag{6.17}$$

where E_h is the measured voltage in a cell or the redox potential (volts). The term RT/nF can be simplified for 25°C as follows:

$$\frac{(0.001987 \text{ kcal °A}^{-1})(298.15°A)(2.303)}{n(23.06 \text{ kcal V-g equivalent}^{-1})} = \frac{0.0592}{n}$$

so that

$$E_h = E° - \frac{0.0592}{n} \log Q \tag{6.18}$$

where

$$Q = \frac{(C)^c(D)^d}{(A)^a(B)^b}$$

Remember, Q is the equilibrium quotient used where ΔF_r and E do not equal zero (i.e., nonequilibrium conditions). Equation 6.18 is commonly called the *Nernst equation.*

Example 6.4: Calculation of Redox Potential
The redox potential will be calculated for water at 25°C with pH = 7 and a dissolved oxygen concentration of 8 mg L^{-1}.

Solution:

$$\frac{8 \text{ mg L}^{-1} O_2}{32,000 \text{ mg mole}} = 10^{-3.60} M$$

$$(H^+) = 10^{-7} M \quad \text{at pH 7}$$

The appropriate reaction is

$$O_2(aq) + 4H^+ + 4e^- = 2H_2O$$

for which $E° = 1.27$ V (Table 6.1).

$$E_h = E^0 - \frac{0.0592}{n} \log Q$$

where

$$\log Q = \frac{1}{(O_2)(H^+)^4} \qquad \text{(H}_2\text{O is taken as unity)}$$

So

$$E_h = 1.27 - \frac{0.0592}{4} \log \frac{1}{(10^{-3.6})(10^{-28})}$$

$$= 1.27 - \frac{0.0592}{4} \log \frac{1}{10^{-31.6}}$$

$$= 1.27 - (0.0148)(31.6) = 0.802 \text{ V}$$

The pH has an effect on the redox potential, and E_h is often adjusted to the appropriate value for pH 7. To adjust E_h to pH 7, subtract 0.0592 V for each pH unit below neutrality and add 0.0592 V for each pH unit above neutrality. Many researchers adjust E_h values to pH 7 and report the redox potential using the symbol E_7 instead of E_h.

Example 6.5: Effect of pH on E_h
    What is the E_h at pH 5, 6, 7, 8, and 9 for the reaction

$$O_2(aq) + 4H^+ + 4e^- = 2H_2O$$

when dissolved oxygen is 8 mg L^{-1} ($10^{-3.60}$ M) and water temperature is 25°C?

Solution: The method of calculating E_h for the reaction was shown in Example 6.4:

$$E_h = +1.27 - \frac{0.0592}{4} \log \frac{1}{(10^{-3.6})(H^{+4})}$$

($H^+ = 10^{-5}, 10^{-6}, 10^{-7}, 10^{-8}$, and 10^{-9} M for pH 5, 6, 7, 8, and 9, respectively). Substituting into the preceding equation gives E_h values as follows: pH 5, 0.921 V; pH 6;, 0.862 V; pH 7, 0.802 V; pH 8, 0.743 V; pH 9, 0.684 V. Notice that E_h increased by 0.059 V for each unit increase in pH.

Changes in temperature will cause E_h to change, as can be seen in Equation 6.17. The influence of temperature on E_h is not as great as the influence of pH, and redox potentials sometimes are not adjusted for temperature. The standard temperature for reporting redox potential is 25°C.

Example 6.6: Effect of Temperature on E_h
 E_h will be calculated for water at pH 7 and 0°C that has 8 mg L^{-1} dissolved oxygen.

Solution: Equation 6.7 must be used for temperatures other than 25°C:

$$E_h = 1.27 - \frac{(0.001987)(273.15)}{4(23.06)} \ln \frac{1}{(10^{-3.6})(10^{-28})}$$

$$= 0.842 \text{ V}$$

At a temperature of 25°C, E_h is 0.802 V (Example 6.4). Thus, decreasing temperature from 25°C to 0°C caused E_7 to decrease by 0.040 V.

6.3.3 Practical Measurement of Redox Potential

Although the hydrogen electrode is a standard half-cell against which other half-cells are commonly compared, it is almost never used as the reference electrode in practical redox potential measurements. The most common reference electrode for redox measurement is the calomel electrode:

$$Hg_2Cl_2 + 2e^- = 2Hg^+ + 2Cl^- \qquad (6.19)$$

A KCl-saturated calomel electrode has $E° = 0.242$ V at 25°C.[1] Standard electrode potentials for the popular KCl-saturated calomel electrode are 0.242 V less than tabulated $E°$ values of standard electrode potentials for the hydrogen electrode.

Example 6.7: Standard Electrode Potentials for the Calomel Electrode
 $E°$ values from Table 6.1 will be adjusted to indicate potentials that would be obtained with a KCl-saturated calomel electrode:

(1) $O_2(aq) + 4H^+ + 4e^- = 2H_2O$ $\qquad E° = 1.27$

(2) $Fe^{3+} + e^- = Fe^{2+}$ $\qquad E° = 0.77$

(3) $NO_3^- + 2H^+ + 2e^- = NO_2^- + H_2O$ $\qquad E° = 0.84$

Solution:

Half-Cell	$E°$		Correction Factor		Calomel Potential (V)
(1)	1.27	−	0.242	=	1.028
(2)	0.77	−	0.242	=	0.528
(3)	0.84	−	0.242	=	0.598

The redox potential is a valuable tool in many chemical endeavors: it can be used to follow the progress of chemical reactions and to determine endpoints of titrations; the redox principle has many industrial applications; corrosion of metals is governed by oxidation–reduction reactions; and the concept is useful in explaining many chemical and biological phenomena.

The practical use of redox potential in natural systems is fraught with difficulty. In thoroughly mixed waters without high concentrations of reducing agents, the redox potential is governed by the dissolved oxygen concentration. Although oxygen is reduced as it oxidizes reduced substances in the water, oxygen is continually replaced by diffusion of oxygen from the atmosphere or by oxygen produced in photosynthesis, and the redox potential may remain fairly constant. In the hypolimnion of stratified bodies of water, at the soil–water interface in unstratified water bodies, and in bottom soils and sediment, dissolved oxygen is at low concentration or even absent, and reducing conditions develop. The driving force causing a decrease in the redox potential is the consumption of oxygen by microbial respiration. When molecular oxygen is depleted, certain microorganisms can utilize relatively oxidized inorganic or organic substances as electron acceptors in metabolism (see Chap. 3). Mechanisms of redox reactions in reduced environments are very complicated. They may occur spontaneously, or they may be mediated by microorganisms. There is a wide variety of compounds, concentrations of relatively reduced and relatively oxidized compounds vary greatly both spatially and temporally, and it is usually impossible to isolate specific components of the system. As a result, the best that can be done is to insert the redox probe into the desired place and obtain a reading. This reading can then be interpreted in terms of departure from the potential established by dissolved oxygen in water. The *oxygen potential*

$$O_2(aq) + 4H^+ + 4e^- = 2H_2O \qquad (6.20)$$

regulates the redox potential of well-oxygenated surface water. At 25°C and pH 7, the oxygen potential should be about 0.802 V (802 V) against a standard hydrogen electrode (see Example 6.4). The oxygen potential of well-oxygenated surface water is about 0.560 V (560 V) when measured with a KCl-saturated calomel electrode. Redox potential drops in aerobic waters or soils, and values below -0.250 V have been observed in pond soils.

Commercially available devices for measuring redox potential do not all provide the same reading for the redox potential of well-oxygenated water or for the same sample of reduced water or soil. This makes it difficult to interpret redox potentials measured in pond waters or soils. Nevertheless, a decrease in the redox potential below values obtained in oxygen-saturated water with a particular instrument indicates that reducing conditions are developing, and the reducing ability of the environment increases as the redox potential drops. Care must be taken not to introduce air or oxygenated water into the medium where redox potential is to be measured.

Example 6.8: Effect of Dissolved Oxygen Concentration on Redox Potential
The E_h of water at 25°C and pH 7 will be calculated for dissolved oxygen concentrations of 1, 2, 4, and 8 mg L^{-1}.

Solution: The molar concentration of dissolved oxygen can be computed by dividing milligrams per liter values by 32,000 mg O_2 mole^{-1}: 1 mg $L^{-1} = 10^{-4.5}\ M$; 2 mg $L^{-1} = 10^{-4.2}\ M$; 4 mg $L^{-1} = 10^{-3.9}\ M$; 8 mg $L^{-1} = 10^{-3.6}\ M$. Use the molar concentrations of dissolved oxygen to solve the expression

$$E_h = +1.27 - \frac{0.0592}{4} \log \frac{1}{(O_2)(10^{-28})}$$

E_h are as follows: 1 mg L^{-1} = 0.789 V; 2 mg L^{-1} = 0.793 V; 4 mg L^{-1} = 0.798 V; 8 mg L^{-1} = 0.802 V. The corresponding readings for a calomel electrode are 1 mg L^{-1} = 0.547 V; 2 mg L^{-1} = 0.551 V; 4 mg L^{-1} = 0.556 V; 8 mg L^{-1} = 0.560 V.

It can be seen from Example 6.8 that only a small amount of dissolved oxygen in water will maintain E_h near 0.8 V (near 0.56 V for a calomel electrode). Natural systems do not become strongly reducing until dissolved oxygen is depleted.

6.4 Anaerobic Respiration

In aerobic soil, oxygen is the terminal electron acceptor and carbon dioxide is the end carbon product for microbial respiration. In anaerobic soil, certain microorganisms called *anaerobes* can decompose organic matter in the absence of oxygen. Oxygen is actually toxic to some of these microorganisms (*obligate anaerobes*). Others can adapt their respiratory pathways and function in either the presence or absence of oxygen (*facultative anaerobes*). Terminal electron acceptors for anaerobic respiration may be organic or inorganic compounds. Carbon dioxide may be produced in anaerobic respiration, but end carbon compounds include alcohols, formate, lactate, propionate, acetate, methane, and other substances.

6.4.1 Fermentation

Fermentation is one type of anaerobic respiration. Organisms capable of fermentation hydrolyze complex organic compounds to simpler ones that can be used in a process identical to glycolysis to produce pyruvate. In the absence of the Krebs cycle, pyruvate cannot be oxidized to carbon dioxide and water with molecular oxygen serving as the terminal electron and hydrogen acceptor. In fermentation, hydrogen ions removed from organic matter during pyruvate formation are combined with one of the products of carbohydrate metabolism.

Ethanol production from glucose is a well-known fermentation in which each glucose molecule is converted to two ethanol molecules and two carbon dioxide molecules:

$$\underset{\text{(Glucose)}}{C_6H_{12}O_6} \rightarrow \underset{\text{(Pyruvate)}}{2C_3H_4O_3} + 4H^+ \tag{6.21}$$

$$2C_3H_4O_3 \rightarrow \underset{\text{(Acetaldehyde)}}{2C_2H_4O} + 2CO_2 \tag{6.22}$$

$$2C_2H_4O + 4H^+ \rightarrow \underset{\text{(Ethanol)}}{CH_3CH_2OH} \tag{6.23}$$

In the conversion of glucose to pyruvate, two ATP molecules are formed per molecule of glucose. Hydrogen ions released in pyruvate formation combine with acetaldehyde to yield ethanol.

Ethanol is only one of many organic compounds that can be produced by fermentation. For example, some microorganisms convert glucose to lactic acid:

$$\underset{\text{(Glucose)}}{C_6H_{12}O_6} \rightarrow 2C_3H_4O_3 + 4H^+ \tag{6.24}$$

$$2C_3H_4O_3 + 4H^+ \rightarrow \underset{\text{(Lactic acid)}}{2C_3H_6O_3} \tag{6.25}$$

In lactic acid production, no carbon dioxide is released as in ethanol production.

Enzymes catalyze fermentation reactions. In ethanol production, carboxylase catalyzes the conversion of pyruvate to acetaldehyde and alcohol dehydrogenase catalyzes the formation of ethanol from acetaldehyde. Lactic acid dehydrogenase is responsible for catalyzing the reaction in which lactic acid is formed from pyruvate. Other enzymes function to catalyze the formation of pyruvate from glucose and other simple organic compounds.

Fermentation reactions also produce hydrogen gas. The hydrogen gas is apparently formed when NADH is oxidized at low hydrogen pressure by liberation of hydrogen:[2]

$$NADH + H^+ = NAD^+ + H_2 \tag{6.26}$$

6.4.2 Inorganic Electron Acceptors

Many microorganisms are capable of using oxidized inorganic substances instead of molecular oxygen as electron acceptors. Bacteria that use nitrate as electron acceptors can hydrolyze complex compounds and oxidize the hydrolytic products to carbon dioxide. Nitrate is reduced to nitrite, ammonia, nitrogen, or nitrous oxide (Eqs. 3.21–3.27). Many of the hydrolytic bacteria also are capable of fermentation.[3] In the zone where nitrate-reducing bacteria occur, a part of the organic carbon is oxidized completely to carbon dioxide and a portion is converted to organic fermentation products.

The iron- and manganese-reducing bacteria utilize oxidized iron and manganese as oxidants in the same manner as nitrate-reducing bacteria use nitrate. They attack organic fermentation products and oxidize them to carbon dioxide. Ferrous iron (Fe^{2+}) and manganous manganese (Mn^{2+}) are released as by-products of respiration.

Sulfate-reducing bacteria and methane-producing bacteria cannot hydrolyze complex organic substances or decompose simple carbohydrates and amino acids from hydrolytic activity. They can only utilize short-chain fatty acids and simple alcohols produced by fermentative bacteria as organic carbon sources.[3,4] Fermentation products are transported downward into the zone of sulfate reduction, and the bacteria use sulfate as an oxidant to oxidize the fermentation products to carbon dioxide. Sulfide is released as a by-product (Eqs. 3.29–3.35).

Fermentation products also move downward into the zone of methane production. In the most common reaction for methane formation, a simple organic molecule is fermented and carbon dioxide is utilized as the electron (hydrogen) acceptor. Carbon dioxide may be obtained from the environment or produced in fermentation:

$$CH_3COOH + 2H_2O \rightarrow 2CO_2 + 8H^+ \qquad (6.27)$$

$$8H^+ + CO_2 \rightarrow CH_4 + 2H_2O \qquad (6.28)$$

Subtraction of the two reactions gives

$$CH_3COOH \rightarrow CH_4 + CO_2 \qquad (6.29)$$

Some bacteria also can use carbon dioxide as an oxidant to oxidize hydrogen:

$$4H_2 + CO_2 \rightarrow CH_4 + 2H_2O \qquad (6.30)$$

Methane production is an important process because the hydrogen that accumulates from fermentation must be disposed of to prevent it from inhibiting the fermentation process.

In fermentation, organic carbon is incompletely metabolized and intermediary organic substances may accumulate. High concentrations of nitrate, ferric iron, sulfate, and other oxidized inorganic compounds in pore water favor decomposition, because they can be used by many microorganisms as alternative electron acceptors (oxidants) in the absence of oxygen. The energy yield in anaerobic respiration (fermentation or with alternative inorganic electron acceptors) is much

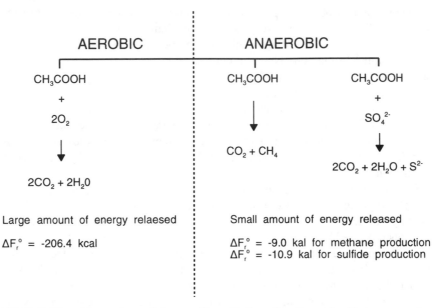

FIGURE 6.2. Comparison of the aerobic oxidation of acetic acid with two examples of anaerobic respiration.

lower than for aerobic respiration (Fig. 6.2). However, both types of anaerobic respiration occurs in pond soils, and organic matter is degraded efficiently in most cases, because aquaculture ponds normally do not have large accumulations of organic matter. For example, Munsiri and Boyd (unpublished) measured organic carbon concentrations in 2.5-cm-long segments cut from cores extending from the soil surface into the original pond soil in 2-, 23-, and 52-year old ponds on the Fisheries Research Unit at Auburn University. Irrespective of pond age, the surface layer of sediment (0–10 cm) contained 3–4% organic carbon, underlaying sediment was 2–3% organic carbon, and original pond soil contained less than 0.5% organic carbon. Only 8–15% of the organic carbon in sediment was associated with the light fraction organic matter pool (see Chapter 5, Section 5.4.2). These observations suggest an increase in organic carbon in pond soil over that of the original pond bottom, but the increase in percentage organic carbon was moderate. The accumulated organic matter consisted primarily of mineral-associated organic carbon (humus). Of course, there was a greater total amount of organic carbon in older ponds than in newer ones, because the sediment layer was thicker in older ponds.

The idea that aquaculture pond soils have high concentrations of organic matter because they are flooded and oxygen-depleted is a misconception. This misconception results from three observations that are misinterpreted. First, wetlands often contain a lot of soil organic matter, and organic soils form in wetlands. Organic matter in wetlands originates in large wetland plants that are high in structural components and low in nitrogen content. Residues of these plants accumulate in wetlands because they have a low C:N ratio and decompose slowly. Residues often are acidic because of organic acids, and this further retards microbial decomposition. Organic residues in aquaculture ponds have little structural material and a high C:N ratio. Acidic ponds are limed to counteract acidity. Second, anaerobic pond soils often are dark, and organic matter makes soil dark. Although this statement is true, the dark color in most anaerobic pond soils results from ferrous iron formed when microorganisms use ferric iron as an alternative electron acceptor. Third, measurements of aerobic respiration of pond soils is much slower when they are flooded than when they are allowed to dry between crops as shown in Fig. 5.12. Anaerobic respiration was no doubt much higher than aerobic respiration when ponds were full of water, and data in Fig. 5.12 do not indicate the actual rate that pond soil organic matter was mineralized. Nevertheless, drying stimulates aerobic respiration, and drying between crops can greatly reduce the amount of organic matter in surface layers in preparation for the next crop.

A problem occurs in aquaculture ponds when the rate of organic matter delivery to the soil is great and the demand for oxygen is too high for the aerobic surface layer of soil to be maintained by movement of dissolved oxygen from pond water into pore water. When this happens, toxic, reduced inorganic substances can enter the pond water. The oxidized soil surface in pond soils is analogous to the thermocline in thermally-stratified lakes. It separates the zone of aerobic processes from the zone of anaerobic processes.

6.5 Reduced Soils

Dissolved oxygen, carbon dioxide, and other gases move very slowly in water-saturated soils. Decomposition of organic matter by aerobic microorganisms depletes the oxygen supply and lowers the redox potential. Once the soil becomes anaerobic, decomposition of organic matter by anaerobic microorganisms drives the redox potential downward even more. Therefore, organic matter and microbial respiration is the source of reducing power (H^+) that causes the redox potential to decline in pond soils.

The most important oxidation–reduction reactions taking place in aquaculture pond soils are aerobic respiration, fermentation, nitrate reduction, reduction of iron and manganese, sulfate reduction, and methane production. Half-cells for these reactions are listed:

Aerobic respiration (oxygen potential is related to presence of dissolved oxygen for aerobic respiration)

$$O_2(aq) + 4H^+ + 4e^- = 2H_2O; \qquad E° = 1.27 \text{ V} \qquad (6.31)$$

Anaerobic respiration (begins when dissolved oxygen is depleted; it also is related to the oxygen potential)

Nitrate reduction and denitrification

$$NO_3^- + 2H^+ + 2e^- = NO_2^- + H_2O; \qquad E^\circ = 0.85 \text{ V} \qquad (6.32)$$

$$NO_3^- + 10H^+ + 8e^- = NH_4^+ + 3H_2O; \qquad E^\circ = 0.88 \text{ V} \qquad (6.33)$$

$$2NO_3^- + 12H^+ + 10e^- = N_2(g) + 6H_2O; \qquad E^\circ = 1.24 \text{ V} \qquad (6.34)$$

$$NO_2^- + 8H^+ + 6e^- = NH_4^+ + 2H_2O; \qquad E^\circ = 0.89 \text{ V} \qquad (6.35)$$

Reduction of iron and manganese

$$Fe^{3+} + e^- = Fe^{2+}; \qquad E^\circ = 0.77 \text{ V} \qquad (6.36)$$

$$Fe(OH)_3(s) + 3H^+ + e^- = Fe^{2+} + 3H_2O; \qquad E^\circ = 1.06 \text{ V} \qquad (6.37)$$

$$MnO_2(s) + 4H^+ + 2e^- = Mn^{2+} + 2H_2O; \qquad E^\circ = 1.23 \text{ V} \qquad (6.38)$$

Sulfate reduction

$$SO_4^{2-} + 10H^+ + 8e^- = H_2S(g) + 4H_2O; \qquad E^\circ = 0.31 \text{ V} \qquad (6.39)$$

Methane production

$$CO_2 + 8H^+ + 8e^- = CH_4(g) + 2H_2O; \qquad E^\circ = 0.17 \text{ V} \qquad (6.40)$$

Reactions indicated by Equations 6.32–6.40 have E° less than the oxygen potential. Some reductions begin before the dissolved oxygen is depleted entirely. For example, conversion of nitrate to nitrite begins when there is still 2–3 mg L^{-1} of dissolved oxygen. Once anaerobic respiration begins, the sequence in the use of terminal electron acceptors by microorganisms is in order of decreasing E°; divalent manganese appears before nitrite, nitrite appears before ferrous iron, ferrous iron before hydrogen sulfide, and hydrogen sulfide before methane. Also, the presence of a particular inorganic electron acceptor will poise the redox potential until the acceptor is depleted. For example, in a system containing dissolved oxygen, nitrate, and ferric iron, nitrate will not be used as an electron acceptor until the oxygen supply is depleted to the point that the redox potential is low enough for reduction of nitrate to nitrite. As long as nitrate is present, it will poise the redox potential. When all of the nitrate has been reduced to nitrite, the redox potential falls, and ferric iron is reduced to ferrous iron. In pond soils the supply of dissolved oxygen and redox potential decrease with depth. Electron acceptors change with increasing depth in the order oxygen, nitrate, nitrite, ferric iron, sulfate, and carbon dioxide. Therefore, the bottom soil in ponds is stratified into thin layers, each of which has a characteristic redox reaction.

The measured redox potential at which a given substance will appear is not the same as $E°$ for the controlling reaction. This results because $E°$ is for the standard hydrogen electrode connected to another half-cell at standard conditions. In practical redox potential measurements, concentrations of oxidants and reductant are much less than $1M$, temperature is seldom 25°C, and a calomel electrode is used instead of a standard hydrogen electrode. Nevertheless, as the measured redox potential falls, the electron acceptors used by anaerobic microorganisms change to compounds that accept electrons at lower and lower potentials.

6.5.1 Redox in Aquatic Environments

The most thorough investigation of redox in aquatic environments was that of Mortimer.[5,6] He used a calomel electrode in measuring redox potential (E_7) in English lakes and in laboratory mud–water systems. He found that the E_7 of well-oxygenated water was around 0.5 V. The redox potential was about 0.5 V throughout the water volume in lakes until thermal stratification developed. The hypolimnion of lakes had E_7 values as low as 0.03 V (Fig. 6.3). Nitrate reduction was usually observed at $E_7 = 0.4$ V and dissolved oxygen of 2 mg L^{-1}. Ferrous iron was usually detected in lake water when E_7 fell below 0.25 V. This corresponded to a dissolved oxygen concentration of about 0.5 mg L^{-1}. The concentrations of total iron, ammonia, and phosphate increased drastically in waters of the hypolimnion during the period of thermal stratification as these substances were released from anaerobic bottom soil by microbial activity. When the lake destratified in the fall, dissolved oxygen concentrations increased, E_7 increased,

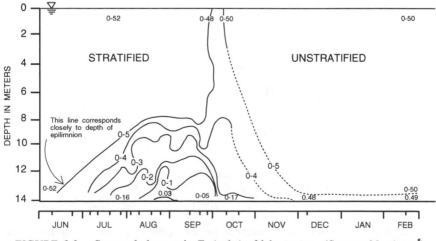

FIGURE 6.3. Seasonal changes in E_7 (volts) of lake waters. (*Source: Mortimer.*[5])

and concentrations of phosphate and reduced substances declined. Phosphate accumulation in the hypolimnion resulted from conversion of iron(III) in iron phosphate compounds to iron(II), which increased phosphate solubility.

The typical winter distributions of E_7 in muds from several lakes, as reported by Mortimer, are provided in Figure 6.4.[6] The water was oxidized, but the redox potential quickly fell in the soil and reducing conditions existed at soil depths of 1 to 4 cm. None of the lakes was highly eutrophic, so most lake or pond bottom soils apparently are anaerobic below a thin, aerobic surface layer. Values for E_7 tended to increase again below a depth of about 5 cm in three of the lakes. This may be the result of less organic matter and microorganisms at these depths.

Mortimer also measured redox potential in laboratory soil–water systems open to the atmosphere or sealed to prevent input of atmospheric oxygen. Redox potentials of sealed and unsealed systems are shown in Figure 6.5. High concentrations of iron, manganese, phosphate, ammonia, and nitrite were observed in

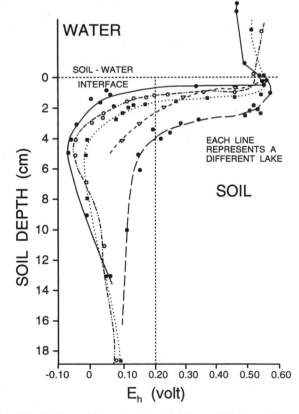

FIGURE 6.4.   Distribution of E_7 (volts) in five lake muds with oxidized surface layers of soil. (*Source: Mortimer.*[6])

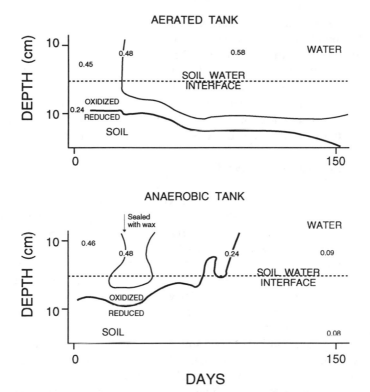

FIGURE 6.5. Changes in E_7 (volts) over time in soils of aerobic and anaerobic laboratory soil–water systems. (*Source: Mortimer.*[5])

water of the sealed system as compared to the open one. Sulfate concentrations also declined in the sealed tank as sulfate was reduced to sulfide. These changes in water quality resulted from the release of substances from anaerobic soil.

Based on results of lake and laboratory studies, Mortimer gave the following E_7 values for the potential below which none of the oxidized forms of the following pairs could be detected: NO_3^- to NO_2^-, 0.45–0.40 V; NO_2^- to NH_4^+, 0.40–0.35 V; Fe^{3+} to Fe^{2+}, 0.30–0.20 V; SO_4^{2-} to S^{2-} (H_2S), 0.10–0.06 V.[6] Dissolved oxygen concentrations associated with these ranges were 4, 0.4, 0.1, and 0 mg l^{-1}, respectively. Mortimer suggested that the appearance of ferrous iron was a good indicator of reducing conditions. Ferrous iron is a good indicator of reducing conditions in aquaculture pond soils because ferrous compounds color the soil gray to deep black. Oxidized soils usually have a brown crust or retain their native color. In pond water the dissolved oxygen concentration is the best indicator of the oxidation–reduction status of the water.

Ranges in the redox potential for the reduction of nitrate, ferric iron, nitrite, and sulfate reported by Mortimer are somewhat different from those of some other authors. According to Takai and Kamura,[7] Turner and Patrick,[8] and Chien,[9] the reduction of inorganic compounds in bottom soils is sequential as follows: NO_3^- to NO_2^- and Mn^{4+} to Mn^{2+} at 0.2 to 0.3 V; Fe^{3+} to Fe^{2+} at 0.05 V; SO_4^{2-} to S^{2-} (H_2S) at -0.15 to -0.20 V; CO_2 to CH_4 at -0.25 V. Patrick and Mahapatra[10] found that aerobic muds had redox potentials of 0.4 to 0.7 V and anaerobic muds had values as low as -0.25 to -0.30 V—much lower than any E_7 values reported by Mortimer (1941, 1942).[5,6] These differences are probably related to types of reference electrodes and measurement techniques.

6.5.2 Redox Potential in Pond Soils

There is little published information on redox potential in pond soils, and it is difficult to interpret reported data. Shigeno measured redox potentials of 0.14 to -0.46 V 2 cm above the soil in shrimp ponds in Japan.[11] He reported that the black color of ferrous iron was visible in soil at -0.215 V. Chien measured redox potentials in shrimp pond soils in Taiwan.[12] Average values were as follows: 0.5-cm depth, -0.113 V; 1-cm depth, -0.162 V; 1.5-cm depth, -0.180 V. The lowest value reported was -0.250 V. Chien found that concentrations of sulfide, ammonia, and nitrite in pond sediment increased as redox potential declined. He suggested that redox potential measurement might be a useful tool for evaluating the condition of pond soil. Chien categorized pond soil according to redox potential as follows: oxidized, 0.40–0.70 V; moderately reduced, 0.10–0.40 V; reduced, -0.10–0.10 V; and highly reduced, -0.10–-0.30 V.[9] Masuda and Boyd measured redox potential in freshwater ponds; they reported that surface, well-oxygenated waters had redox potentials of 0.049–0.151 V, and water about 2 cm above the soil–water interface had redox values of -0.023–0.135 V.[13] Redox potential at 5-cm depth in the sediment ranged from -0.113 to -0.173 V. Masuda and Boyd also removed pore water from 5-cm depth in pond soil and found 30–60 mg L^{-1} ferrous iron, 2–25 mg L^{-1} ammonia N, 0.1–0.2 mg L^{-1} nitrite, and 0.03–0.1 mg L^{-1} sulfide.[13] Concentrations of these substances were 20–80 times greater in pore water of pond soils with redox potentials below -0.100 V than in overlaying pond water with redox potentials of 0.050 V and higher. The failure of reduced substances to diffuse into the overlaying pond water resulted from the aerobic surface layer (Fig. 6.6). Reduced substances were oxidized in the aerobic layer and did not enter the pond water. In shrimp ponds in Ecuador, redox potentials measured at 5-cm depth in soils of 36 ponds were 0.00–0.050 V when ponds were initially filled, and redox potentials then generally declined (Fig. 6.7) with values as low as -0.400 V occurring after three to four months of feed applications to ponds to promote shrimp growth.

Patrick and Delaune studied redox potential in laboratory soil–water systems.[14] They plotted the change in E_7 with depth as shown in Figure 6.8 and concluded that the midpoint between the two points of maximum change in slope of the

A. BOTTOM SOIL WITH OXIDIZED SURFACE LAYER

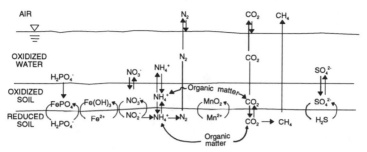

B. COMPLETELY REDUCED BOTTOM SOIL

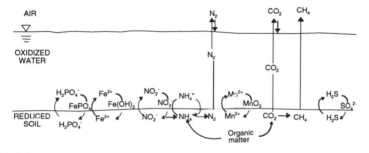

FIGURE 6.6. Transfer of substances between pond soils and water: (A) soils with oxidized surface layer, (B) soils with reduced surface layer.

line depicting E_7 was the dividing point between the surface aerobic layer and the underlaying anaerobic layer. They found this division to occur at about 0.200 V in most soils. Patrick and Delaune also reported that in a silt–loam soil of apparently low organic matter concentration exhibited reduced manganese at 8 mm depth, reduced iron at 12 mm depth, and reduced sulfide at 14 mm depth.[14] The depth of $E_7 = 0.200$ V corresponded closely with the depth of the appearance of reduced iron. This suggests that a substance with a high $E°$ value will be reduced at lesser depth into the soil than a substance with a low $E°$ value.

6.5.3 Use of Redox Potential in Pond Management

Several problems are associated with the use of redox potential as an indicator of soil condition in aquaculture pond management:

1. Different redox potential meters may give different readings for the same sample or location in the pond bottom.

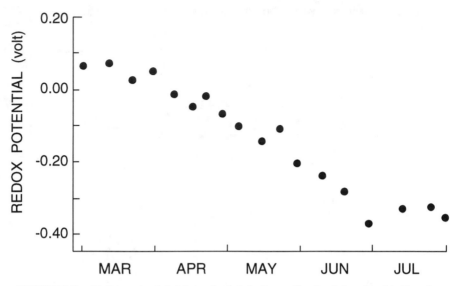

FIGURE 6.7. Redox potential at 5-cm depth in bottom soils of a shrimp pond in Ecuador. (*Source: Gilberto Escobar, Guayaquil, Ecuador.*)

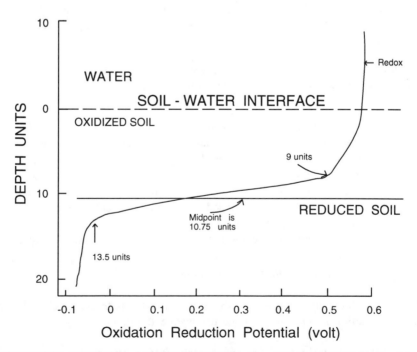

FIGURE 6.8. Illustration of technique for locating the plane between the surface layer of aerobic soil and the underlying anaerobic soil.

2. The redox potential declines rapidly with depth in the soil (Figs. 6.4 and 6.5), and it is difficult to insert the electrode of commercial devices to a precisely known depth.

3. Water containing dissolved oxygen may contaminate the soil when the redox probe is inserted.

4. A decline in redox potential with depth is a normal phenomenon in a water-saturated soil. If an oxidized surface layer exists at the mud–water interface, diffusion of reduced substances from the underlaying anaerobic soil into the water is not likely. Therefore, the existence of an anaerobic layer below the soil–water interface is normal, and it usually has no influence on the quality of water standing above the soil.

5. There are few data correlating redox potential of soil with survival and growth of aquaculture species.

Although the routine measurement of redox potential is not recommended as a technique for assessing condition of aquaculture pond soils, oxidation–reduction reactions occurring in pond soils do influence aquacultural production. If bottom soils are highly reduced because of high-organic-matter input, the anaerobic layer may rise to the surface of the soil. The entire soil–water interface is seldom anaerobic, but there may be anaerobic microenvironments.[15] Retardation of fish growth in ponds was noted when anaerobic microenvironments were detected in surface soil layers by a technique for measuring nitrate reduction. Boyd reported that organic matter and soil particles eroded from shrimp ponds are sometimes deposited in the central areas of shrimp ponds.[16] These deposits often have an anaerobic surface exposed to the pond water, and reduced substances may enter the pond water. Aeration of catfish ponds also disturbs pond bottoms, and nitrite concentrations in aerated ponds often are higher than those in unaerated ponds.[17] This results because disturbance of the pond bottom by water currents created by aeration brings nitrite from the anaerobic layers of the soil into the water.

The pond water volume in aquaculture ponds is normally well oxygenated from natural sources of oxygen or from aeration. This maintains a high redox potential in the water, and reduced substances are quickly oxidized. If the rate of delivery of reduced substances from anaerobic soils to the water exceeds the rate of oxidation of these substances, undesirably high concentrations of reduced substances may be found in oxygenated pond waters. In the culture of shrimp and other crustaceans, animals spend most of their time at or near the pond bottom or burrowed into the bottom soil. Crustaceans are thought to be much more sensitive to soil condition than fish.

References

1. Pagenkopf, G. K. *Introduction to Natural Water Chemistry*. 1978. Marcel Dekker, New York.

2. Wolin, M. J. 1974. Metabolic interactions among microorganisms. *Ann. J. Clin. Nutr.* **27**:1320–1328.

3. Blackburn, T. H. 1987. Role and impact of anaerobic microbial processes in aquatic systems. In *Detritus and Microbial Ecology in Aquaculture,* D. J. W. Moriarty and R. S. V. Pullin, eds., pp. 32–53. ICLARM Conference Proceedings 14, International Center for Living Aquatic Resources Management, Manila, Philippines.

4. Alexander, M. *Introduction to Soil Microbiology.* 1961. John Wiley & Sons, New York.

5. Mortimer, C. H. 1941. The exchange of dissolved substances between mud and water in lakes, I. *J. Ecol.* **29**:280–329.

6. Mortimer, C. H. 1942. The exchange of dissolved substances between mud and water in lakes, II. *J. Ecol.* **30**:147–201.

7. Takai, Y., and T. Kamura. 1966. The mechanism of reduction in waterlogged paddy soils. *Folia Microbiol. (Prague)* **11**:304–313.

8. Turner, F. T., and W. H. Patrick, Jr. 1968. Chemical changes in waterlogged soil as a result of oxygen depletion. *Trans. 9th Internat. Congress Soil Sci.* **4**:53–65.

9. Chien, Y. H. 1989. The management of sediment in prawn ponds. In *Proceedings III, Brazilian Prawn Culture Symposium Proceeding,* pp. 219–243, MCI Aquacultura, Joao Pessoa—PB, Brazil.

10. Patrick, W. H., Jr., and I. C. Mahapatra. 1968. Transformation and availability of nitrogen and phosphorus in waterlogged soils. *Advan. Agron.* **20**:323–359.

11. Shigeno, K. *Problems in Prawn Culture.* 1978. Amerind Publishing Co., New Delhi, India.

12. Chien, Y. H. 1989. Study on the sediment chemistry of Tiger prawn, Kuruma prawn, and Red Tail prawn ponds in I-Lan Hsien. In *Studies on the Environment Improvement and the Control of the Off-Flavor in Fish II,* pp. 257–275. Fisheries Series No. 16, Council of Agriculture, Taipei, Republic of China.

13. Masuda, K., and C. E. Boyd. 1994. Chemistry of sediment pore water in aquaculture ponds built on clayey, Ultisols at Auburn, Alabama. *J. World Aquaculture Soc.* **25**:396–404.

14. Patrick, Jr., W. H., and R. D. Delaune. 1972. Characterization of oxidized and reduced zones in flooded soil. *Soil Sci. Soc. Amer. Proc.* **36**:573–575.

15. Avnimelech, Y., and G. Zohar. 1986. The effect of local anaerobic conditions on growth retardation in aquaculture systems. *Aquaculture* **58**:167–174.

16. Boyd, C. E. 1992. Shrimp pond bottom soil and sediment management. In *Proceedings of the Special Session on Shrimp Farming,* J. Wyban, ed., pp. 166–181. World Aquaculture Society, Baton Rouge, La.

17. Hollerman, W. D., and C. E. Boyd. Nightly aeration to increase production of channel catfish. *Trans. Amer. Fish. Soc.* **109**:446–452.

18. Snoeyink, V. L., and D. Jenkins. *Water Chemistry.* 1980. John Wiley & Sons, New York.

7

Sediment

7.1 Introduction

Most ponds are static water systems, but some are operated with water exchange. Water exchange seldom exceeds 20–30% of pond volume per day, and it is usually much less. A common expression of the average residence time of water in ponds is the *hydraulic residence time*, which is equal to pond volume/inflow rate. Ponds usually have hydraulic residence times of weeks or even months, and water does not even flow rapidly through ponds with water exchange. Long hydraulic residence time favors sedimentation of suspended particles, and ponds behave as sediment basins.

Ponds have external and internal sediment loads. Suspended particles that enter ponds in the water supply are the source of external sediment. When water enters ponds, turbulence is reduced and suspended particles settle. Examples of high external sediment loads are freshwater ponds filled by runoff from denuded watersheds and brackish-water ponds filled from estuaries laden with suspended soil particles. Internal sediment loads result from turbulence that erodes pond bottom soils and levees and suspends solid particles. Sources of turbulence are wind action, aquatic animal activity, mechanical aeration, and harvest operations. Organic sediment in ponds originates primarily from plankton. Other sources are manure applications, uneaten feed, aquatic animal feces, and higher aquatic vegetation.

There usually is not enough turbulence in ponds to maintain high concentrations of soil particles in suspension, and plankton is the main source of turbidity. Some ponds are "muddy" for long periods because of high external sediment loads or strong turbulence. Erosional and depositional processes can reduce pond volume, alter shape of pond bottoms, erode levees, change characteristics of the soil–water interface, influence the structure and function of pond ecosystems, and diminish aquacultural production. An understanding of sediment sources and the principles of sedimentation has practical value to pond managers.

7.2 Sources of Sediment

A wide array of solid particles contribute sediment to aquaculture ponds as follows: rock, sand, silt, clay, finely divided organic matter, and remains of plants and animals. These materials originate from both external or internal sources (Table 7.1). In woodland ponds and in aquatic macrophyte-infested ponds, vascular plant residues accumulate on bottoms. Aquaculture ponds normally are constructed in open areas, and aquatic macrophytes infestations are not permitted to develop. The major sources of sediment in aquaculture ponds are suspended soil particles in inflowing water, manures and feed added to ponds to promote aquatic animal production, and sediment produced internally by biological activity and management procedures. Major sources of sediment will be discussed.

7.2.1 Runoff

Aquaculture ponds filled by direct runoff usually have watershed areas 10–20 times larger than pond surface areas. Rain drops and surface runoff suspend plant residues and other debris and erode soil from watersheds. These materials are washed into ponds and settle.

Erosion by rainfall and runoff varies with several factors: rainfall intensity,

Table 7.1 Major Sources and Types of Sediment in Aquaculture Ponds

Source	Description
External	
Natural inputs	
Wind	Dust, pollen, leaves
Runoff	Rocks, sand, silt, clay, unrecognizable organic fragments, recognizable plant and animal remains
Management inputs	Animal manure, green manure, feed (uneaten fraction), and suspended solids in intentional water additions
Internal	
Natural turbulence and wave action	Suspension and redeposition of particles from pond bottom, erosion of pond edges and levees
Biological activity	Dead organic matter produced in pond, animal feces, erosion and redeposition of bottom soil particles by fish and other animals
Management	Redeposition of bottom soil particles suspended by mechanical aeration, induced water circulation, pond draining, and seining operations
Rainfall	Erosion of upper slopes of empty pond with redeposition of material in deeper areas

amount of rainfall, soil moisture concentration before storm events, soil texture, soil chemical properties, slope, and cover. Large raindrops in intense storms strike the ground with much energy and dislodge more soil particles than small raindrops in less intense storms. If rain is intense and of long duration, sheet flow occurs over the watershed surface. Sheet flow can transport particles dislodged by raindrops. Energy of flowing water can cause sheet erosion or gully erosion. Moist soil generates more runoff than dry soil, but raindrop impact dislodges more particles in dry soil. Water infiltrates readily in sandy or silty soils. They produce less runoff and are less subject to erosion than soils with lower rates of infiltration. Soils that contain large amounts of humus are more absorptive and less productive of runoff than soils with little humus. Velocity of runoff increases with increasing watershed slope. The ability of flowing water to suspend and transport solid particles increases as a function of increasing velocity. Vegetative cover breaks the force of falling raindrops, plant roots hold soil together and tend to prevent it from washing away, friction between grass and flowing water reduces water velocity, and plant residues are a source of humus that tends to hold soil particles together.

The most severe erosion occurs on bare soils, and two or three of the heaviest rainfall events each year cause most of the annual erosion. In the United States, suspended solid loads in runoff from major river basins average less than 1 ton ha^{-1} annually.[1] Much greater losses of soil may occur on small watersheds. Representative erosion rates for various land uses are provided in Table 7.2. Assuming a 10:1 ratio of watershed area to pond surface area, data in Table 7.2 suggest that typical rates of external sediment input to watershed ponds (ton ha^{-1} yr^{-1}) in runoff would differ with land use as follows: forest, 0.34; grassland, 3.4; cropland, 6.8; harvested forest, 170; construction, 680.

Example 7.1: Annual Sediment Input to Watershed Pond
  Estimate sediment depth for a 1-ha pond with a 10-ha watershed on which the woodland has been recently harvested.

  Solution: For a typical sediment input to the pond of 17.0 ton ha^{-1} yr^{-1} of watershed area, 170 tons of sediment would enter the pond in one year. The

Table 7.2   Representative Rates of Erosion for Selected Land Uses

Land Use	Erosion (tons ha^{-1} yr^{-1})	Relative Erosion Rate (forest = 1)
Forest	0.034	1
Grassland	0.34	10
Cropland	6.8	200
Harvested forest	17.0	500
Construction	68.0	2,000

Source: U.S. Environmental Protection Agency.[17]

average weight of soil often is given as 150 tons cm^{-1} of soil depth. The sediment depth is

$$170 \, ton \div 150 \, ton \, cm^{-1} = 1.1 \, cm$$

Of course, the sediment would not be spread uniformly over the pond bottom. It would be much deeper in some places than others.

The universal soil loss equation incorporates most of the factors affecting erosion.[2,3,4] It may be used to estimate how much soil is lost from erosion of individual watersheds. The equation is

$$E = RSK_LCP \tag{7.1}$$

where E = amount of erosion
K = soil-erodibility factor
R = rainfall erosion potential factor or erosion index
S_L = topographic factor (degree of slope and length of slope)
C = crop management factor
P = conservation practice factor

Nomographs and tables for obtaining values of the five factors necessary to solve Equation 7.1 are presented by Schwab et al.[5] The soil-erodibility factor (K) is the only factor that is strictly soil related. When other factors are equal, K is the only factor needed in comparing erodibility among different locations. A value for K can be obtained from Figure 7.1. This value represents the average soil loss in English tons per acre per unit of erosion index for a soil in cultivated continuous fallow with an arbitrarily selected slope length of 73 ft and slope of 9%. The K values may be evaluated as follows: <0.1, slight erosion potential; 0.1–0.2, moderate erosion potential; >0.3, severe erosion potential. A K factor in metric tons per hectare for most soils can be selected directly from Table 7.3.

7.2.2 Water Supply

Water to fill levee ponds is taken from wells or natural bodies of water. Freshwater streams and estuaries may have heavy sediment loads. When water is transferred from streams or estuaries to ponds, its turbulence is greatly reduced and suspended particles quickly settle out.

The suspended solid load in natural water can be measured by filtering a known volume through a tared glass–fiber filter or a membrane filter and weighing the residue on the filter. The weight of residue in milligrams per liter is called *total suspended solids*. Another way of assessing the sediment load is to measure *settleable solids* with an Imhoff cone (Fig. 7.2). The volume of sediment may be measured directly in the cone and expressed as milliliters of settleable solids

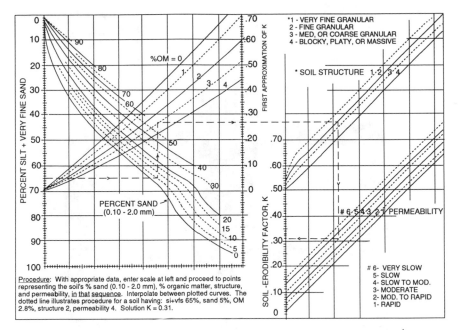

FIGURE 7.1. Soil erodibility nomograph. (*Source: Soil Survey Staff.*[4])

per liter. Settleable solids also may be expressed on a weight basis. The difference between the initial total suspended solid concentration in the sample and the total suspended solid concentration in the supernatant of the Imhoff cone after the 1-h settling period is the settleable solids in milligrams per liter.

Heavy rainfall and associated overland flow erodes large amounts of soil and

Table 7.3 Soil-Erodibility Factor, K (ton ha^{-1}), by Soil Texture Organic Matter Concentration

Textural Class	Organic Matter Content (%)		
	0.5	2	4
Fine Sand	0.36	0.31	0.22
Very fine sand	0.94	0.81	0.63
Loamy sand	0.27	0.22	0.18
Loamy very fine sand	0.99	0.85	0.67
Sandy loam	0.60	0.54	0.43
Very fine sandy loam	1.05	0.92	0.74
Silt loam	1.06	0.94	0.74
Clay loam	0.63	0.56	0.47
Silty clay loam	0.83	0.72	0.58
Silty loam	0.56	0.52	0.43

Source: U.S. Department of Agriculture and Environmental Protection Agency.[18]

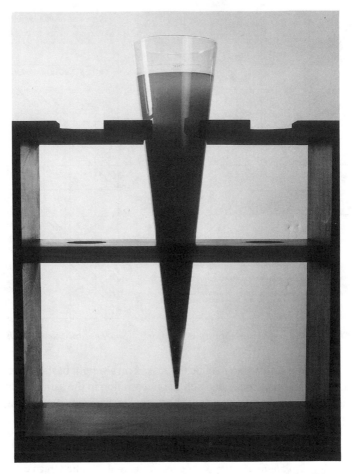

FIGURE 7.2. An Imhoff cone for measuring settleable solids.

causes high concentrations of suspended soils in streams. According to Alabaster and Lloyd, streams may contain 2,000–6,000 mg L^{-1} of suspended solids during floods, but under normal flow conditions values seldom exceed 1,000 mg L^{-1}.[6] High concentrations of suspended solids are common in estuarine reaches of streams. Freshwater streams entering estuaries contain suspended solids, and turbulence created by the tide erodes the bottoms of estuaries and suspends solids. In the Gulf of Guayaquil, Ecuador, settleable solids in water pumped from the Guayas River to supply a shrimp farm averaged 930 mg L^{-1} (460–2,290 mg L^{-1}) over a 12-month period.[7] Dry bulk density of the settleable matter was 1.23 ton m^{-3}.

Example 7.2: Sediment Load for a Shrimp Pond
    The annual sediment load will be calculated for a 1-ha shrimp pond of 1.2-m

average depth. The water supply has an average settleable solid concentration of 1,000 mg L^{-1} (1 kg m^{-3}), and solids have a bulk density of 1.23 ton m^{-3}. Assume that water is exchanged 315 days yr^{-1} at 10% pond volume per day and that ponds are drained and refilled twice per year.

Solution:
  Pond volume = 10,000 m^2 × 1.2 m = 12,000 m^3
  Water use

Filling	12,000 m^3 × 2 times yr^{-1}	= 24,000 m^3 yr^{-1}
Exchange	12,000 m^3 × 0.1 pond volume day^{-1} × 315 day yr^{-1}	= 378,000 m^3 yr^{-1}
Total		402,000 m^3 yr^{-1}

Sediment weight = 402,000 m^3 yr^{-1} × 1 kg m^3 = 402 ton yr^{-1}
Sediment volume = 402 ton yr^{-1} ÷ 1.23 ton m^{-3} = 327 m^3 yr^{-1}
Sediment depth = 327 m^{-3} ÷ 10,000 m^2 = 3.27 cm

Solids do not settle uniformly as one might infer from Example 7.2. The larger particles settle first, and there tends to be gradients in both the quantity and the particle size of sediment. Extensive and semi-intensive shrimp ponds often are constructed with a 5- to 10-m-wide peripheral ditch that is 25–40 cm deeper than the central pond area. The ditch is necessary for draining the pond and facilitating shrimp harvest. Inflowing water follows this ditch because it is deeper and offers less resistance to flow than the central area. Sediment accumulates in the ditch, and the amount of sediment decreases with distance from the inlet.

The calculated sediment volume in Example 7.2 of 327 m^3 was 2.72% of pond volume per year [(327 ÷ 12,000) × 100]. This is a high external sediment load, but even greater amounts of sedimentation may occur in some ponds. Intensive shrimp ponds with high rates of water exchange in Thailand had sediment accumulation up to 8.5 cm over a four-month period.[7] An average sediment depth of 45 cm was observed by the author in a five-year-old shrimp pond in Ecuador. Over a period of several years, sediment can fill ponds and severely interfere with pond operation.

Analysis of sediment from shrimp ponds in Thailand showed that its mineral fraction consisted of 32.2% clay, 21.0% silt, and 45.8% sand. Its organic matter concentration was 3.1%.

7.2.3 Manures and Feeds

Some aquaculture ponds are treated with manure. Application rates normally range from 50 to 100 kg dry weight day^{-1}, but rates as high as 200 kg day^{-1} have been used.[8,9] Manure applications over a 12-month period in tropical ponds may reach 10–20 ton dry weight ha^{-1}. In ponds with feeding, maximum daily feeding rates range from 10 to 250 kg ha^{-1}, and annual inputs of feed may range

Table 7.4 Composition of Selected Manures and a Representative Fish Feed

Material	Dry-Matter Content of Fresh Material (%)	Composition of Dry Matter (%)			
		Organic Matter	Mineral Matter	Nitrogen	Phosphorus
Green Manure					
Alfalfa	24.4	97.8	2.2	0.74	0.06
Mixed grasses	28.1	97.2	2.8	0.78	0.09
Animal manure					
Cattle	15	≈95	≈5	0.70	0.22
Poultry	28	≈93	≈5	1.20	0.57
Swine	18	≈95	≈5	0.50	0.13
Fish feed	92	90	10	5.12	1.10

from 1 to 40 tons. Normal feed applications are 5–10 ton ha^{-1} yr^{-1} for freshwater fish ponds and 12–24 ton ha^{-1} yr^{-1} for intensive shrimp ponds.[9] Feed is normally 90–92% dry matter. The composition of some representative manures and a fish feed is provided in Table 7.4.

Manure is not consumed directly by aquaculture species. It is decomposed by microorganisms to yield mineral nutrients that stimulate plant production. Fine particles of decomposing manure and associated microbial flora serve as food for microscopic animals that may be consumed by plankton-feeding fish. Manure is made of plant material or the unassimilated fraction of animal feed; it has a high fiber content and a low nitrogen content. It decomposes at a comparatively slow rate, and large inputs of manure into ponds can result in the accumulation of organic sediment.

Ingredients for commercial aquaculture feeds are mixed and formed into pellets. Pellets are consumed directly by aquatic animals, and the unassimilated fraction enters the water in feces. Not all of the feed applied to ponds is eaten by the animals. The uneaten portion is particularly large for feeds that contain a high percentage of fines. *Fines* are fragments that break off pellets during shipping and handling. Overfeeding encourages wasted feed. When fish are stressed by disease or poor environmental conditions, they have a poor appetite, and a high percentage of the feed offered may not be consumed.

7.2.4 Primary Productivity

Nutrients added to ponds in inorganic fertilizers, manures, fish excrement, and uneaten feed stimulate phytoplankton productivity. Boyd demonstrated that each ton of fish production resulted in the production of about 2.5 tons of dry organic matter in phytoplankton.[10] The life span of phytoplankton cells is only one to two weeks, and dead cells settle to the pond bottom.

Schroeder made a carbon budget for an intensively manured pond.[11] He found

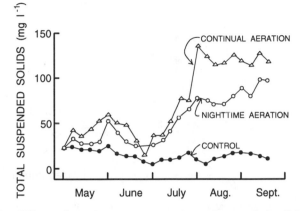

FIGURE 7.3. Effects of aeration on concentrations of suspended solids in waters of channel catfish ponds.

that the carbon accumulation rate was 0.5 g m^{-2} day^{-1}. Organic matter is about 58% carbon, so 0.86 g/m^2 (8.6 kg ha^{-1}) of organic matter accumulated in the bottom soil each day. Zur found accumulations of organic carbon from 2,290–8,900 kg ha^{-1} yr^{-1} (3,950–15,300 kg ha^{-1} yr^{-1} of organic matter) in carp and tilapia ponds.[12]

7.2.5 Erosion and Redeposition in Pond

Aeration can be a major factor in creating sediment in ponds. Aerators produce water currents that erode the pond bottom in areas where water velocities are highest, and the suspended particles are redeposited in other parts of the pond where water velocities are less. Thomforde and Boyd showed that continually aerated ponds had much higher concentrations of total suspended solids than ponds aerated only occasionally (Fig. 7.3), but no observations were made on alterations of pond bottoms by aeration.[13] Mevel and Boyd used an air diffuser system that released air along transects from end to end of ponds.[14] Waters of aerated ponds were more turbid with suspended clay particles than waters of unaerated ponds. At draining, bottoms of aerated ponds had troughs about 2 m wide and 30 cm deep along transects where diffusers operated. Mounds formed along both sides of troughs where eroded particles redeposited (Fig. 7.4).

Very severe incidences of erosion, sedimentation, and reshaping of pond bottoms occur in intensive shrimp ponds. Shrimp ponds in Thailand often have eight or more 0.75-kW paddlewheel aerators positioned to cause circular flow (Fig. 7.5). High-velocity water currents produced by aerators detach soil particles from 10- to 15-m-wide bands around peripheries of ponds and from insides of levees. Suspended particles settle in central areas of ponds where water velocities are less. Deposits are 30–45 cm deep and cover 30–50% of pond areas. Sediment

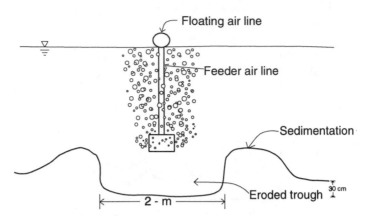

FIGURE 7.4. Cross-sectional view of erosion pattern in pond bottom caused by diffused air aeration. The trough and mound of sediment extended from one end of the pond to the other.

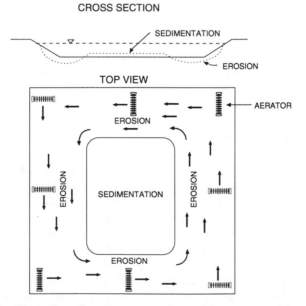

FIGURE 7.5. Illustration of erosion pattern in pond bottom caused by paddlewheel aerators positioned for circular flow.

is mineral soil with 2–4% organic matter. Soil from eroded peripheral areas usually contains less than 1% organic matter.

Aeration equipment is designed for maximum oxygen transfer efficiency, and little consideration has been given to its influence on soils. A great improvement in aeration would be realized if aerators produced uniform, gentle water currents over pond bottoms to suspend fresh organic particles without suspending mineral soil particles. It should be possible to design such equipment, because the particle density of organic matter is much less than for mineral soil particles. Suspended organic particles in the water would enhance the availability of dissolved oxygen for their decomposition. Preventing the deposition of mixtures of organic and mineral particles on the pond bottom would reduce the likelihood of areas with anaerobic conditions at the soil-water interface.

Wave-action erodes edges of ponds, and suspended particles settle over the pond bottom. Erosion also occurs at points of water inflow or outflow. When ponds are drained for harvest, water currents produced by outflow suspend solid particles and remove them from ponds.

7.3 Principles of Sedimentation

7.3.1 Terminal Settling Velocity and Stokes' Law

Settling velocity of a spherical particle is related to its diameter and density. The particle settles in water in response to gravity, but its motion is opposed by buoyant and drag forces of the water (Fig. 7.6). The force balance for a settling particle is

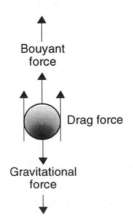

FIGURE 7.6. Forces acting on a particle settling in water.

$$m_p \frac{dv_s}{dt} = G - B - D \tag{7.2}$$

where m_p = mass of settling particle (kg)
v_s = velocity of particle (m sec^{-1})
t = time (sec)
G = gravitation force (N)
B = buoyant force (N)
D = drag force (N)
$m_p(dv_s/dt)$ = acceleration

The *net gravitational force* is the difference in gravitational and buoyant forces:

$$\text{Net gravitational force} = G - B \tag{7.3}$$

or

$$\rho_p g V_p - \rho_w g V_p = G - B$$

and

$$(\rho_p - \rho_w)g V_p = G - B \tag{7.4}$$

where ρ_p = density of particles (kg m^{-3})
ρ_w = density of water (kg m^{-3})
g = gravitation acceleration (m sec^{-2})
V_p = particle volume (m^3)

As the particle accelerates and its velocity increases, the viscous friction force that opposes motion of the settling particle relative to the fluid increases. This force is called the *drag force*. The particle quickly reaches a settling velocity where the drag force is equal to the net gravitational force, and the particle settles at a constant velocity. This is the *terminal settling velocity*. The drag force is related to the cross-sectional area and settling velocity of the particle, density of water, and a drag coefficient:

$$D = C_D \left(\frac{\pi d_p^2}{4} \right) \rho_w \left(\frac{v_s^2}{2} \right) \tag{7.5}$$

where C_D = drag coefficient and d_p = diameter of particle (m).
   At the terminal settling velocity, the acceleration term in Equation 7.2 is zero, and the net gravitational force (Eq. 7.4) equals the drag force (Eq. 7.5):

$$G - B - D = 0 \tag{7.6}$$

thus

$$G - B = D$$

and

$$(\rho_p - \rho_w)gV_p = C_D\left(\frac{\pi d_p^2}{4}\right)\rho_w\left(\frac{v_s^2}{2}\right) \tag{7.7}$$

The particle volume is

$$V_p = \frac{\pi d_p^3}{6} \tag{7.8}$$

and rearrangement of Equation 7.7 gives

$$C_D\left(\frac{\pi d_p^2}{4}\right)\rho_w\left(\frac{v_s^2}{2}\right) = (\rho_p - \rho_w)g\frac{\pi d_p^3}{6}$$

and

$$v_s^2 = \frac{4}{3}\frac{(\rho_p - \rho_w)\,gd_p}{C_D\rho_w} \tag{7.9}$$

The drag coefficient is obtained from the *Reynolds number:*

$$C_D = \frac{24}{R_e} + \frac{3}{\sqrt{R_e}} + 0.34 \tag{7.10}$$

where R_e = Reynolds number and

$$R_e = \frac{v_s d_p \rho_w}{\mu} \tag{7.11}$$

where μ = viscosity of water (kg m·sec^{-1}). At $R_e < 0.3$, only the first term of Equation 7.10 is important, and the drag coefficient is

$$C_D = \frac{24}{R_e} \tag{7.12}$$

Substituting Equation 7.11 into Equation 7.9 gives

$$v_s^2 = \frac{4}{3}\left[\frac{(\rho_p - \rho_w)g d_p}{24}\right]\left[\left(\frac{v_s d_p \rho_w}{\mu}\right)\rho_w\right]$$

and

$$v_s^2 = \frac{g(\rho_p - \rho_w)\,d_p^2}{18\mu} \tag{7.13}$$

Equation 7.13 is commonly known as the *Stokes' law equation*.

Because the drag force was considered to equal the net gravitational force, the settling velocity given by Equations 7.9 and 7.13 is the terminal settling velocity. Direct solution of Equation 7.13 for v_s is possible for cases with $R_e <$ 0.3, but an iterative solution of v_s is necessary with $R_e > 0.3$. This point is illustrated in the following examples. Particle density must be known to calculate v_s. Particle density of organic solids usually is around 1,050 kg m^{-3}; mineral soil particles have a density around 2,500 kg m^{-3}. Particle densities of 1,400–2,000 kg m^{-3} are typical for mineral soil particles flocculated by chemical means.[15] The viscosity and density of water are given in Table 7.5.

Table 7.5 Density (ρ) and Dynamic Viscosity (μ) of Water

Temperature (°C)	Density (kg m^{-3})	Dynamic Viscosity ($\times 10^{-3}$ N·sec m^{-2})	Temperature (°C)	Density (kg m^{-3})	Dynamic Viscosity ($\times 10^{-3}$ N·sec m^{-2})
0	999.8	1.787	21	998.0	0.978
1	999.9	1.728	22	997.8	0.955
2	999.9	1.671	23	997.5	0.932
3	999.9	1.618	24	997.3	0.911
4	1000.0	1.567	25	997.0	0.890
5	999.9	1.519	26	996.8	0.870
6	999.9	1.472	27	996.5	0.851
7	999.9	1.428	28	996.2	0.833
8	999.9	1.386	29	995.9	0.815
9	999.8	1.346	30	995.7	0.798
10	999.7	1.307	31	995.3	0.781
11	999.6	1.271	32	995.0	0.765
12	999.5	1.235	33	994.7	0.749
13	999.4	1.202	34	994.4	0.734
14	999.2	1.169	35	994.0	0.719
15	999.1	1.139	36	993.7	0.705
16	998.9	1.109	37	993.3	0.692
17	998.8	1.081	38	993.0	0.678
18	998.5	1.053	39	992.6	0.665
19	998.4	1.027	40	992.2	0.653
20	998.2	1.022			

Example 7.3: Calculation of Terminal Settling Velocity
A mineral soil particle has a diameter of 0.02 mm and a density of 2,500 kg m^{-3}. The terminal settling velocity will be computed for 30°C.

Solution: From Equation 7.13,

$$v_s = \frac{9.81 \text{ m sec}^{-2}(2500 - 995.7) \text{ kg m}^{-3}(2 \times 10^{-5} \text{ m})^2}{18(0.798 \times 10^{-3} \text{ N·sec m}^{-2})}$$

$$= 4.11 \times 10^{-4} \text{ m sec}^{-1}$$

The value of R_e may be checked:

$$R_e = \frac{4.11 \times 0^{-4} \text{ m sec}^{-1} (2 \times 10^{-5} \text{ m})(995.7 \text{ kg m}^{-3})}{0.798 \times 10^{-3} \text{ N·sec m}^{-2}}$$

$$= \frac{8.18 \times 10^{-6}}{0.798 \times 10^{-3}} = 1.03 \times 10^{-2}$$

Since $R_e < 0.3$, the solution is acceptable.

The particle in Example 7.3 is the size of silt (Table 2.3). In 1 h it will settle 1.18 m. This is a fairly slow settling rate, and water passes the particle in streamlines. This condition is known as *laminar flow*. Larger particles settle faster, and water flow past the particle becomes turbulent. If laminar flow persists, the drag force increases in proportion to the particle's velocity relative to the water, and Equation 7.13 provides a suitable estimate of v_s. When flow is turbulent, the drag force increases in proportion to the square of the particle's speed relative to the water. A value of $R_e > 0.3$ indicates turbulent flow, and Equation 7.9 must be used to estimate v_s. It is not possible to use Equation 7.9 without an estimate of v_s for obtaining R_e and C_D. A first approximation of v_s is made with Equation 7.13, and this estimate allows an iterative solution of Equation 7.9.

Example 7.4: Calculation of Terminal Settling Velocity under Turbulent Flow Conditions
The terminal settling velocity will be computed for a 0.2-mm particle at 30°C. Assume a particle density of 2,500 kg m^{-3}.

Solution: Use Equation 7.13 to get an initial estimate of v_s and R_e:

$$v_s = \frac{9.81 \text{ m sec}^{-2} (2500 - 995.7) \text{ kg m}^{-3} (2 \times 10^{-4} \text{ m})^2}{18 (0.798 \times 10^{-3} \text{ N·sec m}^{-2})} = 0.041 \text{ m sec}^{-1}$$

$$R_e = \frac{0.041 \text{ m sec}^{-1} (2 \times 10^{-4} \text{ m})(995.7 \text{ kg m}^{-3})}{0.798 \times 10^{-3} \text{ N·sec m}^{-2}} = \frac{8.16 \times 10^{-3}}{0.798 \times 10^{-3}} = 10.22$$

$R_e > 0.3$, so Equation 7.9 must be used to obtain v_s. The estimate of R_e can be used to obtain C_D with Equation 7.10:

$$C_D = \frac{24}{10.22} + \frac{3}{\sqrt{10.22}} + 0.34 = 3.63$$

Next, solve Equation 7.9 with the initial estimate of C_D:

$$v_s^2 = \frac{(4/3)(9.81 \text{ m sec}^{-2})(2500 - 995.7) \text{ kg m}^{-3} (2 \times 10^{-4} m)}{3.63 (995.7 \text{ kg m}^{-3})}$$

$$= 0.00109 \text{ m sec}^{-1}$$

$$v_s = 0.033 \text{ m sec}^{-1}$$

Use the v_s estimate from Equation 7.9 to estimate a new R_e and C_D:

$$R_e = \frac{(0.033)(2 \times 10^{-4})(995.7)}{0.798 \times 10^{-3}} = \frac{6.57 \times 10^{-3}}{0.798 \times 10^{-3}} = 8.23$$

$$C_D = \frac{24}{8.23} + \frac{3}{\sqrt{8.23}} + 0.34 = 4.30$$

Use the new estimates of R_e and C_D and solve a second time for v_s:

$$v_s^2 = \frac{(4/3)(9.81)(2500 - 995.7)(2 \times 10^{-4})}{4.30 (995.7)}$$

$$= 0.00092$$

$$v_s = 0.030 \text{ m sec}^{-1}$$

This estimate of v_s is close enough to the first estimate to accept it.

In making iterative solutions, we continue the iterations until two successive iterations give roughly the same answer. This resulted very quickly in Example 7.4 because R_e was not very large. For a larger particle with higher settling velocity, R_e will be greater, and several iterations will be necessary to obtain an acceptable estimate of v_s.

The terminal settling velocity of the 0.2-mm, sand-sized particle was 0.030 m sec^{-1} (1 m in 33 sec). The 0.02-mm, silt-sized particle had a terminal settling of 0.00041 m sec^{-1} (1 m in 40.6 min). The settling rate increases drastically as particle size increases. Clay settles slower than silt, and silt settles slower than sand. The terminal settling velocity of 0.001-mm clay particles at 30°C calculated by Stokes' law is 1.03×10^{-6} m sec^{-1}. It would take this particle more than 11 days to settle 1 m. Many clay particles have diameters less than 0.001 mm and settle at even slower rates. Some colloidal clay particles may remain suspended in water indefinitely.

7.3.2 Coagulation

Colloidal particles in water do not settle rapidly because they are small. They also are electrically charged and, like charged particles, repel each other. Most colloids in natural water are negatively charged. If the charge on colloidal particles in water is neutralized, they coalesce and form a floc large enough to settle.

The source of the charge on colloidal particles was discussed in Chapter 2. In water the surface of a negatively charged colloidal particle attracts a layer of cations, and a layer of anions is attracted to the cations. An electrical double layer forms around a charged particle (Fig. 7.7). Ions in the layers are diffuse, and they become more diffuse with distance from the solid surface. The repelling force that keeps particles from coming together is the zeta potential:

$$\zeta = \frac{4\pi\sigma q}{DE_w} \tag{7.14}$$

where ζ = zeta potential
σ = thickness of zone of influence of particle charge
DE_w = dielectric constant of water
q = charge on the particle

The repelling force is ζ. Intermolecular or interparticle attraction occurs through van der Waals forces, but if the zeta potential is greater than the van der Waals forces, particles cannot coalesce. Coagulation and precipitation of colloidal particles from water can be affected by processes that reduce the zeta potential.

In chemical coagulation, electrolytes reduce the zeta potential. Electrolytes have two effects. Monovalent salts such as NaCl reduce σ, and by compressing the electrical double layer the effect of the surface charge of the particle is greatly reduced. This causes attractive forces to exceed repelling forces, and particles

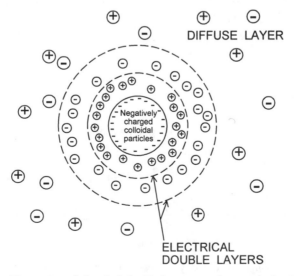

FIGURE 7.7. Illustration of electrical double layer around a negatively charged particle suspended in water.

coalesce. Multivalent cations have a greater attraction for negative charges, and they penetrate the zone of influence to neutralize part of the charge on the particle. This process also permits particles to coalesce. Sometimes changing the pH of a solution can neutralize enough of the charge on clay particles to allow them to coalesce.

Water chemistry influences the stability of colloidals. In freshwater, a high concentration of ions, especially calcium and magnesium, favors rapid coagulation and precipitation. Clay particles settle from brackish water much faster than from freshwater (Fig. 7.8). Chemical coagulation may be affected by applying electrolytes. The order of coagulating power is as follows: monovalent ions < divalent ions < trivalent ions. The most common chemical coagulant is aluminum sulfate $[Al_2(SO_4)_2 \cdot 14H_2O]$, or alum.

An example of chemical coagulation of suspended particles in pond water using alum is shown in Figure 7.9. To determine the alum treatment rate, one applies alum to aliquots of a turbid water sample at different concentrations and ascertains the concentration necessary to flocculate and precipitate the turbidity. Alum forms sulfuric acid in water:

$$Al_2(SO_4)_3 + 6H_2O \rightarrow 2Al(OH)_3 + 6H^+ + 3SO_4^{2-} \qquad (7.15)$$

Therefore, alum treatment reduces total alkalinity and pH. The reaction of alum and bicarbonate (alkalinity) can be written as

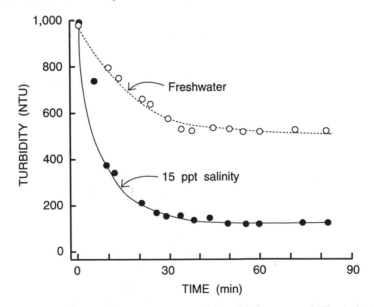

FIGURE 7.8. Changes in turbidity expressed in nephelometer turbidity units (NTU) caused by settling of suspended soil particles in samples of brackish water and freshwater. (*Source: Boyd.*[9])

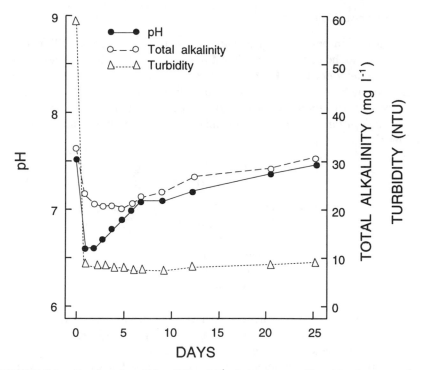

FIGURE 7.9. Results of applying 25 mg L^{-1} of aluminum sulfate (alum) to a pond to remove suspended soil particles on turbidity, pH, and total alkalinity. (*Source: Boyd.*[9])

$$Al_2(SO_4)_3 \cdot 14H_2O + 3Ca(HCO_3)_2 \rightarrow 2Al(OH)_3 + 3CaSO_4 + 6CO_2 + 14H_2O$$

$$(7.16)$$

One molecular weight of alum neutralizes three molecular weights of $Ca(HCO_3)_2$, and three molecular weights of $Ca(HCO_3)_2$ equal six equivalents of alkalinity. In other words, 600 g of alum neutralizes 300 g of $CaCO_3$, or 1 mg L^{-1} of alum neutralizes 0.5 mg L^{-1} of total alkalinity. Effects of alum treatment on pH, total alkalinity, and turbidity of a pond on the Auburn University Fisheries Research Unit are illustrated in Figure 7.9. Alum treatment causes a sudden reduction in pH, total alkalinity, and turbidity, but the pH and total alkalinity gradually returned to pretreatment levels.

Alum does not cause efficient flocculation and turbidity removal at pH values below 5. There must be residual alkalinity for alum treatment to be effective. In the alum treatment test, care must be taken to adjust the alkalinity of water samples so that the pH does not fall below 5. The same precaution is necessary in alum treatments of settling basins or ponds. Lime is applied to low-alkalinity waters just before or simultaneously with alum treatment to prevent low pH.

7.3.3 Sedimentation

In sedimentation basins or aquaculture ponds, particles become sediment only
if they settle to the bottom before they flow out of the pond. In many aquaculture
ponds, overflow occurs only after heavy rains. Even in ponds with daily water
exchange, hydraulic residence time is usually at least one week. The settling
velocity necessary for a particle to settle before flowing out of the pond is called
the *critical settling velocity,* and it is related to the hydraulic residence time as
follows:

$$v_{cs} = \frac{D_p}{\text{HRT}} = \frac{D_p}{V/Q} = \frac{1}{A/Q} = \frac{Q}{A} \tag{7.17}$$

where v_{cs} = critical settling velocity (m min^{-1})
   HRT = hydraulic residence time (V/Q) (min)
   D_p = pond depth (m)
   V = volume of pond (m^3)
   Q = inflow rate (m^3 min^{-1})
   A = surface area of pond (m^2).

This equation assumes that water enters in a homogeneous manner with the same
particle-size distribution at all depths. Particles with $v_s > v_{cs}$ are removed even
if they enter at the surface (Fig. 7.10). Particles with $v_s < v_{cs}$ are removed in
the proportion v_s/v_{cs}. Information on v_s and v_{cs} can be used in the design of settling
ponds or in predicting the sedimentation characteristics of a basin or pond.

Example 7.5: Analysis of Pond Sedimentation
   The water supply for a pond is expected to have an average suspended solid
   load of 0.5 kg m^{-3}. The pond has a water-exchange rate of 5% of pond volume
   per day. It is 1 ha in area and 1 m deep. The solids consist of 35% clay
   (average particle diameter = 0.001 mm), 40% silt (average particle diameter
   = 0.01 mm), and 25% very fine sand (average particle diameter = 0.05 mm).
   The water temperature is 25°C, and the particle density is 2,500 kg m^{-3}. The

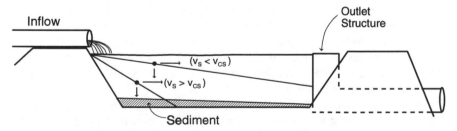

FIGURE 7.10. Illustration of the relationships between critical settling velocity (v_{cs})
and terminal settling velocity (v_s) in a sedimentation pond. Both particles have the same
horizontal velocity, but the terminal settling velocity of particle A is not great enough
for it to settle to the bottom before reaching the outlet structure ($v_s < v_{cs}$). Particle B has
$v_s > v_{cs}$, and it settles before reaching the outlet.

amount of suspended solids and particle characteristics of the pond effluent will be estimated.

Solution: First, the value of v_s will be determined for each particle-size class:

Fine sand

$$v_s = \frac{9.81 \text{ m sec}^{-1}(2,500 - 997.0)\text{kg m}^{-3}(5 \times 10^{-5}\text{ m})^2}{18(0.890 \times 10^{-3})\text{N·sec m}^{-2}}$$

$$= 2.3 \times 10^{-3} \text{ m sec}^{-1}$$

$$R_e = \frac{(0.0023 \text{ m sec}^{-1})(2 \times 10^{-5}\text{ m})(997.0 \text{ kg m}^{-3})}{0.890 \times 10^{-3}\text{ N·sec m}^{-2}}$$

$$R_e = 5.2 \times 10^{-2} \text{ (<0.3, so solution is acceptable)}$$

Silt

$$v_s = \frac{(9.81 \text{ m sec}^{-1})(2,500 - 997.0)\text{kg m}^{-3}(10^{-5}\text{ m})^2}{18(0.890 \times 10^{-3})\text{N·sec m}^{-2}}$$

$$= 9 \times 10^{-5} \text{ m sec}^{-1}$$

There is no need to solve for R_e because it will be even smaller than R_e for the fine sand.

Clay

$$v_s = \frac{(9.81 \text{ m sec}^{-1})(2,500 - 997.0)\text{kg m}^{-3}(10^{-6}\text{ m})^2}{18(0.890 \times 10^{-3})\text{N·sec m}^{-2}}$$

$$= 9 \times 10^{-7} \text{ m sec}^{-1}$$

Next, compute the critical settling velocity:

$$v_{cs} = \frac{Q}{A}$$

The pond has a volume of 10,000 m³ and a 5% daily water-exchange rate, so the inflow is 500 m³ day^{-1} or 0.00579 m³ sec^{-1}, and

$$v_{cs} = \frac{0.00579 \text{ m}^3 \text{ sec}^{-1}}{10,000 \text{ m}^2} = 5.79 \times 10^{-7} \text{ m sec}^{-1}$$

Finally, compare v_s for each particle-size class with v_{cs} of 5.79×10^{-7} m sec^{-1}:

	v_s (m sec^{-1})
Fine sand	2.3×10^{-3}
Silt	9×10^{-5}
Clay	9×10^{-7}

All particles have a terminal settling velocity greater than the critical settling velocity and should be removed from the water by sedimentation before flowing out of the pond. The sediment load is

$$0.5 \text{ kg m}^{-3} \times 500 \text{ m}^3 \text{ day} = 250 \text{ kg day}^{-1}$$

7.4 Sediment Control

High concentrations of suspended solids and associated sedimentation in aquaculture ponds result in loss of pond volume and have adverse effects on production. Erosion is a major source of suspended solids in water supplies for ponds, and it can be prevented through good conservation practices on watersheds or drainage basins. In many instances aquaculturists have no means of solving erosion problems affecting water supplies for their ponds. Sedimentation basins may be constructed to remove suspended solids before water enters ponds. Suspended solids and sediment that originates within ponds can also cause serious problems, and methods for dealing with internal sediment load are discussed in Chapter 9. When ponds are drained for harvest, suspended solids generated by harvest operations and released in effluents can cause sedimentation problems in receiving waters. Sedimentation basins can be valuable for treatment of pond effluents.

7.4.1 Influent Water

Runoff entering watershed ponds can be a major source of sediment. Where it is impossible to control erosion on a watershed, it may be feasible to build a sediment pond through which runoff passes before it enters into one or more aquaculture ponds. Sometimes sediment ponds may have marginal use as production ponds, while in other cases they serve only for sedimentation and water storage. A system of sediment control once used for some ponds on the Auburn University Fisheries Research Unit is shown in Figure 7.11. Turbid runoff from row-cropped watersheds entered ponds S-23, S-25, and S-27. Water from these ponds overflowed into ponds S-22, S-24, and S-26 and finally into ponds S-3 and S-6. Ponds S-23, S-25, and S-27 functioned only as sediment ponds, and waters were always turbid during rainy weather. The other ponds were seldom excessively turbid with suspended soil particles and were suitable for aquaculture. Ponds S-3 and S-6 were protected from high concentrations of suspended soil particles originating on the disturbed watersheds by the intervening ponds. Water held in the upper ponds (S-23, S-25, and S-27) seeped downgrade and stabilized water levels in lower ponds during dry weather.

There are many instances where water for filling ponds is pumped from streams or estuaries with high concentrations of suspended solids. These solids can be prevented from settling in production ponds if water is held in settling basins before it enters production ponds. Data on the terminal settling velocity of particles may be used as an aid in designing settling ponds as illustrated in the following examples.

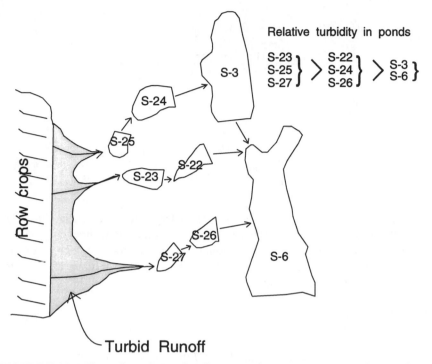

Relative turbidity in ponds

$$\left.\begin{matrix} S\text{-}23 \\ S\text{-}25 \\ S\text{-}27 \end{matrix}\right\} > \left.\begin{matrix} S\text{-}22 \\ S\text{-}24 \\ S\text{-}26 \end{matrix}\right\} > \left.\begin{matrix} S\text{-}3 \\ S\text{-}6 \end{matrix}\right\}$$

Turbid Runoff

FIGURE 7.11. Illustration of use of sediment ponds to protect watershed ponds from suspended soil particles in runoff. Ponds were located on the Auburn University Fisheries Research Unit. Ponds S-23, S-25, and S-27 served as sediment ponds. Fish were cultured in the other ponds.

Example 7.6: Size of Sediment Pond

Shrimp farms often have water supplies with large loads of suspended solids. It is desired to build a sediment pond to remove all particles larger than 0.001 mm in diameter to prevent them from settling in grow-out ponds on a farm with 500 ha of ponds. The water temperature is normally between 25 and 30°C. Water is pumped into the farm at an average rate of 300 m^3 min^{-1}. The settling pond will be 1.0 m deep. How large should it be?

Solution: Calculate v_s using water temperature of 25°C because settling velocity will be lowest at the lowest temperature:

$$v_s = \frac{(9.81 \text{ m sec}^{-1})(2{,}500 - 997.0) \text{ kg m}^{-3}(10^{-6} \text{ m})^2}{18(0.890 \times 10^{-3}) \text{ N·sec}^{-2}}$$

$$= 9.2 \times 10^{-7} \text{ m sec}^{-1}$$

The value of v_{cs} should not be greater than 9.2×10^{-7} m sec^{-1}. The pond will be 1 m deep and have an inflow of 5 m^3 sec^{-1}, and the outflow must equal the inflow. By rearrangement of Equation 7.17, the necessary pond volume is

$$v_{cs} = \frac{D_p}{V/Q}$$

$$V = \frac{D_p Q}{v_{cs}} = \frac{(1 \text{ m})(5 \text{ m}^3 \text{ sec}^{-1})}{9.2 \times 10^{-7} \text{ m sec}^{-1}} = 5.4 \times 10^6 \text{ m}^3$$

The required area is

$$A = \frac{V}{D_p}$$

$$\frac{5.4 \times 10^6 \text{ m}^3}{1 \text{ m}} = 5.4 \times 10^6 \text{ m}^2 = 540 \text{ ha}$$

It is obviously impossible to devote such a large area to a settling basin. In a smaller pond the finest soil particles could not be removed without coagulation. Assuming that coagulation is not economically feasible for such a large amount of water, the problem in Example 7.6 will be solved to remove all particles smaller than 0.005 mm in diameter.

Example 7.6: (continued)
    The terminal settling velocity of a 0.005-mm particle is

$$v_s = \frac{(9.81 \text{ m sec}^{-1})(2,500 - 997.0) \text{ kg m}^{-3}(5 \times 10^{-6} \text{ m})^2}{18(0.890 \times 10^{-3}) \text{ N} \cdot \text{sec}^{-2}}$$

$$= 2.3 \times 10^{-5} \text{ m sec}^{-1}$$

Let $v_{cs} = 2.3 \times 10^{-5}$ m sec^{-1} and

$$V = \frac{(1 \text{ m})(5 \text{ m}^3 \text{ sec}^{-1})}{2.3 \times 10^{-5} \text{ m sec}^{-1}} = 2.17 \times 10^5 \text{ m}^3$$

and

$$A = \frac{2.17 \times 10^5 \text{ m}^3}{1 \text{ m}} = 21.7 \text{ ha}$$

It is feasible to build a 21.7-ha sediment pond on a 500-ha farm. This pond will fill with sediment, so it should be larger than 21.7 ha. To determine the amount of sediment, we must know the amount and particle-size distribution of the suspended solids. For this example, suppose that the water contains an average of 250 mg L^{-1} (0.25 kg m^{-3}) of suspended solids. The particle-size distribution is as follows: 0.005 mm and larger, 70%; 0.002 mm, 10%; 0.001 mm, 5%; 0.0005 mm, 5%; 0.0002 mm, 5%; 0.0001 mm, 5%. These data will be used to determine the sediment load.

Example 7.6: (continued)
Using data on amount and particle-size distribution of the suspended solids (SS), calculate the critical settling velocity for each particle-size fraction, and determine the proportion of each fraction that will become sediment.

Particle-Size Fraction (mm)	Distribution (% of SS)	v_s (m sec^{-1})	v_s/v_{cs}	Proportion Removed
0.0001	5	9.2×10^{-9}	0.0004	0.0004
0.0002	5	3.7×10^{-8}	0.0016	0.0016
0.0005	5	2.3×10^{-7}	0.01	0.01
0.001	5	9.2×10^{-7}	0.04	0.04
0.002	10	3.7×10^{-6}	0.16	0.16
0.005	70	2.3×10^{-5}	1	1.00

The amount of each particle-size fraction that settles out is

Amount of fraction removed = SS concentration × proportion of fraction in SS
× proportion of fraction removed

For the 0.005-m fraction we get

Amount of 0.005-m fraction removed = 0.25 kg m^{-3} × 0.70 × 1.00
= 0.175 kg m^{-3}

The preceding computation was made for each fraction:

Particle-Size Fraction (mm)	Amount Removed (kg m^{-3})
0.0001	0.000005
0.0002	0.00002
0.0005	0.000125
0.001	0.0005
0.002	0.004
0.005	0.175
	0.180 g m^{-3}

Thus, the sediment load is 0.180 kg m^{-3} for each cubic meter of water. In one day the inflow is 432,000 m^3, and 77.8 tons of sediment are produced.

The sediment load is in tons, but it must be expressed on a volume basis in order to determine how fast the sediment pond will fill. The dry bulk density of mineral soil particles is usually 1,250–1,500 kg m^{-3}; a value of 1,400 kg m^{-3} will be assumed for the sediment in this example.

Example 7.6: (continued)
    The sediment pond has a sediment load of 77.8 tons day^{-1}. The volume is

$$77.8 \text{ tons} \div 1.4 \text{ ton m}^{-3} \text{ day}^{-1} = 55.6 \text{ m}^3 \text{ day}^{-1}$$

This is 20,294 m^3 year^{-1}. The volume of the pond is 217,000 m^3, so about
10% of its volume is lost in one year. The pond obviously should be built
larger than the minimum size needed for initial operation to be effective.

    The useful life of a sediment pond depends on its capacity to store sediment
and still maintain adequate hydraulic residence time for effective particle removal.
Suppose it is desired to build the pond of Example 7.6 large enough to provide
a useful life of 10 years. In 10 years the sediment load would be 202,940 m^3.
This would fill a 20.3-ha pond to a depth of 1 m. Thus, if the pond was constructed
20.3 ha larger than the minimum size of 21.7 ha, it would have a total area of
42 ha. Such a pond could possibly be used for 10 years before its performance
would fall below that predicted in Example 7.6. Its actual performance would
be better than that predicted in Example 7.6, because it would have a long
hydraulic residence time. This means that it would fill somewhat faster than
indicated.
    A common method for settling particles from water supplies for large aquacul-
ture farms is to use the water supply canal (Fig. 7.12) for a settling basin. Where
there is a heavy external sediment load, the canal may fill with sediment, and
the resulting loss of retention time renders the canal ineffective for removing
solids from the water before it enters ponds. A water supply canal almost completely filled with sediment is shown in Figure 7.13. Canals can be effective
sediment basins if sediment is removed to prevent its accumulation. The best
procedure is to build two canals. One canal can be used for water supply while
sediment is removed from the other canal. It is difficult to dry the sediment in
an empty sedimentation basin so that it can be removed with a bulldozer or
backhoe. Draglines can be effective for removing sediment, but floating dredges
(Fig. 7.14) are best. Sediment dredged from canals is discharged from a pipe
(Fig. 7.14) and constitutes a disposal problem. The dredged material should not
be discharged back into the water supply, and it should not be allowed to
contaminate natural wetlands. Suitable containment areas for dredged material
must be selected on a site-to-site basis. In many places where shrimp farming
is conducted, it is possible to discharge dredged material so that it contributes
to the natural sediment accumulation. Estuaries are natural sediment basins, and
the dredge material can be diverted into areas that are being filled by natural
processes. In this manner the material simply hastens the growth of small islands
or the extension of the land into the estuary. On small aquaculture farms it is
possible to remove sediment with shovels.
    Construction of a separate settling pond (Fig. 7.15) has advantages over the
canal system. A settling pond can be large enough to permit storage of sediments

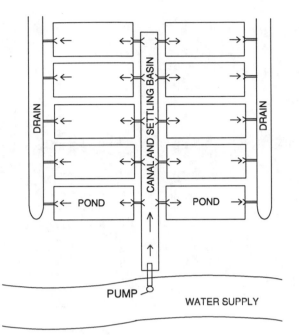

FIGURE 7.12. Canal sedimentation system for aquaculture ponds.

for years without a decline in performance. A dredge can be operated to remove sediment while the sediment pond is still in operation. The large surface area of the sediment pond encourages aeration of the water. Water flows out of the basin at one point, so the concentration of suspended solids is uniform. In the canal system for sedimentation, water near the head of the canal may have a greater concentration of suspended solids than water near the tail.

Baffle levees may be installed in sediment ponds to cause a circuitous flow (Fig. 7.16). This practice prevents water from flowing straight through the pond to the exit. It also is a convenient system for dredging operations.

Where only one large pond is available for aquaculture and there is no space to construct a sediment pond, a small sediment pond can be constructed in the influent end of the large pond (Fig. 7.16). Removal of the sediment to prevent its accumulation on the floor of the grow-out area by devoting 5–10% of the pond area to sedimentation is often a good investment.

7.4.2 Effluent

When ponds are drained for harvest, soil particles are suspended in the water and discharged in the effluent. Schwartz and Boyd showed that an average of 9.4 ton ha^{-1} (range 2.9–16.3 ton ha^{-1}) of suspended solids were discharged during drain harvest of channel catfish ponds.[16] More than half of the solids were

FIGURE 7.13. (Upper) A water supply canal for a shrimp farm in Honduras that has been almost completely filled by sediment. (Lower) Close-up view of sediment in the canal. (*Photographs supplied courtesy of Innovative Material Systems, Inc., Olathe, Kansas.*)

FIGURE 7.14. (Upper) A dredge removing sediment from the water supply canal of a shrimp farm in Honduras. (Lower) Discharge from a dredge. (*Photographs supplied courtesy of Innovative Materials Systems, Inc., Olathe, Kansas.*)

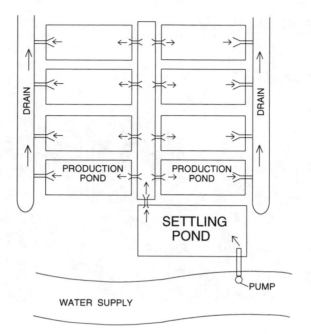

FIGURE 7.15.   Separate sediment pond for aquaculture farm.

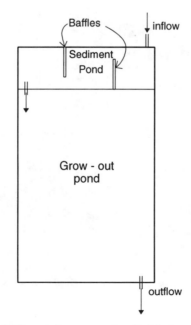

FIGURE 7.16.   Sedimentation system installed in large aquaculture pond.

discharged in the last 5% of water volume drained from the ponds. Where several ponds are available, the last portion of effluent from a pond can be pumped into an adjacent empty pond for settling. Where such an arrangement is not feasible, the effluent can be passed through a sedimentation pond before being discharged into natural waterways.

Aquaculture pond effluents contain both suspended soil particles and plankton. Constructed wetlands have been advocated as a means for treating aquacultural effluents. Shallow ponds are constructed and stocked with marsh plants. The layout of an experimental, two-celled constructed wetland used to treat effluents from a channel catfish pond in Alabama is shown in Figure 7.17. Water depth is maintained at about 45 cm. With a hydraulic residence time of two days or more, the constructed wetland removed large proportions of the suspended solids, biochemical oxygen demand, and plant nutrients from pond effluents (Table 7.6). The disadvantage of such a system is the large surface-area requirement.

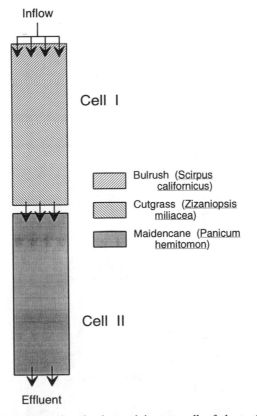

FIGURE 7.17.   A constructed wetland containing two cells of plants. (*Source: Schwartz and Boyd, Auburn University, unpublished data.*)

Table 7.6 Average Concentrations of Water-Quality Variables[a] in Influent and at Outflow of Two Cells in a Constructed Wetland to Which Water from a Channel Catfish Pond Was Applied at Different Hydraulic Residence Times (HRT)

HRT (days)	Sampling Location	TAN (mg L^{-1})	NO_2–N (mg L^{-1})	NO_3–N (mg L^{-1})	TKN (mg L^{-1})	TP (mg L^{-1})	BOD (mg L^{-1})	SS (mg L^{-1})	VSS (mg L^{-1})	Settleable Solids (ml L^{-1})
2	Influent	0.338	0.048	0.310	1.96	0.310	9.25	58.5	27.9	0.09
	Cell I effluent	0.098	0.007	0.225	1.01	0.188	4.87	19.0	11.1	0.01
	Cell II effluent	0.064	0.003	0.151	0.77	0.127	4.28	7.5	6.2	0
4	Influent	1.287	0.132	0.501	5.72	0.208	12.92	45.0	31.3	0.11
	Cell I effluent	1.072	0.003	0.178	3.63	0.078	4.75	8.1	5.7	0.01
	Cell II effluent	0.684	0.003	0.125	2.58	0.033	4.45	5.1	3.1	0

[a]TAN = total ammonia nitrogen; NO_2–N = nitrite–nitrogen; NO_3–N = nitrate–nitrogen; TKN = total Kjeldahl nitrogen; TP = total phosphorus; BOD = five-day biochemical oxygen demand; SS = suspended solids; VSS = volatile suspended solids.

However, Schwartz and Boyd showed that at least 50% of the total complement of potential pollutants in pond effluents was discharged in the last 15–20% of effluent volume.[16] If constructed wetlands were only used for treating the final concentrated effluents from ponds, the space requirements for constructed wetlands would be less, and they might be feasible water treatment systems for aquaculture pond effluents.

References

1. van der Leeden, Frits, F. L. Troise, and D. K. Todd. *The Water Encyclopedia*. 1990. Lewis Publishers, Chelsea, Mich.

2. Smith, D. D., and W. H. Wischmeier. 1957. Factors affecting sheet and rill erosion. *Trans. Amer. Geophys. Union* **38**:889–896.

3. Smith, D. D., and W. H. Wischmeier. 1962. Rainfall erosion. *Advan. Agronomy* **14**:109–148.

4. Soil Survey Staff. *National Soils Handbook*. 1983. USDA Soil Conservation Service, Washington, D.C.

5. Schwab, G. O., R. K. Frevert, T. W. Edminster, and K. K. Barnes. *Soil and Water Conservation Engineering*. 1981. John Wiley & Sons, New York.

6. Alabaster, J. S., and R. Lloyd. *Water Quality Criteria for Freshwater Fish*. 1982. Butterworth Scientific, London.

7. Boyd, C. E. 1992. Shrimp pond bottom soil and sediment management. In *Proceedings of the Special Session on Shrimp Farming*, J. A. Wyban, ed., pp. 166–181. World Aquaculture Society, Baton Rouge, La.

8. Wohlfarth, G. W., and G. L. Schroeder. 1979. Use of manure in fish farming—a review. *Agriculture Wastes* **1**:279–299.

9. Boyd, C. E. *Water Quality in Ponds for Aquaculture*. 1990. Alabama Agricultural Experiment Station, Auburn University, Ala.

10. Boyd, C. E. 1985. Chemical budgets for channel catfish ponds. *Trans. Amer. Fish. Soc.* **114**:291–298.

11. Schroeder, G. L. 1987. Carbon and nitrogen budgets in manured fish ponds on Israel's coastal plain. *Aquaculture* **62**:259–279.

12. Zur, O. 1981. Primary production in intensive fish ponds and a complete organic carbon balance in the ponds. *Aquaculture* **23**:197–210.

13. Thomforde, H., and C. E. Boyd. 1991. Effects of aeration on water quality and channel catfish production. *Bamidgeh* **43**:3–26.

14. Mevel, J., and C. E. Boyd. 1992. Channel catfish production and water quality in ponds with different rates of continuous, diffused-air aeration. *J. Aquaculture Tropics* **6**:275–285.

15. Tchobanoglous, G., and E. Schroeder. *Water Quality*. 1985. Addison-Wesley, Reading, Mass.

16. Schwartz, M., and C. E. Boyd. 1994. Effluent quality during harvest of channel catfish from watershed ponds. *Prog. Fish-Cult*. **56**:25–32.

17. U.S. Environmental Protection Agency. *Methods for Identifying and Evaluating the Nature and Extent of Nonpoint Sources of Pollutants*. 1973. EPA 430/9-73-014, U.S. Government Printing Office, Washington, D.C.

18. U.S. Department of Agriculture and U.S. Environmental Protection Agency. *Control of Water Pollution from Cropland*, Vol. I. 1975. U.S. Department of Agriculture, Agricultural Research Service, Washington, D.C.

8

Relationships to Aquatic Animal Production

8.1 Introduction

Relationships between pond soils or sediment and aquatic animal production are mostly indirect. Soil affects nutrient concentrations in the water, which, in turn, influence plant productivity. In extensive pond culture, aquatic animal production is the culmination of the food web that has its base in plant production (Fig. 5.1). Plants also are the primary source of dissolved oxygen in most ponds, but plant respiration and plant residues also exert an oxygen demand. Pond soils can be a source of toxic metabolites that can enter the water and harm aquatic animals. Suspended soil particles reduce light transmission through the water and adversely impact plant productivity and photosynthetic oxygen production. Sedimentation can smother aquatic organisms and destroy fish nesting sites. Suboptimal soil texture and pH may limit production of benthos, which is the food of many aquatic animals.

Desirable ranges of most water-quality variables have been established, but less is known about optimal ranges of pond soil condition factors. It is generally assumed that acidic soils, anaerobic conditions at the soil–water interface, high concentrations of soil organic matter, high rates of sedimentation, erosion of pond soils, and infestations of macrophytes in the pond bottom are detrimental factors in aquaculture ponds. This chapter attempts to explain relationships among soil condition factors and aquatic animal production, and to define optimal ranges for levels of several soil properties. Knowledge about this subject is poorly developed, and few definitive statements are given.

8.2 Flora and Fauna

Communities of plants and animals that inhabit the bottom of a water body are called *benthos*. These organisms live in the bottom soil, on the soil surface, or attached to the soil surface. Producer (plants), consumer (animals), and decomposer (bacteria) organisms constitute the benthic community of a water body. Organisms may be microscopic (bacteria, small algae, and protozoa) or visible to the naked eye (macrophytic algae, higher aquatic plants, worms, and sessile animals).

A pond or lake may be divided conceptually into several zones. The *littoral zone* extends from the shore to the depth where light becomes too dim for rooted plants, and it includes the water and soil in this area. The well-illuminated surface layer of water that extends beyond the littoral zone is called the *limnetic zone*. The littoral and limnetic zones make up the epilimnion of a stratified lake, and all plant productivity occurs in these two zones. The *profundal zone* is the deeper, poorly illuminated part of a water body. The profundal bottom is covered by the hypolimnion in a thermally stratified lake. The sublittoral zone occurs between the littoral zone and the profundal zone. In a stratified lake the thermocline occurs across the sublittoral zone. The littoral, sublittoral, and profundal bottom areas make up the benthic zone. The zones of a lake or pond are illustrated in Figure 8.1.

The littoral zone is obviously the most productive part of the bottom, for it has enough light for photosynthesis and is normally aerobic at the soil–water interface. The area of the pond bottom contained in each zone is greatly influenced by basin shape and water turbidity. In a shallow, clear body of water, the entire bottom may be within the littoral zone. In a deep body of water with steep basin slopes, the littoral zone is greatly restricted in extent. The boundaries of the littoral zone in a given body of water increase or decrease as turbidity changes and alters the proportion of the bottom that receives adequate light for photosynthesis.

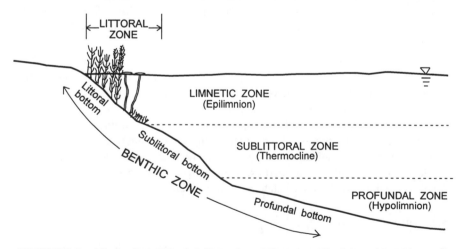

FIGURE 8.1. Illustration of littoral, limnetic, sublittoral, profundal, and benthic zones in a lake.

Sedimentation can fill in deep-water areas and enlarge the littoral zone. The littoral zone is highly productive, and benthic organisms in the littoral bottom provide food for many species of macrocrustaceans and fish. A brief description of benthic flora and fauna will be given.

8.2.1 Vegetation

Littoral zone vegetation typically forms zones as shown in Figure 8.2. The zone of emergent vegetation consists of grasses, rushes, sedges, and other "reed-swamp" plants that are rooted in the bottom and have long leaves that reach above the water surface. In deeper water the zone of emergent vegetation is followed by the zone of floating-leafed plants. These plants are rooted in the bottom and send floating leaves to the water surface. The zone of submerged vegetation occurs beyond the floating-leafed plants. Submerged plants are rooted and completely covered with water. Some submersed plants occur within the zones of emergent and floating-leafed vegetation. Macrophytic algae grow on the bottom or float in mats on the surface anywhere within the littoral zone. Benthic algae grow on the soil surface and on the submerged surfaces of rooted vegetation. Floating vascular plants also may be found in the littoral zone. Table 8.1 provides a list of some common littoral zone vegetation.

The outward extent of macrophytes in the littoral zone is limited by underwater light availability. There normally is insufficient light for aquatic plants at depths greater than twice the Secchi disk visibility. Typical standing crops of different types of aquatic vegetation also are provided in Table 8.1. Benthic algae do not

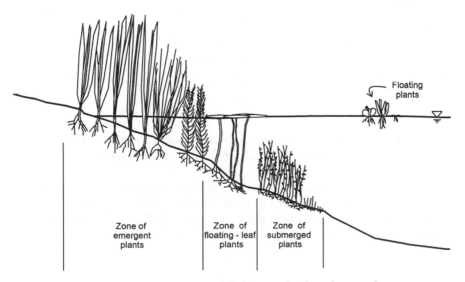

FIGURE 8.2. Zonation of higher aquatic plants in a pond.

Table 8.1   Typical Species and Standing Crops of Vegetation Found in the Littoral Zone of Lakes and Ponds

Scientific Name	Common Name	Typical Standing Crop (g dry weight m^{-2})
Macrophytic Algae		
Spirogyra spp.	—	20–60
Pithophora spp.	—	100–200
Cladophora spp.	—	50–100
Hydrodictyon reticulatum	Water net	50–100
Chara spp.	Stonewort	100–200
Nitella spp.	Muskgrass	100–200
Submersed Plants		
Eleocharis acicularis	Needle rush	200–400
Elodea densa	Water weed	100–200
Cabomba caroliniana	Fanwort	—
Potamogeton crispus	Pond weed	100–200
Najas guadalupensis	Bushy pondweed	100–200
Myriophyllum spicatum	Water milfoil	200–400
Floating-Leafed Plants		
Brasenia schreberi	Water shield	100–200
Nymphaea odorata	Water lily	150–300
Nuphar advena	Spatterdock	100–200
Nelumbo lutea	American lotus	150–300
Emergent Plants		
Sagittaria latifolia	Duck potato	800–1,000
Typha latifolia	Cattail	1,000–1,500
Juncus effusus	Softrush	1,000–2,000
Alternanthera philoxeroides	Alligator weed	500–1,000
Phragmetes communis	Reed grass	1,000–2,000
Floating Plants		
Eichhornia crassipes	Water hyacinth	1,500–3,000
Pistia stratiotes	Water lettuce	500–1,000
Lemna spp.	Duckweed	100–200

have large standing crops, but their turnover time in the community is much more rapid than for macrophytes. Their productivity often is greater than standing crop data suggests.

Some smaller aquatic plants are grazed directly, but larger ones usually die, fall into the water, and become detritus. Because of their high fiber content, they decompose slowly and large amounts of partially decomposed aquatic plant residue may accumulate in littoral environments.

8.2.2 Decomposers

The surface layer of soil in the littoral zone is normally aerobic and inhibited by a rich microbial flora of bacteria, actinomycetes, fungi, and protozoans.

Because of the high rate of oxygen consumption by microorganisms in the littoral bottom, anaerobic conditions may occur at soil depths below a few millimeters. Where algal mats grow on the bottom, water above and inside the mat may contain high dissolved oxygen concentrations because of photosynthesis, but beneath the mat, decay of dead algal material causes strongly anaerobic conditions.

Profundal areas of thermally stratified, eutrophic bodies of water contain little or no dissolved oxygen at the soil–water interface because there is insufficient light for photosynthesis. Large communities of anaerobic bacteria occur in the soil.

8.2.3 Consumers

The bottom fauna of littoral and profundal areas of lakes and ponds is extremely variable. Wetzel states that benthic communities of natural water bodies include protozoans, hydroids, turbellarian flatworms, nematodes, bryozoans, annelid worms, ostracods, mysid shrimp, isopods, decapods, amphipods, mollusks, and a wide variety of aquatic insects.[1] Reid indicated that some species of aquatic animals occur in littoral and profundal areas, and others occur primarily in the littoral or profundal area.[2] In lakes, dipteran larvae and oligochaete worms are common in soils of the profundal zone. Table 8.2 shows that benthic biomass may be just as high in the profundal area as in the littoral area.

There have been many studies of benthic animals in aquaculture ponds, because in some types of fish culture benthic animals are the primary food organisms for fish. For example, in the culture of sunfish and bass in sportfish ponds in the southern United States, aquatic insects and other benthic animals are the primary source of food for sunfish, which in turn, serve as forage for bass.

Selected studies will be summarized to illustrate the kinds and amounts of benthic animals found in aquaculture ponds. Dendy et al. observed many genera of benthic animals in Alabama fish ponds (Table 8.3), but only oligochaete worms and Diptera (chironomids) were abundant.[3] Patriarche and Ball reported that the most common benthic animals in fertilized ponds in Michigan were oligochaete worms, leeches (Hirudinea), and aquatic insects (Diptera and Odonata).[4] In other fertilized ponds in Michigan, aquatic insects (midges, dragonflies,

Table 8.2   Average Number and Weight of Benthic Animals at Different Depths (m) in Lake Simoe, Ontario

Variable	Littoral Zone					Sublittoral and Profundal Zones		
	0–1	1–5	5–10	10–15	15–20	20–25	25–30	30–35
Organisms (m^{-2})	405	788	926	574	844	947	1,068	1,134
Dry weight (g m^{-2})	1.02	1.28	1.48	0.65	1.17	1.42	1.34	1.22

Source: Modified from Reid[2].

Table 8.3 Benthic Animals Found in
Fertilized Ponds at Auburn, Alabama

Coelenterata	Ephemeroptera
Hydra	*Hexagenia*
Turbellaria	*Caenis*
Dugesia	Odonata
Rotifera	*Somatochlora*
Keratella	*Celithemis*
Nematoda	*Neurocordulia*
Ectoprocta	*Ischnura*
Plumatella	*Lestes*
Oliogochaeta	Trichoptera
Nais	*Ochrotrichia*
Pranais	*Leptocella*
Pristina	*Hydroptila*
Naidium	Coleoptera
Chaetogaster	*Dipneustes*
Cladocera	*Berosus*
Eurycercus	Diptera
Copepoda	*Tendipes*
Ectocyclops	*Pentaneura*
Brycocamptus	*Bezzia*
Ostracoda	*Probezzia*
Cyrpis	Gastropoda
Cypridopsis	*Gyraulus*
	Physa
	Ferrissia

mayflies, caddis flies, and beetles), annelid worms, mollusks, and scuds were the most abundant benthic animals (Fig. 8.3).[5,6] Hall et al. reported that the benthic animal communities in fertilized ponds in New York were dominated by amphipods and aquatic insects.[7] The two most common species were *Chironomus tentans* and *Caenis simulans*. In heavily fertilized ponds, chironomid larvae often made up 80% of the biomass. Benthic animals in brackish-water shrimp ponds consisted primarily of polychaetes, harpacticoid copepods, and nematodes.[8] Chironomid larvae were the most common benthic animals in manured carp ponds in Israel.[9,10]

The biomass of aquatic animals in ponds generally increases with increased inputs of nutrients.[3,9,11] Biomass changes over time in response to nutrient inputs and grazing by fish.[10,11] The total dry weight of benthic animal biomass in fertilized ponds was approximately 10–12 g m^{-2}, compared to 3–9 g m^{-2} in unfertilized ponds.[4,5,11]

8.3 Soil Condition and Fish Production

Before considering the influence of specific soil properties on fish and shrimp production, we note some general relationships among soils and fish production.

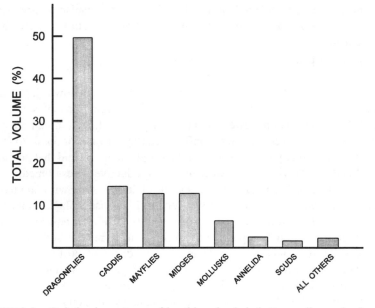

FIGURE 8.3. Relative importance of benthic animals in bottom soil samples from fish ponds in Michigan. Abundance of organisms is expressed as the percentage of the total volume of organisms. (*Source: Ball and Tanner.*[6])

In ponds where aquaculture species depend on natural food organisms for their nutrition, the fertility of the pond soil is a key factor regulating the production of fish and shrimp because it governs the fertility of the water. In ponds with feeding, most of the nutrition for fish and shrimp comes directly from the feed, and the fertility of the soil is not as important. However, even in fed ponds, soil condition is extremely important in determining concentrations of water-quality variables that influence survival and growth of fish and other aquatic animals.

8.3.1 Fertilized Ponds

Early reviews of pond fertilization pointed to the importance of pond bottom soils. Neess stated that in the European literature on pond fertilization, an impressive number of generalizations about fertilizers was made in terms of bottom soils.[12] Neess summarized these generalizations as follows:

1. The mineral composition of pond water is to a large degree a reflection of the mineral composition of the soils of the pond bottom.
2. The colloidal fraction of soils consisting of humic substances, ferric hydroxide gels, and clay is a powerful adsorbent of soluble mineral nutrients and governs their availability.

3. The pond soil is the medium in which live microorganisms responsible for decomposition of organic matter and transformations of chemical elements.

4. The pond soil is a source of food organisms for fish.

Neess concluded that the two most common problems causing low fish production in fertilized ponds were acidic soils and high concentrations of soil organic matter.[12] Mortimer also reviewed literature on pond fertilization and concluded that reactions between soil and water influenced the response of ponds to fertilizers.[13] He stressed that pond soils were sinks for phosphorus, and that ponds with peat soil and ponds with acidic water were unproductive unless limed.

Jhingran discussed the major classification of Indian soils with emphasis on retentivity of water by the soil and soil fertility, because soils for ponds must retain water tightly and they have adequate fertility to support aquatic productivity.[14] Jhingran stressed that pond soils differ from terrestrial soils as follows: (1) pond soils are conglomerated from soils of different profiles, completely waterlogged, and devoid of air-filled spaces; (2) ponds have large catchment areas and dissolved nutrients and soil particles from the catchment area enters ponds in runoff; (3) sedimentation of organic mater on the pond bottom modifies its properties. Jhingran reported that ponds built on different kinds of soil adsorbed phosphate from fertilizers at different rates and that the ability of ponds to produce fish was related to the soil phosphorus concentration.

Boyd and Walley[15] and Arce and Boyd[16] found that the mineral composition of pond waters differed among the major edaphic regions of Alabama and Mississippi. The most striking differences were obtained when sandy soils were compared with fertile, clay-loam soils or when noncalcareous soils were compared with calcareous soils. Pond waters with total alkalinity and total hardness values below 20 mg L^{-1} are usually less productive of phytoplankton following fertilization than ponds with higher total alkalinity and total hardness concentrations.[17,18] The total alkalinity and total hardness of surface waters are governed primarily by the kind of soil with which the water has had contact.[15,19,20] Ponds with low total alkalinity and total hardness typically are found on acid-sulfate soils, peat soils, sandy soils, or soils with a high degree of base unsaturation on cation-exchange sites.

8.3.2 Fed Ponds

Soils have a pronounced influence on water quality even in fed ponds. Nutrients that reach the water in excrement and un-eaten feed stimulate phytoplankton blooms in ponds, and dissolved oxygen dynamics in fed ponds are dominated by the phytoplankton. During the day, dissolved oxygen concentrations are high because photosynthesis produces oxygen faster than oxygen is consumed in respiration, but at night photosynthesis ceases, and dissolved oxygen concentra-

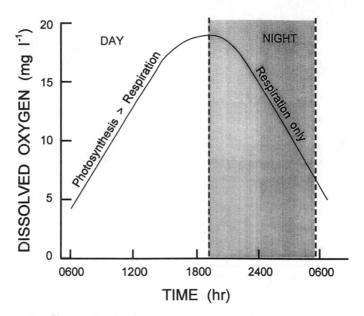

FIGURE 8.4. Changes in dissolved oxygen concentrations over a 24-h period in an aquaculture pond.

tions decline as oxygen is used in respiration (Fig. 8.4). At high feeding rates, heavy phytoplankton blooms develop, and the likelihood of nighttime dissolved oxygen depletion is great. Phosphorus usually is the most important nutrient regulating phytoplankton productivity. For a given feeding rate a pond with bottom soil that strongly adsorbs phosphorus has a lower potential for phytoplankton productivity than a pond with soil that adsorbs phosphorus less strongly.

Ponds with acidic bottom soils and low-alkalinity water are poorly buffered against pH change. Thus, when rates of photosynthesis are high, carbon dioxide removal from the water causes the pH to rise. At night, carbon dioxide is replaced by respiration and the pH falls. Ponds on acidic soils with low-alkalinity water typically have much greater fluctuation in pH over a 24-h period than ponds on neutral or alkaline soils with higher-alkalinity water (Fig. 8.5).

High concentrations of organic matter in bottom soils, either as a result of high soil organic matter in original soils or from high-organic-matter inputs to ponds, favor anaerobic conditions in pond bottoms and can lead to the production of toxic metabolites by microorganisms.

8.3.3 Soil Properties and Pond Aquaculture

Soil scientists and agronomists have correlated crop production with levels of soil properties such as texture, pH, nutrient concentration, and organic matter concentration. With such calibration data available it is possible to interpret

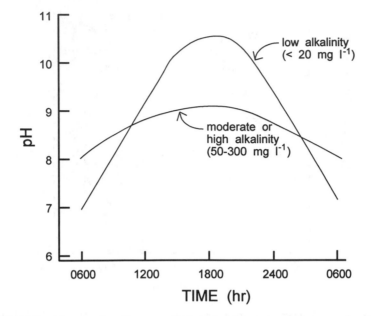

FIGURE 8.5. Changes in pH over a 24-h period in an aquaculture pond with low-alkalinity water compared to a pond with moderate- or high-alkalinity water.

analyses of soil samples to determine if levels of individual soil properties are within optimal ranges for production of specific crop plants. When levels of soil properties are suboptimal, remedial treatments may be applied to bring the levels of individual properties within the desirable ranges for crop growth. Studies also have been conducted to calibrate soil treatment rates against levels of soil properties. These calibration data can be used to estimate remedial treatment rates necessary to mitigate suboptimal soil conditions.

The approach used in crop science can be applied to aquaculture, as demonstrated by Boyd in developing a lime requirement test for pond soils.[18] In this procedure, 20 mg L^{-1} total hardness and total alkalinity were selected as the minimum permissible concentrations for fertilized ponds. It was then found that if pond soil was limed to decrease its base unsaturation to 0.20 or less, the total hardness and total alkalinity would be 20–50 mg L^{-1}. The liming rate is estimated from soil bulk density, depth of lime reaction in soil, total exchange acidity, and the relationship between pH and base unsaturation. Although the methods for developing the lime requirement test were complex, the laboratory aspects of the test are simple. Other soil test procedures have not been developed specifically for pond soils.

The most comprehensive study of relationships between individual soil properties and fish production was done by Banerjea.[21] He was able to obtain data on fish production and levels of soil properties for 80 ponds in India. Most of these ponds had never been fertilized. Fish production was classified as low (<200 kg ha^{-1}), average (200–500 kg ha^{-1}), and high (>500 kg ha^{-1}).

All ponds with soil pH <5.5 or above 8.5 had low production. The optimal soil pH for high production was 6.5–7.5. Average production was achieved in ponds with soil pH in the ranges 5.5–6.5 and 7.5–8.5.

Available phosphorus was extracted from soils with 0.002 N sulfuric acid. Available soil phosphorus concentrations corresponding to low, average, and high fish production were <30 ppm, 30–60 ppm, and >60 ppm, respectively.

Available nitrogen concentrations in soils were determined by measuring the nitrogen extractable in an alkaline potassium permanganate solution. Low production was obtained in ponds with less than 250 ppm available nitrogen, 250–500 ppm available nitrogen provided average production, and high production was achieved at greater available nitrogen concentrations.

Organic carbon concentrations in soils of less than 0.5% and greater than 2.5% resulted in low fish production. Low organic carbon was associated with low productivity of phytoplankton and bottom organisms, whereas high organic carbon caused anaerobic conditions in the pond bottom soils. Average fish production was achieved in ponds with 0.5–1.5% organic carbon, and 1.5–2.5% organic carbon was associated with high fish production.

The C:N ratio was computed using organic carbon and available nitrogen data. Low production was achieved at C:N ratios <5 and >15. A C:N ratio of 5–10 provided average production, and a C:N ratio of 10–15 gave high production. Total Kjeldahl nitrogen is usually employed in computing C:N ratios in soil, so C:N ratios reported by Banerjea are somewhat higher than those reported by many other authors.

Exchangeable calcium concentrations were related to fish production as follows: low, <100 ppm and >300 ppm; average, 100–200 ppm; high, 200–300 ppm.

The pond soils studied by Banerjea had clay contents ranging from 25% to 55%, and all but one was classified as clayey in texture.[21] He considered the soils were generally ideal in texture, because they were not so sandy as to allow excessive leaching of nutrients or so clayey as to adsorb nutrients too strongly.

Organic carbon and pH values provided by Banerjea would be applicable to ponds with feeding. Data on nitrogen, phosphorus, and exchangeable calcium concentrations have little meaning in ponds with feeding because nutrients for fish production would come from the feed rather than natural productivity.

Avnimelech et al.[22] and Avnimelech and Zohar[23] suggested that fish pond soils deteriorate over time as a result of high organic matter input and development of low redox potential. Even through anaerobic conditions and toxic metabolites were not detected in pond water, fish growth declined when reduced conditions developed in the soil (Fig. 8.6). Two methods for detecting reducing conditions in the sediment were developed. In one method, the extent that fish grazed on the sediment was chosen as a criterion of sediment quality. Radioactive phospho-

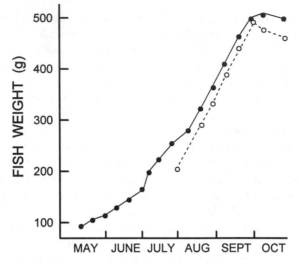

FIGURE 8.6. Fish growth patterns in two ponds in Israel. The cessation of growth in both ponds in September corresponded with an increase in denitrification rates. (*Source: Avnimelech and Zohar.*[23])

rus (P^{32}) was introduced into fresh, aerobic soil enriched with food pellets and into aged, anaerobic soil also enriched with food pellets. Fish were placed into the soil–water systems prepared with the two soils, and after a few hours exposure the P^{32} activity in digestive tracts of the fish was three times higher in fish from the containers with fresh sediment.[22] In the other technique, nitrate was added to ponds and the rate of disappearance of nitrate from the water was taken as an index of the denitrification rate and as evidence of anaerobic conditions in the pond bottom. When fish were growing well in ponds, denitrification rates were $0–0.05$ mg N $L^{-1} h^{-1}$. Later, when fish growth rates declined, denitrification rates averaged 0.2 mg N L^{-1}. Ponds were aerated continually, and low dissolved oxygen concentrations were not detected in the water column. However, the decline in nitrate suggested that denitrification was occurring in microenvironments of the pond soil. Neither procedure has gained acceptance as a tool in practical aquaculture, but each has potential application in research and practical aquaculture.

8.4 Effects of Aquatic Animals on the Soil

Aquatic animals also affect soil properties. Fish stir the bottom by producing water currents when they swim, some species stir the bottom while capturing prey, and many species make shallow depressions for nesting. There are many species of micro- and macro-invertebrate animals that live by burrowing in the bottom and feeding on other benthos or organic matter.

Stirring of the pond bottom can resuspend sediment (bioturbation). Ponds stocked with goldfish (*Carassius auratus*), common carp (*Cyprinus carpio*) or other bottom-feeding fish often are turbid with soil particles suspended by fish feeding activities. Most heavily-stocked ponds have greater than normal turbidity caused by stirring of the bottom by swimming fish. Organisms which burrow in the bottom loosen the soil and make it more susceptible to erosion by water currents, and burrowing activity mixes the sediment. Burrowing also enhances aeration of the upper layers of soil by facilitating the exchange of pore water with pond water. Benthic organisms die and contribute organic mater to the soil. Aquatic animals are influenced by pond soil condition, but they also are factors contributing to the development of pond soils.

References

1. Wetzel, R. G. *Limnology*. 1975. W. B. Saunders Co., Philadelphia.

2. Reid, G. K. *Ecology of Inland Waters and Estuaries*. 1961. Reinhold Publishing Co., New York.

3. Dendy, J. S., V. Varikul, K. Sumawidjaja, and M. Potaros. 1968. Production of *Tilapia mossambica* Peters, plankton, and benthos as parameters for evaluating nitrogen in pond fertilizers. Proceedings World Symposium on Warm-Water Pond Fish Culture, FAO United Nations, Fisheries Report No. 44, 3:226–240, Rome, Italy.

4. Patriarche, M. H., and R. C. Ball. 1949. *An Analysis of the Bottom Fauna Production in Fertilized and Unfertilized Ponds and Its Utilization by Young-of-the-Year Fish*. Technical Bulletin 207, Michigan State College Agricultural Experiment Station, East Lansing, Mich.

5. Ball, R. C. 1949. *Experimental Use of Fertilizer in the Production of Fish-Food Organisms and Fish*, Technical Bulletin, 210, Michigan State College Agricultural Experiment Station, East Lansing, Mich.

6. Ball, R. C., and H. A. Tanner. 1951. *The Biological Effects of Fertilizer on a Warm-Water Lake*. Technical Bulletin 223, Michigan State College Agricultural Experiment Station, East Lansing, Mich.

7. Hall, D. J., W. E. Cooper, and E. E. Werner. 1970. An experimental approach to the production dynamics and structure of freshwater animal communities. *Limnol. Oceanogr*. 15:839–928.

8. Rubright, J. S., J. L. Harrell, H. W. Holcomb, and J. C. Parker. 1981. Responses of planktonic and benthic communities to fertilizer and feed applications in shrimp mariculture ponds. *J. World Mariculture Soc*. 12:281–299.

9. Rappaport, U., S. Sarig, and Y. Begerano. 1977. Observations on the use of organic fertilizers in intensive fish farming at the Genosar station in 1976. *Bamidgeh* 29:57–70.

10. Zur, O. 1981. Primary production in intensive fish ponds and a complete organic carbon balance in the ponds. *Aquaculture* 23:197–210.

11. Howell, H. M. 1942. Bottom organisms in fertilized and unfertilized fish ponds in Alabama. *Trans. Amer. Fish. Soc.* **71**:165–179.

12. Neess, J. C. 1946. Development and status of pond fertilization in central Europe. *Trans. Amer. Fish. Soc.* **76**:335–358.

13. Mortimer, C. H. *Fertilizers in Fish Ponds.* 1954. Fisheries Publication No. 5, Her Majesty's Stationery Office, London.

14. Jhingran, V. G. *Fish and Fisheries of India.* 1975. Hindustan Publishing Corp., New Delhi, India.

15. Boyd, C. E., and W. W. Walley. *Total Alkalinity and Hardness of Surface Waters in Alabama and Mississippi.* 1975. Bulletin 465, Alabama Agricultural Experiment Station, Auburn University, Ala.

16. Arce, R. G. . and C. E. Boyd. *Water Chemistry of Alabama Ponds.* 1980. Bulletin 522, Alabama Agricultural Experiment Station, Auburn University, Ala.

17. Thomaston, W. W., and H. D. Zeller. 1961. Results of a six-year investigation of chemical soil and water analysis and lime treatment in Georgia fish ponds. *Proc. Annual Conf. S.E. Assoc. Game Fish Comm.* **15**:236–245.

18. Boyd, C. E. *Lime Requirements of Alabama Fish Ponds.* 1974. Bulletin 459, Alabama Agricultural Experiment Station, Auburn University, Ala.

19. Moyle, J. B. 1956. Relationships between the chemistry of Minnesota surface waters and wildlife management. *J. Wildlife Manag.* **20**:303–320.

20. Mairs, D. F. 1966. A total alkalinity atlas for Maine lake waters. *Limnol. Oceanogr.* **11**:68–72.

21. Banerjea, S. M. 1967. Water quality and soil condition of fish ponds in some states of India in relation to fish production. *Indian J. Fish.* **14**:113–144.

22. Avnimelech, Y., M. Lacher, A. Raveh, and O. Zur. 1981. A method for the evaluation of conditions in a fish pond sediment. *Aquaculture* **23**:361–365.

23. Avnimelech, Y., and G. Zohar. 1986. The effect of local anaerobic conditions on growth retardation in aquaculture systems. *Aquaculture* **58**:167–174.

9

Pond Bottom Management

9.1 Introduction

Soil properties and relationships among pond soils, water quality, and aquacultural production have been discussed. It has been shown that initial soil condition in new ponds sometimes is unsatisfactory, and the condition of pond soils may deteriorate over time in response to demands imposed on them by aquaculture. Some pond management schemes cause less degradation of soil condition than others, and a few treatments are available for improving soil condition. This chapter discusses pond management and soil treatment techniques for improving or maintaining the condition of pond soils. Application of these techniques can enhance soil condition and water quality, improve production efficiency, and prolong the useful life of ponds.

9.2 Liming

9.2.1 Pond Soils with Exchangeable Acidity

Soil acidity is a management concern because both total alkalinity and total hardness of pond water are closely related to bottom soil acidity. Total alkalinity and total hardness should be above 20 mg L^{-1} in aquaculture ponds to provide enough buffering capacity to maintain pH of 6–9.5 and sufficient dissolved inorganic carbon to support good phytoplankton growth.[1] Crustaceans have a requirement for calcium and bicarbonate in molting, and total alkalinity and hardness should be 50 mg L^{-1} or more in ponds for their culture. Liming can be used to increase total alkalinity and hardness, and in established ponds the need for liming can be readily determined by water analysis. In new ponds that have never been filled with water, the combination of soil pH <6.5 and a low-alkalinity water supply suggests that liming will be required.

Exchangeable acidity is caused by exchangeable aluminum ion on soil colloids, and the amount of soil acidity depends on pH and cation-exchange capacity (see Chap. 2). The lime dose should be based on the lime requirement of the pond soil rather than total alkalinity and total hardness of the water. The amount of lime needed to neutralize soil acidity is much greater than the quantity of lime required to initially raise the total alkalinity and total hardness of the water to a desirable concentration. The soil acidity must be neutralized, or it will quickly react with total alkalinity and total hardness and lower their concentrations.[1]

In humid inland areas, soils often are highly leached and contain a high proportion of exchangeable aluminum, water usually is not highly mineralized, and ponds commonly need to be limed. Ponds on organic soils also tend to have acidic water. In arid regions salts accumulate in soils, waters are highly mineralized, and ponds normally do not need to be limed. Brackish water or seawater used to fill coastal ponds has moderate concentrations of total alkalinity (75–150 mg L^{-1}) and high concentrations of total hardness (500–6,000 mg L^{-1}). Water often is exchanged in coastal ponds at 2–10% of pond volume per day, and alkalinity and hardness removed from water by reaction with acidic soil is replenished by water exchange. Unless coastal ponds have extremely acidic, organic, or acid-sulfate soils, liming is not needed during the grow-out period. When coastal ponds are drained between crops, bottoms of empty ponds are often limed.

Procedures for liming ponds will be provided. The best liming material for general use is agricultural limestone.[2] Liming should be done two to three weeks before fertilizer applications are initiated, because an immediate effect of liming is to reduce dissolved inorganic phosphorus and carbon dioxide concentrations. However, after the liming material has reacted with water and soil for a few days, there will be an improvement in phosphorus and carbon dioxide availability.[1]

Total alkalinity and total hardness should be measured at 6- to 12-month intervals after liming, because a single lime treatment seldom has a residual life of more than three to five years or about 10 pond volumes of water exchange.[3] After heavy rains or large intentional water exchange, total alkalinity and hardness decline in recently limed ponds, but they will return to normal within a few weeks.

New Ponds

Remove soil samples (15-cm-deep cores) from the pond bottom and determine pH. If pH is below 7, measure the lime requirement. Methods for sampling and for determining the lime requirement are given in Chapter 10. The indicated dose of agricultural limestone should be spread uniformly over the bottom and mixed into the soil by tilling to a depth of 10–15 cm with a rototiller (Figs. 9.1 and 9.2) or disk harrow (Fig. 9.3). A turning plow (Fig. 9.4) should not be used. The pond may be filled with water as soon as liming has been completed.

FIGURE 9.1. Small, self-powered rototiller.

FIGURE 9.2. Tractor-powered rototiller.

FIGURE 9.3. Disk harrow.

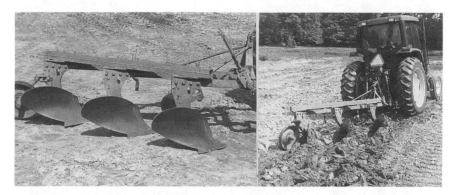

FIGURE 9.4. Turning plow.

Established Ponds

If the total alkalinity of water is below 20 mg L^{-1} in fish ponds or 50 mg L^{-1} in crustacean ponds, soil samples (15-cm-deep cores) should be collected and the lime requirement determined (see Chap. 10). It often is more convenient to collect samples and apply agricultural limestone to empty ponds after bottoms have been allowed to dry between crops.

Soil in deeper parts of a pond normally has a higher lime requirement than soil from shallower areas (Table 9.1). It is not necessary to map the lime requirement of the pond bottom and apply liming material accordingly. The total amount of agricultural limestone needed for a pond is calculated from the average lime requirement of several soil samples from the bottom area. The pond can be divided by visual inspection into areas less than or more than 1 m deep. Agricul-

Table 9.1   Lime Requirements of Muds from Different Depths in Ponds

Depth (m)	Lime Requirement (kg ha^{-1})		
	Pond 1	Pond 2	Pond 3
0.01	600	700	900
0.33	800	500	900
0.66	1,300	600	1,500
1.00	2,800	2,100	1,800
1.33	3,000	3,300	2,600
1.66	3,100	3,600	3,500
2.00	3,100	3,800	4,000
2.33	—	3,800	4,300
2.66	—	3,700	4,300
3.00	—	3,600	4,100

Source: Boyd.[1]

tural limestone should be applied two to three times heavier in deep areas than in shallow ones. In ponds less than 1 m deep, spread limestone uniformly over the entire bottom. Mechanical lime spreaders mounted on trucks or pulled by tractors are convenient for liming large ponds with dry bottoms. Manual application is necessary in small ponds or in large ponds that cannot be dried enough to support liming equipment. When pond soils are dry enough to permit tilling, agricultural limestone should be incorporated into the surface 10–15-cm soil layer with a rototiller or disk harrow.

Limestone must be spread from a boat in ponds that cannot be drained and dried. Two techniques are widely used. Bagged limestone is poured from a boat as the boat moves over the pond surface. Bulk agricultural limestone is cheaper than bagged material. Bulk limestone may be loaded on a platform built on a boat or barge and spread manually with a shovel (Fig. 9.5) or washed into the pond by a jet of water as the vessel moves over the pond surface (Fig. 9.6). Limestone should be spread two to three times as heavily over areas more than 1 m deep as for areas less than 1 m deep.

9.2.2 Ponds with Acid-Sulfate Soils

Identification

Potential acid-sulfate soils may be identified by a high total sulfur concentration (>0.75%) or a low pH (2–3) upon drying for several days. Field identification of potential acid-sulfate soils is made by mixing a few grams of fresh soil with a few milliliters of fresh, 10–30% hydrogen peroxide (H_2O_2). If pyrite is present, there will be a vigorous reaction with release of gas bubbles as the hydrogen peroxide oxidizes FeS_2 to H_2SO_4. Other reduced substances can react with H_2O_2 with release of gas bubbles. The "foolproof" test is to place a piece of Universal pH paper (pH 0–6) in the soil–peroxide mixture after it has reacted for about 10 min. Insert only the tip of the pH paper in the edge of the mixture and allow the liquid to migrate along the length of the strip of indicator paper. If the color change of the indicator paper suggests a pH of 3 or less, the soil is a potential acid-sulfate soil.

Old coastal soils may not contain pyrite in surface layers, because it may have already oxidized and the resulting acidity leached out. Subsurface soils in coastal areas often are poorly drained and anaerobic because of a high water table. Thus, surface soil may not be acidic, but subsurface horizons at depths of 0.5–2 m may be potential acid-sulfate soil. When an area suspected to have potential acid-sulfate soils is considered as a site for aquaculture ponds, core samples should be taken to a depth of at least 0.5 m more than the anticipated depth of excavation, and soil from different segments of the cores should be examined for potential acid-sulfate conditions.

FIGURE 9.5. Manual application of limestone from a small boat.

FIGURE 9.6. Hydraulic application of limestone from a liming barge.

Mitigation of Acidity

As long as soils containing pyrites are submerged and anaerobic, they remain reduced and the pyrite does not oxidize to yield acidity. When ponds are drained and pond soils exposed to air, oxidation results and sulfuric acid forms. Specific chemical reactions occurring in pyrite oxidation are found in Chapter 2; the summary reaction for sulfuric acid formation from iron pyrite is

$$FeS_2 + 3.75O_2 + 3.5H_2O \rightarrow Fe(OH)_3 + 2SO_4^{2-} + 4H^+ \qquad (9.1)$$

The $Fe(OH)_3$ crystallizes as a reddish brown material in the sediment. After draining, a sediment containing pyrite is called an acid-sulfate soil.

Singh provided information on acid-sulfate soils at the Brackishwater Aquaculture Center, Iloilo, Philippines.[4] Drying the soil resulted in decreased soil pH because of oxidation of pyrite upon exposure to the air. Lime requirements calculated from total potential acidity values were 40–70 tons $CaCO_3$ ha^{-1}. Acid-sulfate soils at Iloilo resulted in low pH in pond water. Pyrite in levees oxidized during dry weather, and the problem was especially severe when highly acidic runoff entered ponds after rains.[5]

Simpson et al. evaluated the influence of acid-sulfate soils on brackish-water aquaculture in Malaysia.[6] Ponds on acid-sulfate soils had low-pH, low-alkalinity water, and aquacultural production was low because of the acidity. They suggested the following: use soil surveys to identify areas of acid-sulfate soils; design ponds to minimize the amount of levee soil exposed to the atmosphere; select acid-tolerant culture species; monitor pH and total alkalinity of the water; dry bottom soil periodically and then flush ponds with water to remove acidity; lime to counteract acidity. Hajek and Boyd also stressed the importance of soil survey data in identifying areas with potential acid-sulfate soils.[7] They provided a systematic procedure for rating limitations of acid-sulfate soils in pond construction and management. Brinkman and Singh developed a method for rapid reclamation of brackish-water ponds with acid-sulfate soils.[8] The procedure is as follows:

Step 1. In the early part of the dry season, dry the pond bottom and harrow.

Step 2. Fill with brackish water. Measure the pH of the water frequently. The pH will drop from that of seawater (7 to 9) to below 4. Once the pH has stabilized, drain the pond. Repeat this procedure until the pH stabilizes at a pH above 5. Three or more drying and filling cycles may be required.

Step 3. At the same time the pond bottom is being reclaimed, acid should be removed from levees. To achieve this, level the levee tops and build small bunds along each side of the levee tops to produce shallow basins. Fill basins with brackish water. When the pond is drained for drying, also drain the small basins on the levee tops for drying. Repeat if necessary. Finally, remove the bunds and broadcast agricultural limestone over the tops and sides of levees at $0.5–1.0$ kg m^{-2}.

Step 4. Once the last drying and refilling cycle is complete, broadcast agricultural limestone over the pond bottom at 500 kg ha^{-1}. Fertilize the pond with manures or chemical fertilizers as necessary to promote phytoplankton blooms.

Step 5. To prevent seepage of acid from levees from killing fish, check pH frequently and apply agricultural limestone if necessary.

If potential acid-sulfate soils are drained and exposed to the air, oxidation and leaching by rainfall will gradually reduce acidity. Over the years, surface soil pH will increase and vegetation will develop. Deep soil layers will still be potential acid-sulfate material. When ponds are dug into such soils, potential acid-sulfate soils will be exposed. Many times, potential acid-sulfate soils are unknowingly used to construct levees. The iron pyrite in levee surfaces oxidizes and runoff entering ponds is highly acidic. The pond bottom also may be of potential acid-sulfate soil, but water contains little oxygen compared to air, and iron pyrite in bottoms of ponds does not oxidize quickly as long as ponds are full of water.

Gaviria et al. suggested that most problems with acidity can be solved through lowering pyrite oxidation rates.[9] They suggested that ponds be maintained full of water at all times. After harvest, ponds should be refilled immediately to prevent bottoms from drying. Ponds should be no deeper than necessary, and levees should have the smallest possible surface areas. Grass cover should be established on levees to provide a barrier between air and soil sulfide and to minimize the contact of runoff with raw soil. Levees should be treated with agricultural limestone at 0.05–0.1 kg m^{-2} or more and planted with native, acid-resistant grass. Some experimentation may be required to determine the best species of grass. The grass should be fertilized and, during the dry season, watered. Even with these measures, some acidity will be produced in pond bottoms and runoff will transport acidity from levees into ponds. Total alkalinity should be monitored, and if it drops below 20 mg L^{-1}, agricultural limestone should be applied to ponds at 1,000 kg ha^{-1}. The procedure suggested by Gaviria et al. is especially useful in freshwater sites with acid-sulfate soil where there is inadequate water to flush ponds.[9]

Lin gave the following recommendations for construction of freshwater ponds on acid-sulfate soils of Thailand's central plain.[10] Digging should be avoided in pond construction. Levees should be built on top of the original terrace with the top soil saved from digging the drainage canals. Reduce the amount of runoff from levees relative to pond volume by reducing the size of levees, increasing pond size, and maintaining a high water level. The top surface of levees should be built in a V shape with a central depression to collect excess rainwater and empty it into the drainage canal.

In summary, there are three ways of dealing with acid-sulfate soils:

1. Drain soils and wait until natural oxidation and leaching removes the acidity.

2. Lime to neutralize the acidity.

3. Prevent the oxidation of the iron pyrite so that acidity is not expressed.

It takes many years for the acidity to naturally leach from acid-sulfate soil, but it is not economically feasible to build ponds and wait for several years to use them. Lime requirements are 25–150 ton ha^{-1}, and such large lime applications normally are infeasible. Most programs for controlling acidity rely on a combination of techniques.

Example 9.1: Liming Rate for a Potential Acid-Sulfate Soil Material
Levees of a pond are made of potential acid-sulfate soil with a potential acidity of 41,200 mg CaCO$_3$ kg^{-1}. The soil has a dry bulk density of 1.42 ton m^{-3}. How much agricultural limestone should be incorporated into the upper 5-cm layer of this soil to counteract the potential acidity. Calculate the answer in kilograms of CaCO$_3$ per square meter.

Solution:
Soil weight

$$1 \text{ m}^2 \times 0.05 \text{ m} \times 1420 \text{ kg m}^{-3} = 71 \text{ kg}$$

Lime dose

$$71 \text{ kg soil} \times 0.0412 \text{ kg CaCO}_3 \cdot \text{kg soil}^{-1} = 2.93 \text{ kg CaCO}_3$$

Soils containing iron pyrite also contain exchange acidity. However, where the acid-sulfate problem is severe enough to warrant liming, exchange acidity constitutes a small proportion of the total acidity and it usually can be ignored. If the total lime requirement of an acid sulfate soil is desired, the lime requirement resulting from both pyrite acidity and exchange acidity must be determined separately and summed.

Mine spoils, particularly coal mine spoils, contain iron pyrite. Mine spoils behave exactly like acid-sulfate soils.[11] Waller and Bass found that conversion of depleted coal strip mines to productive lakes required that all of the acid-sulfate spoils on the watershed, including lake banks, be treated to prevent acidic seepage.[12] Next, agricultural limestone was applied to lakes at the rate of 25 mg L^{-1} of calcium carbonate for each milligram per liter of boiling point acidity. Several strip-mine lakes treated as described maintained pH values near 7, measurable alkalinity, and satisfactory fish production.[13]

Construction techniques can sometimes be modified to minimize problems with acid-sulfate soils. If surface soils have already oxidized and been leached free of sulfuric acid, but deeper layers of soil are potential acid-sulfate soils, surface soils may be stockpiled and later used for a surface layer on levees and pond bottoms. In some areas near Chanthaburi, Thailand, nonacidic soil material was brought by truck and used to cover pond bottoms and levees.

It is important not to use total hardness data when working with ponds on acid-sulfate soils. I analyzed a sample of water from a pond excavated into acid-sulfate soil in southern Georgia (USA). The pond had been limed heavily, but total alkalinity was undetectable. The water had a pH of 3, total acidity of 350 mg L^{-1}, and a total hardness of 250 mg L^{-1}. The high hardness resulted because the acidity of the pond water neutralized all of the limestone. The carbonate from the limestone was converted to carbon dioxide and water, and the calcium and magnesium from the limestone remained in the water to increase total hardness.

9.3 Bottom Soil Sterilization

It is sometimes desirable to sterilize pond bottom soils to prevent pathogens from surviving between aquaculture crops and infecting fish or crustaceans stocked for the next crop. The three most common sterilants are of burnt lime (CaO), hydrated lime [Ca(OH)$_2$], or chlorine compounds. Burnt and hydrated lime raise soil pH above 10 long enough to kill pathogenic organisms. Chlorine is directly toxic to pathogenic organisms. These sterilants kill pathogens and normal, non-pathogenic microbial flora as well. The residual time of high pH and chlorine compounds is short, and after a few days the soil will be quickly reinoculated with microbial organisms from the air. Nevertheless, after pathogens are eliminated, they will not be reinoculated unless diseased animals or pathogen-contaminated water are introduced into ponds.

9.3.1 Burnt and Hydrated Lime

Calcium oxide (CaO) is more effective per unit weight than calcium hydroxide [Ca(OH)$_2$] in increasing soil pH. The gram molecular weight of CaO is 56 and that of Ca(OH)$_2$ is 74. Therefore, 1.32 times more Ca(OH)$_2$ than CaO (74/56) must be applied to give the same concentration of base. Usually, 1 ton ha^{-1} of CaO or 1.32 ton ha^{-1} of Ca(OH)$_2$ is adequate to increase soil pH above 10. The residual time for high pH is brief (Fig. 9.7), and a treatment rate of 2 ton ha^{-1} is more efficient in killing pathogens. Commercially available burnt and hydrated lime are seldom 100% pure; use of the higher treatment rate also ensures that adequate lime is applied to provide a high pH. Agricultural limestone will not raise soil pH high enough to kill pathogenic organisms (Fig. 9.7).

Burnt or hydrated lime should be applied uniformly over the pond bottom so that the entire soil surface is covered. There should be adequate moisture in the soil for the lime to dissolve and move down into the entire volume of the upper 10-cm soil layer. If the soil is very dry, it should be irrigated. Best results will be achieved if lime is spread uniformly over the soil surface with a mechanical lime spreader, the upper 10-cm layer of soil is tilled, and enough water is applied to thoroughly moisten the soil. In addition to killing pathogenic organisms, high pH resulting form liming causes movement of ammonia from soil to air according to the reaction

$$NH_4^+ + OH^- = NH_3 \uparrow + H_2O \tag{9.2}$$

Calcium oxide applied to pond soils reacts with water and is converted to calcium hydroxide:

$$CaO + H_2O = Ca(OH)_2 \tag{9.3}$$

Carbon dioxide from moist soil and air neutralizes hydroxide ion resulting from dissolution of Ca(OH)$_2$:

$$Ca(OH)_2 + 2CO_2 = Ca^{2+} + 2HCO_3^- \tag{9.4}$$

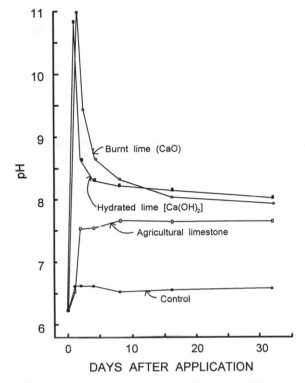

FIGURE 9.7. Effects of liming materials on soil pH in dry pond bottom soils. Materials were applied at 1 ton ha^{-1}. (*Source: after Boyd and Masuda.*[2])

Ponds may be refilled and stocked with fish or other aquatic animals 10–14 days after applying burnt or hydrated lime. Nevertheless, it is wise to check the pH of the water to ascertain that the pH is suitable before stocking aquatic animals.

9.3.2 Chlorination

Chlorine also is used to disinfect water and soil. Chlorine is applied as free chlorine (Cl_2), sodium hypochlorite or bleach ($NaOCl$), or calcium hypochlorite [$Ca(OCl)_2$], also known as high-test hypochlorite (HTH). When applied to soil or water, the form of chlorine (Cl_2, $HOCl$, or OCl^-) depends on pH:

$$Cl_2 + H_2O = HOCl + H^+ + Cl^- \tag{9.5}$$

and

$$HOCl = H^+ + OCl^- \tag{9.6}$$

Almost no Cl_2 can exist above pH 2, and at pH 7.48, the proportions of HOCl and OCl^- are equal (Fig. 9.8). The disinfecting value of HOCl is about 100 times greater than that of OCl^-, so the amount of chlorine that must be applied to effect disinfection increases rapidly above pH 6 because the proportion of HOCl decreases (Fig. 9.8). For safety and convenience it is better to apply bleach or HTH instead of free chlorine to pond soils; HTH is used most frequently. Pond soils usually can be disinfected by application of 100 ppm HTH. In computing the HTH dose, assume that the soil has a dry bulk density of 1.5 ton m^{-3}. The effectiveness of HTH treatment will depend on soil pH, and at pH values above 7.5–8 it is probably better to treat pond bottoms with lime instead of chlorine compounds. Lime and chlorine compounds should not be applied at the same time, because lime increases pH and lowers the effectiveness of chlorination.

High-test hypochlorite can be applied in the same manner as lime. The residual effect of HTH lasts for 7 to 10 days. After refilling ponds, test the chlorine concentration before stocking aquatic animals, because chlorine is extremely toxic.

Example 9.2: Application Rate for HTH to Pond Soil
A pond soil is to be treated with HTH at 100 ppm and the HTH incorporated into the surface 10-cm layer by tilling. Calculate the treatment rate per hectare. Assume dry bulk density is 1.5 ton m^{-3}.

Solution:
Soil weight

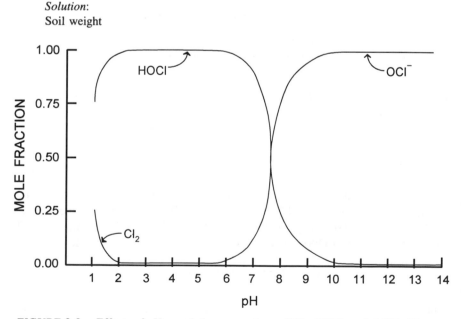

FIGURE 9.8. Effects of pH on relative proportions of Cl_2, HOCl, and OCl^-. (*Source: Boyd.*[1])

$$1,500 \text{ kg m}^{-3} \times 10,000 \text{ m}^2 \text{ ha}^{-1} \times 0.10 \text{ m} = 1,500,000 \text{ kg ha}^{-1}$$

HTH dose

$$100 \text{ ppm} = 100 \text{ mg kg}^{-1} = 0.1 \text{ g kg}^{-1} = 0.0001 \text{ kg chlorine kg soil}^{-1}$$
$$1,500,000 \text{ kg soil ha}^{-1} \times 0.0001 \text{ kg HTH kg soil}^{-1} = 150 \text{ kg HTH ha}^{-1}$$

9.4 Oxidation of Bottom Soil

Techniques that enhance dissolved oxygen concentrations at the pond bottom increase the redox potential in the surface layer of soil. This lessens the possibility for migration of toxic metabolites from deeper, anaerobic soil into the pond water. Aeration, destratification, and water circulation increase dissolved oxygen concentrations in bottom waters. Mechanical disturbance of the soil surface improves contact with oxygenated water. However, as already discussed, disturbances of bottom soil can cause excessive turbidity in the water and sediment accumulation. Chemical oxidants have been applied to ponds to poise the redox potential of bottom soils. Various treatments are applied to empty ponds between crops to accelerate organic matter decomposition and to lower the potential oxygen demand of soil.

9.4.1 During Grow-Out Period

pH Control

Low pH retards decomposition of organic matter and permits it to accumulate and cause a high oxygen demand in the surface layer of soil. Acidic ponds should be treated with agricultural limestone as discussed earlier.

Nitrogen Fertilization

Low concentrations of nitrogen (wide C:N ratio) retard the decomposition of organic mater. Acidic ponds with high concentrations of soil organic matter should be treated with nitrogen fertilizer in addition to agricultural limestone. Nitrogen fertilization should be applied to static ponds at 10 kg N ha^{-1} every two to four weeks. Any nitrogen fertilizer may be used, but sodium nitrate is probably most effective because it is both a source of nitrogen and an oxidant. Soil microorganisms can use oxygen from nitrate when molecular oxygen is depleted. Granular fertilizer should be broadcast over the pond surface, or liquid fertilizer released in the propeller wake of an outboard motor as the boat is driven back and forth across the surface. Aquaculture ponds usually obtain plenty of nitrogen in feeds or fertilizers, but sometimes in manured ponds large amounts of organic matter accumulate and nitrogen fertilization can be beneficial. Liming and nitrogen fertilization are more effective in enhancing aerobic microbial activity when aeration is used to supplement the oxygen supply.

Bacterial Amendments

Bacteria capable of fixing nitrogen and mineralizing phosphorus have been widely applied to agricultural soils in the former Soviet Union. They were claimed to increase nutrient concentrations and crop yields.[14,15] Research in other nations has failed to show benefits of bacterial amendments on crop yield.[14,16,17] There also has been interest in applying bacterial suspensions to wastewater to augment treatment,[18,19] but the value of this practice has not been established.

Recently, bacterial amendments have been advocated for use in aquaculture ponds. Claims for bacterial amendments include prevention of off-flavor; reduction in proportion of blue–green algae; less nitrate, nitrite, ammonia, and phosphate; more dissolved oxygen; and more rapid organic matter degradation. One such bacterial preparation, which was claimed to contain a mixture of live *Bacillus, Nitrobacter, Pseudomonas, Enterobacter, Cellulomonas,* and *Rhodopseudomonas*, was applied to channel catfish ponds as recommended by the distributor, and other ponds served as untreated controls. Concentrations of inorganic nitrogen, total phosphorus, chemical oxygen demand, biochemical oxygen demand, and chlorophyll *a*; abundance of bacteria and phytoplankton; and percentage of blue–green algae did not differ between ponds treated with bacteria and control ponds on any sampling dates.[20] On three dates there were higher dissolved oxygen concentrations in bacteria-treated ponds, but at other times dissolved oxygen concentrations did not differ or were higher in control ponds. Fish production was not influenced by bacterial treatment. Almost identical results were obtained by Tucker and Lloyd, who also compared water quality and channel catfish production in ponds treated with a bacterial amendment and in control ponds.[21]

In Asia several companies sell photosynthetic bacterial preparations that can convert hydrogen sulfide to sulfate as follows:

$$2H_2S + CO_2 + 4H_2O \rightarrow 2H_2SO_4 + 4(CH_2O) \qquad (9.7)$$

There is no experimental evidence that these bacteria lower concentrations of hydrogen sulfide in ponds.[22] Shrimp farmers in Asia often apply bacterial amendments and enzyme preparations with the purpose of improving water quality in ponds. However, there is no conclusive evidence that these products are of benefit.

It is not surprising that bacterial and enzyme amendments are ineffective. Commercial suspensions of bacteria contain common bacteria that already are common in aquaculture ponds. These bacteria normally will increase abundantly if appropriate substrate is supplied. Native bacteria produce copious supplies of extracellular enzymes. It is not necessary to add bacteria or enzymes; they already are present. To enhance organic matter decomposition, provide optimal conditions for microbial activity: pH above 7; adequate nitrogen; plenty of oxygen; favorable temperature.

Aeration

Aerators are commonly used in aquaculture ponds. The most common types are paddlewheel aerators (Fig. 9.9), vertical-pump aerators (Fig. 9.10), propeller–aspirator–pump aerators (Fig. 9.11), and diffused-air aerators (Fig. 9.12). When a large amount of aeration is applied with any type of aerator, water currents produced by the aerator mix pond water, prevent thermal stratification, and provide better oxygenation at the soil surface. In Chapter 7 it was noted that aeration can become excessive and cause serious erosion and redeposition patterns on the pond bottom. Where it is desired to aerate ponds specifically to enhance the decomposition of organic mater in the bottoms by preventing anaerobic conditions, propeller–aspirator–aerators or diffused-air aeration systems are probably the most effective. This is especially true for ponds deeper than 2 m. Diffused-air systems release air bubbles at the pond bottom, and propeller–aspirator–aerators can be positioned at a 30° to 60° angle with the horizontal to

FIGURE 9.9. Paddlewheel aerator.

FIGURE 9.10.   Vertical pump aerator.

FIGURE 9.11. Propeller–aspirator–pump aerator.

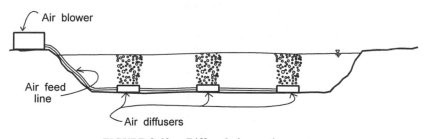

FIGURE 9.12. Diffused-air aeration system.

propel water with entrained air bubbles toward the bottom. In most cases, aeration should be applied at 2–3 kW ha^{-1} to improve conditions at the soil–water interface. In intensive aquaculture much higher rates of aeration are applied.

Destratification and Water Circulation

Mechanical destratifiers and water circulators enhance water circulation, but they provide less aeration than conventional aerators. They direct more energy to producing water currents to improve circulation than to splashing water to effect aeration. Surface water that contains a high dissolved oxygen concentration during daylight hours is blended with bottom waters of lower dissolved oxygen concentration. A more uniform dissolved oxygen profile is maintained, and dissolved oxygen concentrations in bottom water are enhanced in mechanically circulated ponds (Fig. 4.11). Providing more dissolved oxygen in bottom waters means there is less likelihood of anaerobic conditions in the upper layer of bottom soil.

Airlift pumps (Fig. 9.13) are simple water circulation devices. Air bubbles released into the vertical tube of airlift pumps rise and lift water so that it is discharged at the surface. Enough airlift units can mix the entire water column.[23] Downflow destratifiers direct water currents toward the bottom and cause destratification.[24] They consist of an axial flow, low-head pump mounted on floats (Fig. 9.14). The pump discharges water downward and perpendicular to the water surface. Fast et al. designed a portable device that discharges water at an angle to the water surface (Fig. 9.15).[25] Howerton et al. fabricated and tested a horizontal, axial-flow water circulator that can be used in water no more than 1 m deep (Fig. 9.16).[26] This device can discharge a large volume of water at low head and velocity; it causes little erosion of the pond bottom. Commercial manufacturers provide horizontal, axial-flow water circulators similar to the one shown in Figure 9.16 but mounted on floats.

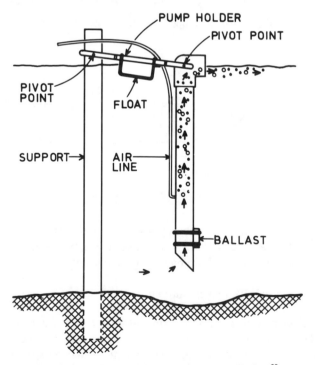

FIGURE 9.13.   Airlift pump. (*Source: Parker.*[38])

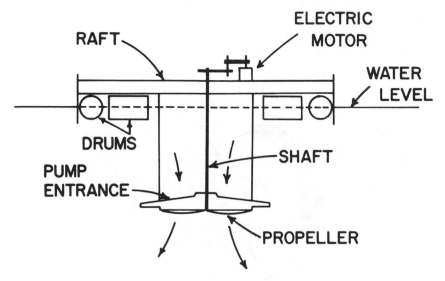

FIGURE 9.14. Schematic of a downflow destratifier.

Bottom Disturbance

Oxygenated water can move into the upper layer of pond soil if the soil surface is disturbed. Borrowing animals living within the soil and fish and other animals feeding on benthos disturb bottom soil and improve aeration by encouraging movement of oxygenated water within the soil mass. This process is called *bioturbation*. Aerators and water circulators enhance dissolved oxygen concentrations in bottom waters, but they tend to cause lateral, surface erosion rather than vertical disturbance within the soil mass. Underwater tilling can supply vertical disturbance with a minimum of lateral erosion. Heavy chains have been dragged over pond bottoms, rakes used to scarify surface soil, and water jets used to drive oxygenated water into surface soil. Studies on benefits of induced bottom soil disturbance on water quality and aquatic animal production and on the best techniques for creating bottom soil disturbance have not been conducted.

Chemical Oxidants

Ripl suggested applying sodium nitrate to eutrophic lakes to prevent highly reduced conditions in sediment.[27] Nitrate functions as an oxidant and poises the oxidation–reduction potential:

$$NO_3^- + 2H^+ + 2e^- = NO_2^- + H_2O \qquad (9.8)$$

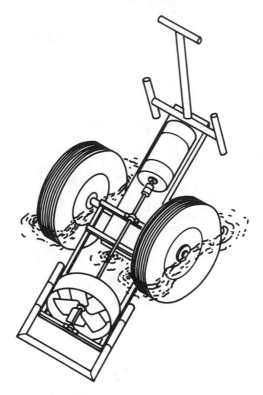

AXONOMETRIC VIEW

FIGURE 9.15. Water circulator that discharges water at about a 45° angle to the pond bottom. (*Source: Fast et al.*[25])

Only when nitrate has been expended by microbial respiration can the redox potential decrease. Avnimelech and Zohar recommended sodium nitrate for the same purpose in fish ponds.[28]

According to Chamberlain, researchers on eel production in Japan applied calcium peroxide (CaO_2) granules to pond bottoms at 25–100 g m^{-2} at monthly intervals to supply oxygen in the surface layer of bottom soil.[29] Calcium peroxide releases oxygen as follows:

$$2CaO_2 + 2H_2O = 2Ca(OH)_2 + O_2 \qquad (9.9)$$

Hydrogen peroxide also releases oxygen when it is placed in water:

$$2H_2O_2 \rightarrow 2H_2O + O_2 \qquad (9.10)$$

Sodium nitrate is cheaper than peroxides and less hazardous to handle. The nitrite released when bacteria remove oxygen from nitrate would probably be

FIGURE 9.16. A horizontal, axial-flow water circulator: (A) Side view. Wooden structure is a baffle used in testing. (B) Intake showing bell-shaped entrance section. (C) Discharge showing baffles to streamline flow. (D) Fan blade impellers. (*Source: Howerton et al.*[26])

quickly oxidized back to nitrate if it diffused into the pond water. However, nitrite is potentially toxic to fish and other aquatic animals. Peroxides would poise the redox potential at a higher value than nitrate. High concentrations of peroxides are toxic to aquatic organisms. Because of their high solubilities, these oxidants would dissolve and be diluted by pond water. Some type of coating could possibly be applied over pellet surfaces so that they dissolve more slowly and provide a continuous source of oxidizing power in the soil. Technology used to produce controlled released fertilizers might be applicable. Obviously, much research needs to be done before oxidants can be recommended for general use in pond management.

Example 9.3: Oxygen Availability from Sodium Nitrate
    The amount of oxygen that microorganisms could obtain by reducing 100 kg
    of sodium nitrate to sodium nitrite will be estimated.

Solution:

$$
\begin{array}{ccc}
100\,kg & & x \\
NaNO_3 \rightarrow & NaNO_2 & + \; O \\
85 & & 16
\end{array}
$$
$$x = 18.8 \text{ kg oxygen}$$

Potassium Permanganate and Ferrous Oxide

Hydrogen sulfide released from anaerobic pond soils can be harmful to shrimp and fish. Mathis et al. suggested applying potassium permanganate ($KMnO_4$) to ponds to oxidize H_2S according to the following reaction:[30]

$$4KMnO_4 + 3H_2S \rightarrow 2K_2SO_4 + S + 3MnO + MnO_2 + 3H_2O \quad (9.11)$$

This treatment has not been successful, because potassium permanganate is highly soluble and only the water column can be treated.

A possible method for preventing hydrogen sulfide release from pond muds is iron application. Ferrous iron reacts with hydrogen sulfide and precipitates it as highly insoluble ferrous sulfide:

$$Fe^{2+} + S^{2-} = FeS \quad (9.12)$$

Shigeno stated that powdered ferrous oxide was applied to pond bottoms in Japan at 1 kg m^{-2} to prevent hydrogen sulfide release from anaerobic soils.[31] Even though the application of iron compounds is logical, it seems more expedient to apply treatments that reduce the possibility of anaerobic conditions and hydrogen sulfide production in the surface layer of soil.

9.4.2 During Fallow Period between Crops

Drying

The rate of soil organic mater decomposition by aerobic processes increases when ponds are drained and their bottoms exposed to air (Fig. 5.12). The optimal length of the fallow period between crops depends on drying conditions and temperature. Empty ponds often cannot be dried during the wet season, and high water tables or seepage from adjacent, full ponds can prevent drying. Ponds that cannot be dried should be refilled at once. In the dry season, pond bottoms may become so dry after two to four weeks that microbial activity is repressed by lack of moisture. Unless the soil can be irrigated to replenish soil moisture, there is little value in extending the fallow period of moisture-depleted soils. The minimum moisture concentration for rapid microbial processing of organic matter is 10–40% for different soils.

It is difficult to base the length of the fallow period on the soil moisture concentration. Simple methods for determining the minimum moisture concentration in individual ponds are not available, and even if the minimum moisture concentration for a pond was known, most pond managers have no means for determining soil moisture concentration. A suitable alternative is to observe the pond soil, and when the surface 5-cm layer appears too dry for the growth of seedling plant or dust is produced when a handful of soil is tossed into the air, soils are too dry to support high rates of microbial activity.

In regions with cold winters, ponds often are left fallow during winter months.[32] In warmer climates a fallow period of two to four weeks is usually adequate. Long fallow periods expose the pond bottom to erosion by rainfall. Terrestrial vegetation may develop in ponds left fallow during warm weather. Drying of pond soils containing iron pyrite can cause extremely low pH. Acidity must be washed out of soils before ponds are filled and restocked.

Tilling

When soils which contain expandable clay dry, cracks form in the surface and extend to depths of 10–100 cm into the soil (Fig. 9.17). Pettry and Switzer studied cracks that formed during dry weather in alligator clay soil in Leflore County, Mississippi, over a three-year period.[33] Cracks had an average width of 7.4 cm, an average depth of 66.8 cm, and made up 18.2% of the soil surface. Air can penetrate cracks and improve oxygenation, but air will not readily penetrate the soil mass comprising columnar blocks between cracks. Surfaces of a columnar soil block are oxidized (usually of brownish color), but if the block is broken the newly exposed soil surfaces usually are reduced and have a darker color (often gray or black) than the outside surfaces, which had been exposed to the air longer. Also, surface soil dries quickly compared to deeper

FIGURE 9.17. Cracks in a dry soil.

soil, because there is better exchange of water vapor with the atmosphere. A dry surface layer acts as a barrier to evaporation. In shrimp ponds heavy aeration often causes mounds of sediment in the centers of ponds. Surfaces of the mounds may by dry enough to walk on, but a few centimeters below the soil surface the sediment is saturated with water. The dry surface crust is an effective barrier to evaporation.

Plowing breaks up surface crusts and columnar blocks that form during drying and greatly improves conditions for drying and aeration in deeper layers of soil. The most efficient devices for plowing pond soils are tractor-powered rototillers or disk harrows (Figs. 9.2 and 9.3), which break the soil into relatively fine particles. A turning plow (Fig. 9.4) turns the furrow slice over and does not pulverize soil as well as a rototiller or disk harrow. A tooth harrow (Fig. 9.18) may be pulled over recently plowed soil to smooth the surface and break up larger clods. Small, self-powered rototillers are excellent devices for pulverizing bottoms of small ponds (Fig. 9.1), but rakes and other hand implements may be used.

Tilling is valuable in fine-textured soils that do not crack upon drying or crack and form large blocks. Tilling is of questionable value in coarse-textured soils that dry readily without forming large blocks. A tillage depth of 5–10 cm is normally adequate. After the bottom has been tilled, it should be left fallow for one to two weeks. Tilling makes the soil more vulnerable to erosion by water currents created by aerators and mechanical water circulators. Compaction of tilled soil with heavy rollers before refilling reduces soil erodibility in such ponds.

Liming

The objective of a fallow period is to aerate soil and encourage oxidation of reduced inorganic compounds and organic matter. The optimal pH range for organic matter decomposition is 7.5–8.5. If soils have a pH below 7.0, as measured in a 1:1 mixture of soil and water, application of agricultural limestone can stimulate microbial respiration. The lime requirement procedure may be used to determine the liming rate, but the following guide may be used when the lime requirement is not known:

pH	Agricultural Limestone (kg ha^{-1})
4.5–5.0	4,000
5.0–5.5	2,000
6.0–7.0	1,000

Pond soils with a pH less than 4.5 usually are acid-sulfate soils. Methods for mitigating acidity in such soils were presented earlier in this chapter.

FIGURE 9.18. Tooth harrow.

Agricultural limestone should be spread evenly over the soil surface, but uniform application is less critical than when spreading burnt or hydrated lime for soil sterilization. Agricultural limestone should be incorporated into the upper 5–10-cm soil layer by tilling. Limestone will not dissolve unless the soil contains plenty of moisture.

Burnt or hydrated lime is not recommended as a routine soil treatment between crops. These materials cause an initial high pH that kills microorganisms and retards the decomposition rate. Agricultural limestone is safer to apply and will not cause soil pH to rise above 8.

Fertilization

Decomposition is slow in soils with a wide C:N ratio. The C:N ratio is between 5 and 10 in most pond soils, and organic residues readily decompose if other factors affecting soil respiration are favorable. Organic soil in ponds needs to be decomposed to provide a more stable bottom. The C:N ratio is wide in organic soil, and fertilization to increase nitrogen availability accelerates decomposition. Application rates of 25–100 kg N ha^{-1} have been used by shrimp farmers. A rate of 100 mg N kg^{-1} was necessary to significantly enhance decomposition of soil organic matter in laboratory studies.[34] Any common nitrogen fertilizer can be used, but sodium nitrate appears to be the most effective. Nitrate is a source of oxygen for microorganisms when oxygen is depleted. Fertilizer should be broadcast uniformly over the pond bottom and incorporated by tilling.

Example 9.4: Calculation of Soil Fertilization Rate
A pond bottom consists of an organic soil with a dry bulk density of 0.6 ton m^{-3}. It is desired to treat the upper 15-cm layer of soil with 150 mg N kg^{-1} using sodium nitrate fertilizer (16% N). Calculate the sodium nitrate application rate per hectare.

Solution:
Soil weight:

$$600 \text{ kg m}^{-3} \times 10,000 \text{ m}^2 \text{ ha}^{-1} \times 0.15 \text{ m} = 900,000 \text{ kg ha}^{-1}$$

Sodium nitrate dose

$$(900,000 \text{ kg soil ha}^{-1} \times 0.00015 \text{ kg N} \cdot \text{kg}^{-1} \text{ soil}) \div 0.16 \text{ kg N} \cdot \text{kg sodium nitrate}$$
$$= 844 \text{ kg sodium nitrate ha}^{-1}$$

Bacterial and Enzyme Amendments

Comments regarding the application of bacterial amendments to pond waters are equally true for pond soils. Boyd and Pippopinyo applied a commercial bacterial preparation to soils in laboratory experiments at 50 and 200 ppm.[34] No increase in organic matter decomposition was measured. Enzyme preparations and bacteria have been applied widely to soils of shrimp ponds in Southeast Asia, but there is no reason to suspect that these treatments enhance organic matter decomposition.

There is much interest in the bioremediation of soils contaminated with toxic chemicals by use of microbial products.[35] Soil microorganisms are capable of degrading toxic chemicals into simple, nontoxic forms. It has long been known that many pesticides were gradually detoxified by soil microorganisms. However, in recent years, special strains of microorganisms capable of rapidly degrading specific classes of toxic chemicals have been developed and commercialized as a technique for remediating soils containing toxic compounds. Mixed results have been obtained, and bioremediation is still a controversial subject. Nevertheless, some companies are advertising microbial preparations made for soil bioremediation as a means of improving soil condition in aquaculture ponds. Aquaculture ponds seldom contain toxic chemicals. The major problem in pond soils is high concentrations of ordinary organic matter and low dissolved oxygen concentrations. It is not logical to assume that commercial bacterial preparations used to remediate toxic chemical spills in terrestrial soils would be beneficial in ponds.

9.5 Sediment

Ponds should be managed to avoid sediment problems. Vegetative cover on watersheds will minimize erosion by rainfall and runoff. Where wave action causes serious erosion of levees, trees or shrubs can be planted to make windbreaks and riprap can be placed on vulnerable expanses of levees. Excessive aeration should not be applied, and in aerated ponds, bottoms should be hardened in areas with high water velocities. Erosion resistance may be enhanced by compacting soil, applying a layer of rock, or by installing a cover over vulnerable areas. Tilling pond bottoms is an aid to drying and organic matter decomposition, but it encourages erosion in aerated ponds. Sediment basins can remove suspended particles from a turbid water supply. Ponds should be quickly refilled during rainy weather to prevent erosion in shallow areas and deposition in deeper areas. In spite of best efforts, deeper areas of most ponds gradually fill with sediment.

9.5.1 Removal

Sediment is sometimes removed from ponds and water supply canals. In the case of external sediment entering ponds in the water supply or runoff, the use of settling basins and erosion protection on the watershed is recommended to prevent suspended solids from entering ponds. However, for various reasons, this is not always done, and ponds fill with sediment. Canals are often constructed around inside edges of shrimp ponds to facilitate draining and shrimp harvest

(Fig. 9.19). Canals frequently fill with sediment, and it is common practice to remove this sediment and place it on levees or dispose of it outside of pond areas. Where sedimentation is not provided for shrimp pond water supplies with high concentrations of suspended solids, the entire bottoms of shrimp ponds may become covered by a layer of sediment. This sediment eventually has to be removed at considerable expense, as illustrated for a shrimp pond in Ecuador (Fig. 9.20). Large amounts of organic matter sometimes are applied to freshwater ponds in green manures and animal manures. Residues of manures accumulate in pond bottoms, and they must be removed periodically. In China, sediment from manured ponds is disposed of on agricultural land where it serves as fertilizer.

Even if ponds are protected from high external sediment loads, the internal sediment load tends to fill in the deeper areas of ponds over the years. It is well known in commercial pond aquaculture that maintenance of water quality within

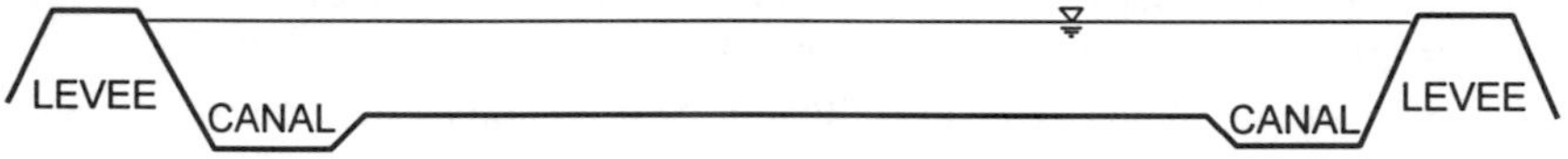

FIGURE 9.19. Drainage canal around the edge of a shrimp pond to facilitate draining and harvest.

FIGURE 9.20. Sediment removal from the bottom of a shrimp pond in Ecuador.

adequate ranges for optimal aquatic animal production becomes more difficult as ponds age. The accumulation of sediment enriched with nutrients and organic mater is thought to be a major factor causing the intensification of management problems in old ponds. Therefore, techniques for removing sediment from old ponds should be evaluated. Development of dredges that can remove sediment from full ponds during the grow-out period might be a valuable tool for use in aquaculture.[36]

In intensive shrimp farming and in other types of intensive aquaculture, farmers may remove sediment from ponds after each crop. For example, in some areas of Southeast Asia, shrimp farmers use high-pressure water jets to resuspend sediment and wash it out of ponds. This is not a good practice, because resuspended sediment contains readily decomposable organic mater. It can cause an oxygen demand and sedimentation problems in receiving water bodies, which also are the water supply for nearby ponds.

An example from southern Thailand will be used to show how sediment removal after each crop is probably unnecessary and counterproductive. In southern Thailand, shrimp ponds have low external sediment loads because they are supplied by seawater with low concentrations of suspended solids. Shrimp are produced intensively (3–8 ton ha^{-1} per crop), and ponds are aerated heavily. Water currents caused by aerators erode and suspend soil particles from insides of levees and peripheral areas of pond bottoms. Organic matter from uneaten feed, shrimp feces, and dead plankton also is resuspended by the currents. These suspended particles of mineral soil and organic matter are an internal sediment load, because they settle in central areas of ponds where water currents are weak. When ponds are drained, large mounds of sediment with volumes of 50–100 m^3 or more are visible. Shrimp farmers believe that the sediment is highly organic with toxic properties. They usually dry pond bottoms and excavate sediment between crops. This practice is expensive, alters bottom slopes, and creates a solid-waste disposal problem. Also, benefits of sediment removal are unproven.

Soil and sediment samples were collected from a shrimp farm in southern Thailand as follows: (1) bottoms of new ponds never filled with seawater; (2) sediment mounds in recently harvested ponds; (3) piles of excavated sediment exposed outdoors for 2–3 months; (4) piles of excavated sediment exposed outdoors for 6–18 months (Boyd, Munsiri, and Hajek).[37] Results from analyses of these samples are summarized in Table 9.2. Soils from new ponds were slightly basic, high in clay content, and low in concentrations of organic matter, nitrogen, sulfur, and phosphorus. An x-ray defraction analysis indicated that clay minerals were kaolinite and clay-sized particles of quartz and mica. The CEC was only 14.7 meq 100 g^{-1} soil. Basic cation concentrations (Ca, Mg, K, and Na) were quite high, but the SAR was only 7.92. Saline soils have SAR values of 13 or above. Concentrations of micronutrients (Fe, Mn, Zn, Cu, B, Co, and Mo) and other metals (Pb, Cr, and Ba) were low to moderate. The

Table 9.2   Average Chemical and Physical Characteristics of New Shrimp Pond Soil, Sediment in Shrimp Ponds, and Excavated Sediment Piles in Southern Thailand[a]

Variable	New Ponds	Sediment in Recently Drained Ponds	Excavated Sediment Piles (age in months)	
			2–3	6–18
Number of samples	3	7	4	7
pH (standard units)	7.42 a	7.56 a	7.56 a	7.64 a
Organic matter (%)	1.12 a	1.90 b	1.53 b	0.98 a
Clay (%)	64.1 a	56.9 a	66.3 a	62.9 a
Silt (%)	34.7 a	38.6 a	30.6 a	32.5 a
Sand (%)	1.1 a	8.0 a	2.6 a	3.7 a
Soil texture	Clay loam	Loam	Clay loam	Clay loam
Clay minerals	Kaolinite, quartz, and mica	Kaolinite, quartz, and mica	—	Kaolinite, quartz, and mica
Cation-exchange capacity (meq 100 g^{-1})	14.7 a	12.0 a	12.7 a	13.4 a
Sodium adsorption ratio	7.92 a	32.9 b	30.93 b	8.34 a
Total nitrogen (ppm)	500 a	890 b	1,100 c	630 a
Total sulfur (ppm)	490 a	1,710 b	1,000 c	420 a
Acid-extractable phosphorus (ppm)	6.4 a	22.7 b	15.6 b	11.4 b
Calcium (ppm)	4,712 a	6,851 b	4,607 a	4,065 a
Magnesium (ppm)	1,448 a	2,122 b	1,651 a	1,245 a
Potassium (ppm)	48 a	306 b	279 b	80 a
Sodium (ppm)	1,545 a	12,177 b	9,606 c	2,371 a
Iron (ppm)	9.9 a	16.0 a	12.0 a	17.2 a
Manganese (ppm)	33 a	100 b	113 b	47 a
Zinc (ppm)	0.53 a	1.66 b	0.95 a	0.98 a
Copper (ppm)	0.13 a	0.16 a	0.12 a	0.17 a
Boron (ppm)	2.0 a	8.1 b	7.8 b	2.6 a
Cobalt (ppm)	0.27 a	0.37 a	0.37 a	0.33 a
Molybdenum (ppm)	0.25 a	0.31 a	0.24 a	0.24 a
Lead (ppm)	1.03 a	1.54 a	1.25 a	1.67 a
Chromium (ppm)	0.43 a	0.61 a	0.65 a	0.51 a
Barium (ppm)	1.33 a	0.61 a	0.58 a	1.49 a

[a]Entries indicated by the same letter were not significantly different ($P<0.1$) as determined by Duncan's New Multiple Range Test (horizontal comparisons only).

milliequivalents of extractable cations per 100 g soil exceeded the CEC, so a large proportion of the cations was present in sediment as soluble salts or other easily dissolved minerals.

Samples of sediment mounds in recently drained ponds had greater concentrations of all chemical substances than new pond soils. The increase in organic matter, nitrogen, and phosphorus resulted from residues of feed, shrimp feces, dead plankton, and, possibly, manure and chemical fertilizer. Concentrations of other substances resulted from inputs of dissolved ions in seawater. The accumulation of sodium was particularly great, and SAR was 32.9. Shrimp farmers often think that sediment in intensive shrimp ponds contains large amounts of organic matter, but the data (Table 9.2) are not in agreement. Sediment from recently drained ponds contained only 1.90% organic matter; it was a mineral soil with loamy texture and a clay content of 56.9%. Clay minerals were unchanged from those in the new ponds. This is contrary to the popular view that flooding with seawater alters the composition of the clay minerals in shrimp pond soils. When the ponds were drained, pore spaces in bottom soils remained saturated with seawater. When the soil and sediment dried, a residue of salts was left that caused the accumulation of metals, boron, and sulfur.

Salts in the sediment are highly soluble as indicated by a test in which 10 g of pond sediment were leached with 200 mL of distilled water. Leaching reduced the concentrations of chemical substances as follows: calcium, 28%; magnesium, 61%; potassium, 93%; sodium, 94%; iron, 68%; manganese, 6%; zinc, 61%; copper, 31%; boron, 82%. Concentrations of organic matter, nitrogen, phosphorus, and sulfur were not measured in leachates, and concentrations of other substances were below limits of detection. Thus, rain should quickly leach large amounts of salts from the piles of excavated sediment.

Analyses of samples from piles of excavated sediment showed that substances had, in fact, been leached by rainfall. When compared to sediment in ponds, there were noticeable declines in concentrations of most substances in 2- to 3-month-old sediment piles, and samples from 6- to 18-month-old sediment piles were similar in composition to the soil collected from ponds that had never been filled with seawater.

Sediment in intensive shrimp ponds contained more organic matter than original site soils, but it was largely mineral soil. Its properties had been altered primarily by addition of salts from seawater. The chemical composition of sediment does not suggest properties toxic to marine or brackish-water plants and animals, and the sediment does not contain enough organic matter to cause an extremely high oxygen demand. There does not appear to be a valid reason for removing sediment from ponds. The practice is expensive and destructive to ponds. Sediment removal is also an environmental hazard. The usual method of disposal is to dump truckloads of sediment in the vicinity of shrimp farms. These piles take up useful space, they are unsightly, and large amounts of salt will be leached from sediment piles by rainfall. This process can increase the salinity of local soils and kill vegetation. The leachate also can contaminate both surface and subsurface bodies of freshwater by increasing their salinity.

The most reasonable approach is to use a bulldozer or tractor-drawn scraper to spread the dry sediment back over the areas of the pond bottoms from which it was eroded. The loose soil should be thoroughly compacted to reduce erodibility by aerator-induced water currents. Of course, there are areas, such as Ecuador, where water supplies for many shrimp ponds have tremendous concentrations of suspended soil particles. Unless these solids are removed in settling basins before waters enter grow-out ponds, sediment rapidly accumulates. In such situations, frequent sediment removal is necessary to reclaim pond volume.

9.5.2 Bottom Reshaping

If adequate precautions are taken to reduce the external and internal sediment loads, ponds can be operated for many years before it is necessary to remove sediment. If most of the sediment load is internal sediment, sediment may be removed from the deeper areas and placed in eroded areas of the bottom and on the insides of levees. Of course, in some instances, sediment must be removed from pond areas, and should be disposed of in an environmentally safe manner.

9.6 Summary

The following actions should be considered by pond managers who wish to maintain good soil condition in the bottoms of aquaculture ponds:

1. Avoid highly intensive aquaculture operations. High nutrient and organic matter inputs to ponds lead to soil deterioration.
2. Use conservative fertilization, manuring, and feeding techniques. Excessive inputs of nutrients and organic matter are wasteful and harmful to pond ecosystems.
3. Reduce external sediment loads by maintaining good vegetative cover on watersheds and using sediment ponds where the water supply is contaminated with suspended solids.
4. Moderate mechanical aeration and water circulation enhance dissolved oxygen concentrations in bottom waters and in surface layers of soil.
5. Reduce internal sediment loads by protecting banks from wave erosion and avoiding excessive aeration and water circulation.
6. Prevent infestations of rooted aquatic macrophytes by maintaining water depths greater than 0.5 m in all areas of ponds. Encourage plankton growth so that visibility into the water is not more than 30–45 cm. Do not allow infestations of emergent macrophytes along levees and in shallow water areas or mats of floating macrophytes.
7. An aggressive liming program in acidic waters will maintain pH, total alkalinity, and total hardness in desirable ranges.
8. Identify areas with potential acid-sulfate soils. Use pond construction and management techniques that minimize exposure of potential acid-sulfate soils to the air.

9. Where possible, allow a fallow period to dry soils between aquacultural crops. Tilling, liming, and fertilization enhance organic matter decomposition between crops in some ponds. Compact tilled soils before refilling ponds to reduce erosion in aerated ponds.

10. Pond soils can be sterilized to destroy fish pathogens. Application of burnt or hydrated lime is probably more reliable than chlorination.

11. Further research is needed to determine if chemical oxidants can be effective in maintaining aerobic conditions in surface layers of bottom soils.

12. Positive benefits of bacterial and enzyme amendments have not been demonstrated.

13. Ponds should be renovated periodically, with removal of sediment from deep areas to reshape bottoms and levees.

References

1. Boyd, C. E. *Water Quality in Ponds for Aquaculture.* 1990. Alabama Agricultural Experiment Station, Auburn University, Ala.

2. Boyd, C. E., and K. Masuda. 1994. Characteristics of liming materials used in aquaculture ponds. *World Aquaculture* **24**:76–79.

3. Boyd, C. E., and M. L. Cuenco. 1980. Refinement of the lime requirement procedures for fish ponds. *Aquaculture* **21**:293–299.

4. Singh, V. P. 1980. Management of fishponds with acid-sulfate soils. *Asian Aquaculture* **5**:4–6.

5. Potter, T. 1976. The problems of fish culture associated with acid sulfate soils and methods for their improvement. Association of South East Asian Nation, Manila, The Philippines, Shrimp Culture Conference 2, Paper 1.

6. Simpson, H. J., H. W. Ducklow, B. Deck, and H. L. Cook. 1983. Brackish water aquaculture in pyrite-bearing tropical soils. *Aquaculture* **34**:333–350.

7. Hajek, B. F., and C. E. Boyd. 1994. Rating soil and water information for aquaculture. *Aquacultural Eng.* **13**:115–128.

8. Brinkman, R., and V. P. Singh. 1982. Rapid reclamation of brackish water fishponds in acid sulfate soils. In *Proceedings of the Bangkok Symposium on Acid Sulfate Soils*, H. Dost and N. van Breeman, eds., pp. 318–330. Publication 18, International Institute of Land Reclamation and Improvement, Wageningen, The Netherlands.

9. Gaviria, J. I., H. R. Schmittou, and J. H. Grover. 1986. Acid sulfate soils: Identification, formation, and implications for aquaculture. *J. Aquaculture Tropics* **1**:99–109.

10. Lin, C. K. 1986. Acidification and reclamation of acid sulfate soil fishponds in Thailand. In *The First Asian Fisheries Forum*, J. L. Maclean, L. B. Dizon, and L. V. Hosillos, eds., pp. 71–74. Asian Fisheries Society, Manila, Philippines.

11. United States Department of Agriculture. *Restoring Surface-Mined Land.* 1973. Misc. Publ. No. 1082, U.S. Govt. Printing Office, Washington, D.C.

12. Waller, W. T., and J. C. Bass. *Pre- and Post-Improvement Limnological Analyses of Certain Strip-Mine Lakes in Southeast Kansas.* 1967. Kansas Forestry, Fish and Game Commission, Topeka.

13. Scheve, J. W. 1971. Limnological and fisheries investigations of strip-mine lakes in southeast Kansas. M.S. thesis, Kansas State College of Pittsburgh.

14. Cooper, R. 1959. Bacterial fertilizers in the Soviet Union. *Soils Fert.* **22**:327–333.

15. Brown, M. E. 1974. Seed and root bacterization. *Annual Rev. Phytopathol.* **12**:181–197.

16. Smith, J. H., F. E. Allison, and D. A. Soulides. 1961. Evaluation of phosphobacteria as a soil inoculant. *Soil Sci. Soc. Amer. Proc.* **25**:109–111.

17. Weaver, R. W., E. P. Dunigan, J. R. Parr, and A. E. Hiltbold. 1974. *Effect of Two Soil Activators on Crop Yields and Activities of Soil Microorganisms in the Southern United States.* Southern Cooperative Series Bulletin 189, Texas Agricultural Experiment Station, College Station, Tex.

18. Horsfall, F. L. 1979. Bacterial augmentation of wastewater treatment. *J. New England Water Pollution Control Assoc.* **13**:158.

19. Schuetzle, D., J. R. Kiskinen, and F. L. Horsfall. 1982. Chemometric modeling of wastewater treatment processes. *J. Water Pollution Control Fed.* **54**:457–465.

20. Boyd, C. E., W. D. Hollerman, J. A. Plumb, and M. Saeed. 1984. Effect of treatment with a commercial bacterial suspension on water quality in channel catfish ponds. *Prog. Fish-Cult.* **46**:36–40.

21. Tucker, C. S., and S. W. Lloyd. 1985. Evaluation of a commercial bacterial amendment for improving water quality in channel catfish ponds. *Miss. Agr. For. Exp. Sta., Miss. State Univ., Miss. State, Miss. Res. Rep.* **10**:1–4.

22. Chien, Y. H. 1989. The management of sediment in prawn ponds. In Proceedings *III, Brazilian Prawn Culture Symposium Proceedings,* pp. 219–243. MCI Aquaculture, Joao Passoa—PB, Brazil.

23. Parker, N. C., and M. A. Suttle. 1987. Design of airlift pumps for water circulation and aeration in aquaculture. *Aquacultural Eng.* **6**:97–110.

24. Quintero, J. E., and J. E. Garton. 1973. A low energy lake destratifier. *Trans. Amer. Soc. Agr. Eng.* **16**:973–978.

25. Fast, A. W., D. K. Barclay, and G. Akiyama. *Artificial Circulation of Hawaiian Prawn Ponds.* 1983. UNIH-SEAGRANT-CR-84-01, University of Hawaii, Honolulu.

26. Howerton, R. D., C. E. Boyd, and B. J. Watten. 1993. Design and performance of a horizontal, axial-flow water circulator. *J. Applied Aquaculture* **3**:163–183.

27. Ripl, W. 1976. Biochemical oxidation of polluted lake sediment with nitrate—a new lake restoration method. *Ambio* **5**:132–135.

28. Avnimelech, Y., and G. Zohar. 1986. The effect of local anaerobic conditions on growth retardation in aquaculture systems. *Aquaculture* **58**:167–174.

29. Chamberlain, G. 1988. Rethinking shrimp pond management. Texas Agricultural Extension Service, College Station, Tex., Coastal Aquaculture, Vol. No. 2.

30. Mathis, W. P., L. E. Brady, and W. J. Gilbreath. 1962. Preliminary report on the use of potassium permanganate to produce oxygen and counteract hydrogen sulfide gas in fish ponds. *Proc. Annual Conf. S.E. Assoc. Game Fish Comm.* **16**:357–359.

31. Shigeno, K. *Problems in Prawn Culture.* 1978. Amerind Publishing Co., New Delhi, India.

32. Wurtz, A. G. 1960. *Methods of Treating the Bottom of Fish Ponds and Their Effects on Productivity.* General Fisheries Council Mediterranean Studies and Reviews No. 11, FAO, Rome, Italy.

33. Pettry, D. E., and R. E. Switzer. *Expansive Soils in Mississippi.* Bulletin 986, 1993. Mississippi Agriculture and Forestry Experiment Station, Mississippi State University, Miss. State, Miss.

34. Boyd, C. E., and S. Pippopinyo. 1994. Factors affecting respiration in dry pond bottom soils. *Aquaculture* **120**:283–294.

35. Skladany, G. J., and F. B. Metting, Jr. 1993. Bioremediation of contaminated soil. In *Soil Microbial Ecology*, F. B. Metting, Jr., ed., pp. 483–513. Marcel Dekker, New York.

36. Hopkins, J. S., P. A. Sandifer, and C. L. Browdy. 1994. Sludge management in intensive pond culture of shrimp: Effects of management regime on water quality, sludge characteristics, nitrogen extinction, and shrimp production. *Aquacultural Eng.* **13**:11–30.

37. Boyd, C. E., P. Munsiri, and B. J. Hajek. 1994. Composition of sediment from intensive shrimp ponds in Thailand. *World Aquaculture* **25**:53–55.

38. Parker, N. C. 1983. Air-lift pumps and other aeration techniques in water quality in channel catfish ponds. In *Water Quality in Channel Catfish Ponds*, C. S. Tucker, ed., pp. 24–27. Bulletin 290, Agriculture and Forestry Experiment Station, Mississippi State University, Miss. State, Miss Southern Cooperative,

10

Pond Bottom Soil Analyses

10.1 Introduction

Soil testing laboratories can make most of the common soil analyses. They are operated by universities, government agencies, or private business and can be found in most countries. Analyses usually are done on a fee basis. Although analyses can be made in soil testing laboratories, aquaculturists must know how to collect and prepare samples for delivery to laboratories. Some aquacultural workers may want to make their own soil analyses. Researchers have access to analytical laboratories, and they can conduct soil physical and chemical analyses by methods found in Jackson,[1] Page et al.,[2] Klute,[3] and Soil Survey Investigations Staff.[4] Some common and important pond soil analyses, such as moisture content, texture, bulk density, organic matter, soil respiration, pH, and lime requirement, can be made with little equipment. Many extension workers, consultants, and farm managers may find it convenient to conduct these analyses themselves. Apparatus and reagents for these analyses can be purchased for a reasonable price, and the analytical expertise for conducting these analyses is not great. Soil analysis kits may be purchased for making some key soil analyses.

This chapter provides data on soil sampling and sample preparation and instructions for a few basic soil analyses. The best sources of detailed information on the principles and methods of soil analysis are Page et al.[2] and Klute.[3]

10.2 Soil Samples

Sampling is a critical aspect of soil analysis. Unless a sample adequately represents conditions in a pond soil, useful information will not be obtained even if laboratory analyses are conducted properly. Several factors must be considered, including time of sampling, number of samples per pond, location of samples in the pond, thickness of soil layer to be sampled, type of sampler, drying technique, method of pulverizing the sample, and sample storage. Unless these issues are properly addressed, one cannot expect to obtain data useful for answering questions posed in research projects or for making management decisions.

10.2.1 Samplers

Dredges or core samplers normally are used for securing soil samples from ponds filled with water. Dredges provide grab samples of known area, but the depth cannot be regulated and depends on the hardness of the bottom. Core samples provide an intact core of known area, and the core can be segmented to give samples from specified depth layers. It is easier to sample empty ponds than full ones. Samples can be obtained from empty ponds with core tube liners or removed with augers or spades.

Ekman Dredge

The Ekman dredge is a lightweight, metal sampler for taking samples from soft, vegetation-free bottoms. It consists of a sample chamber with a cross-sectional area of 15.2×15.2 cm (Fig. 10.1). The body of the dredge is a square box of sheet brass. The lower end of the box can be closed by a pair of jaws. The jaws can be opened to expose the entire cross section of the box. Two strong external springs, when released by a messenger, forcefully close the jaws. The top of the box is fitted with two thin doors. These doors open in response to water flowing through the box when the sampler is falling through the water to the pond bottom. After the jaws have been released to enclose the soil sample and the sampler is being retrieved, water pressure holds the doors shut as the sampler is lifted.

An Ekman dredge is ideally suited for ponds with soft bottoms. It will not penetrate hard clay bottoms, and rocks or sticks prevent the jaws from closing completely. The dredge normally is attached to a rope for use in deep water. A modification for shallow water is available in which the dredge is attached to the end of a 2-m-long rod.

Petersen Dredge

The Petersen dredge is designed for sampling hard bottoms that cannot be penetrated by the Ekman dredge. The dredge (Fig. 10.2) is constructed of steel, and weight is added as necessary. The dredge is dropped to the bottom, and its weight causes it to cut into the soil. The leverage mechanism is activated when the cable is tightened to retrieve the sampler; this causes the sampler to cut into the bottom as it closes. Because of its heavy weight, the Petersen dredge is operated from a boat-mounted, mechanical hoist. The Petersen dredge is rarely used in sampling aquaculture pond soils.

FIGURE 10.1. Ekman dredge with carrying case. *(Photo courtesy Wildlife Supply Co., Saginaw, Michigan.)*

Core Samplers

There are several types of core samplers for obtaining relatively undisturbed soil cores from lakes, streams, and ponds. Most core samplers consist of a head assembly, a core tube, and a core tube liner (Fig. 10.3). The core tube penetrates the bottom soil, and the core tube liner is filled with soil (sediment) to provide a soil core that has not been greatly disturbed by the sampling process. For deep-water sampling the sampler is heavily weighted so that it falls through the water with enough velocity and force to penetrate the bottom. The upper end of the core tube and liner is fitted with a valve that opens during descent to permit free flow of water through the open tube. Stabilizing fins can be attached to ensure that the tube descends vertically. When the device strikes and penetrates the

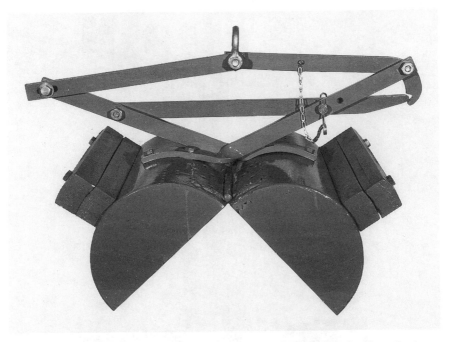

FIGURE 10.2. Petersen dredge. *(Photo courtesy Wildlife Supply Co., Saginaw, Michigan.)*

bottom, the valve closes and creates a partial vacuum that holds the sample in the core tube liner when the sampler is extracted and pulled up by a cable. The closed valve also prevents water from mixing into the core tube during ascent. The core tube liner with core sample is removed from the core tube, and the soil core pressed out of the liner with a plunger inserted into the bottom of the core tube liner (Fig. 10.4).

Before the core is removed, it is desirable to take out the overlaying water with a siphon or large syringe. The upper few centimeters of the soil core have a high water content. This material cannot be pressed out directly, for it will flow or fall down the outside of the liner. A piece of core tube liner can be cut to provide a core segment holder of the desired length (Fig. 10.5). The core segment holder can be placed on the end of the core tube liner, the soil core pushed upward until its top is even with the top of the core segment holder, and the soil core segment cut off by inserting a broad spatula between the bottom of the core segment holder and the core tube liner (Fig. 10.5).

Aquaculture ponds are not deep, and hand-operated core samplers normally can be used (Fig. 10.6). The head assembly has handles for use in very shallow water (Fig. 10.6). Handles on the head assembly may be replaced by an extension handle (Fig. 10.7) for use in deeper water.

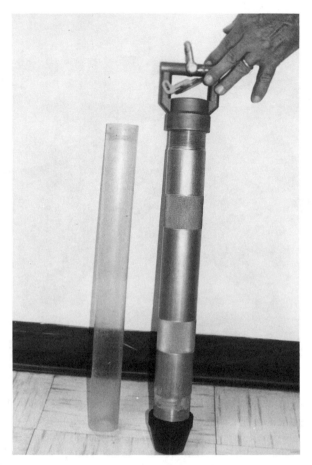

FIGURE 10.3. Core sampler and core tube liner. Note valve and swivel for attaching cable or rope at top.

Other Devices

Soil samples sometimes are taken with an empty can attached to the end of a wooden stick (Fig. 10.8). Soil samples from shallow edges of ponds may be taken by manually digging surface soil with an open container or inserting a core tube liner or other suitable tube into the bottom. Wading or diving is required to reach other locations in a pond. When ponds are empty between crops, soft, moist soil can be removed from the surface with a large spatula or a small garden shovel. Cores can be obtained by pressing core liners into the soil by hand. After the soil dries and hardens, a shovel or soil auger (Fig. 10.9) is required for removing samples.

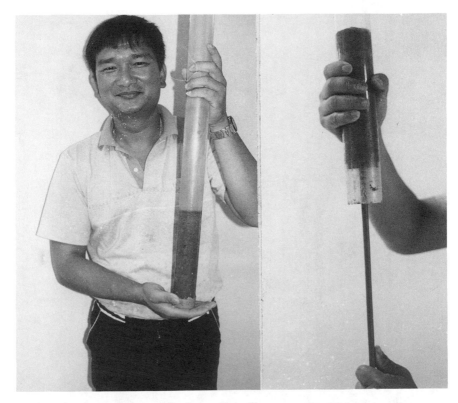

FIGURE 10.4. (Left) Core tube liner with sediment sample. (Right) Removing a core sample from a core tube liner.

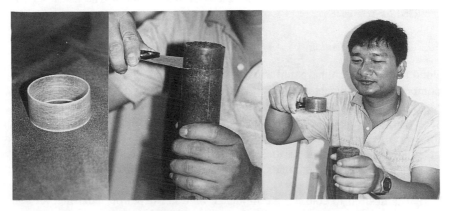

FIGURE 10.5. (Left) Core segment holder. (Center and right) Using the core segment holder and wide spatula to remove a core segment.

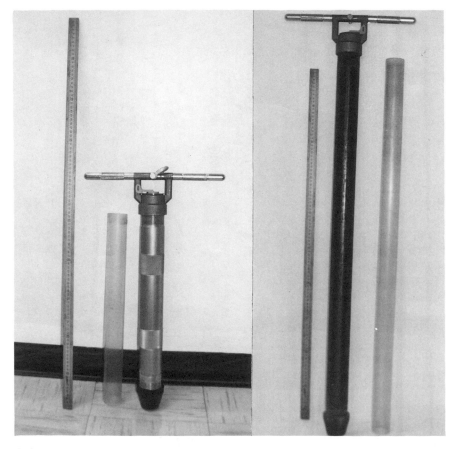

FIGURE 10.6. Hand-operated core samplers for shallow water. The meter stick is provided for reference.

10.2.2 Sample Location

Pond bottoms do not have uniform depths of sediment, soil texture, or chemical composition. Sediment depth increases from shallow to deep-water areas, soil texture is finer in deep water and coarser in shallow water, and soil properties such as organic matter, cation-exchange capacity, and organic nitrogen usually increase in concentration from edges to centers of ponds. Even along a contour of equal water depth, intensities of individual soil properties exhibit random variation among different points. Intensities of soil properties also change with depth into the soil.

If it is desired to consider the influence of water depth on intensities of soil properties, samples should be taken along transects from shallow to deep water. Samples should be taken at 5–10 equally spaced depth intervals along transects

FIGURE 10.7. Core sampler with extension handle for deep water.

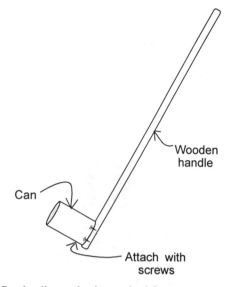

FIGURE 10.8. Pond soil sampler improvised from wooden stick and empty can.

from edges to centers of ponds. Variation in samples from the same water depth may be included by sampling along several transects. It is necessary to take samples of identical soil depths at all locations to avoid the effects of variation within the soil profile.

If it is only desired to obtain average concentrations of pond soil variables, samples can be collected randomly. There are several ways of accomplishing this feat, as illustrated in Figure 10.10. Several samples may be taken from the pond bottom at randomly selected points. Sampling along an S-shaped pattern is frequently recommended.[5] The sampling points along the pattern may be selected randomly. A system of nine quadrats often is employed for sampling soils in agronomic studies,[6] and it is suitable for application in aquaculture ponds. The sampling point in each quadrat should be randomly selected. As mentioned, samples must be taken to the same soil depth.

A study of variability in soil organic carbon concentrations in 1,000-m^2 ponds in Honduras suggests that intensive sampling is required to show differences in this variable among treatments.[7] A two-stage nested (hierarchical) linear statistical model was used to estimate components of variance for carbon data from ponds.[8,9] These variance component allowed the relationship among detectable change in soil carbon concentration ($P < 0.05$), number of samples per pond, and number of ponds per treatment to be defined (Fig. 10.11). From Figure 10.11 it may be determined that in order to detect a change of 0.2–0.3% in soil carbon—for example, a change from 1.5% to 1.7% or 1.8%—at least three ponds per treatment and eight samples per pond are required.

In practical pond management it is not necessary to obtain estimates of varia-

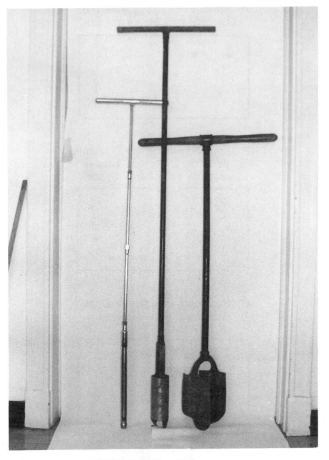

FIGURE 10.9. Soil augers.

tion, but it is necessary to obtain representative and reliable samples. One method illustrated in Figure 10.10 may be used to obtain a random set of samples from a pond. Instead of analyzing all samples to obtain the mean concentration of a variable, one may composite and thoroughly mix equal volumes or weights of each sample to provide a composite sample for analysis. Analysis of the composite sample provides an average concentration of the soil variable(s) for use in pond management. Because of variation resulting from water depth, location, and soil depth, it is critical to include at least 10–12, and preferably more, random samples in the composite sample.

10.2.3 Sample Preparation

Some soil analyses are conducted on moist samples, but most are made on dry samples. Soils should be dried soon after sampling to stop microbial activity.

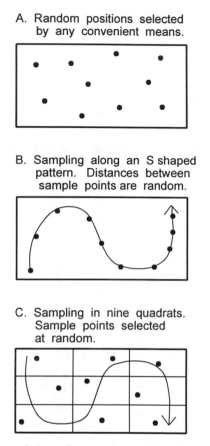

FIGURE 10.10.   Three techniques for selecting positions for taking soil samples from a pond bottom.

The best drying procedure is to place samples in a mechanical convection oven at 60°C or 105°C, as dictated by the type of analysis. Samples may be placed in porcelain dishes, small aluminum cans, or heat-resistant plastic containers. Samples of 50–100 g usually can be dried in 24–48 h. A longer drying period may be necessary in a gravity convection oven. Soil samples can be spread in thin layers and air-dried outdoors or indoors, but the soils must not be contaminated by foreign material.

Dry soils must be pulverized, with a wooden rolling pin, a mortar and pestle, or a mechanical soil crusher (Fig. 10.12). Depending on the analysis, soils are crushed to different particle sizes. Many analyses require soil that has passed a Number 10 (2.00-mm openings), a Number 20 (0.85-mm openings), or a Number 60 (0.25-mm openings) sieve. The sizes of the openings in sieves are provided in Table 10.1. Sieving can be done by hand or with a mechanical sieve shaker (Fig. 10.13).

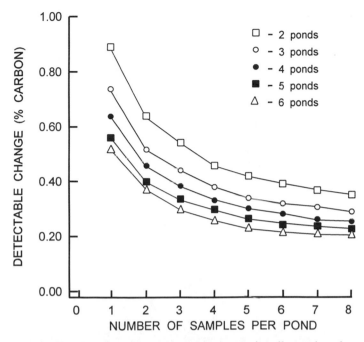

FIGURE 10.11. Relationships among detectable changes in soil organic carbon concentrations ($P<0.05$), number of ponds per treatment, and number of samples per pond. (*Source: Ayub et al.*[7])

FIGURE 10.12. Mechanical soil crusher.

Table 10.1   USA Standard Wire Cloth Sieve Number and Opening Designations

Sieve No.	Opening (mm)	Sieve No.	Opening (mm)
10	2.00	50	0.30
14	1.40	60	0.25
18	1.00	80	0.18
20	0.85	100	0.15
25	0.71	140	0.106
30	0.60	200	0.075
35	0.50	270	0.053
40	0.425		

FIGURE 10.13.   Mechanical sieve shaker.

After sieving, soils can be stored in paper or plastic bags, small cardboard boxes, glass vials, small metal cans, or other convenient containers. Samples must be kept dry. Soil samples absorb moisture from the air during grinding, sieving, and storage. Analyses often are reported on an oven-dry basis. To do this, samples may be dried at 105°C and desiccated before analysis. Alternatively, analyses may be conducted on air-dried samples, and moisture determinations made on separate portions of samples used to correct analytical data to an oven-dry basis.

10.3 Soil Moisture: Gravimetric Analysis of Water Content

There are direct and indirect methods for determining the moisture content of soil. These methods include field and laboratory procedures, and they are discussed by Gardner.[10] For pond aquaculture applications the simple, oven-drying technique is adequate.

10.3.1 Principle

The wet sample is weighed, water is removed by heating at 105°C, and the amount of water removed is determined by weight loss. Results usually are expressed as the percentage of water in a sample on an oven-dry soil basis.

10.3.2 Apparatus

Metal cans or weighing bottles with lids, gravity or mechanical convection drying oven, desiccator with desiccant, and balance.

10.3.3 Procedure

Place 5–10 g of wet soil or 1–2 g of air-dried soil in tared weighing containers (tare must include lid weight), close lids, and weigh to the nearest milligram. Remove lids, place samples in the oven, and dry them to a constant weight. Samples should be dried for at least 24 h in a mechanical convection oven. Longer drying time may be needed in a gravity convection oven. Once drying has started, additional wet samples should not be placed in the oven. Remove samples from the oven, replace lids, and cool to room temperature in a desiccator. Weigh the cool samples. Compute the moisture content as follows:

$$\text{Moisture } (\%) = \frac{\text{Initial weight} - \text{final weight}}{\text{Final weight} - \text{tare weight}} \times 100 \qquad (10.1)$$

This equation is the same as

$$\left[\frac{\text{Weight of wet soil}}{\text{Weight of dry soil}} - 1 \right] \times 100$$

10.4 Particle Size and Texture: Hydrometer Method

Sieve analysis separates soil particles of 0.050–2.00 mm. Particles larger than 0.05 mm are sand and gravel, so sieve analysis can be used only to separate soil samples into gravel, sand, and silt plus clay fractions. Sieve analysis data cannot be used to separate silt from clay, and are not useful in naming soil texture. The standard method for particle-size analysis is the pipet method.[11] In this method, samples are pretreated to remove carbonates, soluble salts, and organic matter. Gravel- and sand-sized particles are removed and determined by sieve analysis. Silt- and clay-sized particles are dispersed in a hexametaphosphate solution, and the soil dispersion is allowed to stand in a 1-L cylinder. A pipet is used to subsample the dispersion at a depth and time calculated to ensure that all silt-sized particles have been eliminated by sedimentation but that all clay-sized particles remain in suspension. This method is too complex for use in most aquaculture projects, but it should be used in research on pond soils where rigorous estimates of particle size are required.

A relatively simple hydrometer method provides particle-size data accurate enough for most aquaculture projects.[12]

10.4.1 Principle

American Society of Testing Materials (ASTM), Model 152H, Bouyoucos-style hydrometers are calibrated to indicate the suspended soil particle concentration in grams of soil per liter at 20°C (Fig. 10.14). The concentration of suspended soil is read directly from the stem of the hydrometer, but the reading must be corrected for the hydrometer reading in a control solution containing no soil and at the same temperature as the test suspension. Soil is dispersed in hexametaphosphate solution, and the dispersion is held in a settling cylinder. Hydrometer readings are taken at time intervals that correspond to elimination of sand-sized and silt-sized particles by sedimentation. Percentages of sand, silt, and clay can be calculated from hydrometer readings, and soil texture can be obtained from a soil triangle (Fig. 2.5).

10.4.2 Apparatus

Wrist action shaker (Fig. 10.15) or rotating table shaker (Fig. 10.15), drying oven, balance, and ASTM 152H (Bouyoucos style) hydrometer.

10.4.3 Reagents

Hexametaphosphate Solution. Dissolve 50 g $(NaPO_3)_6$ in distilled water and dilute to 1,000 mL.

10.4.4 Procedure

Air-dry the soil sample and gently pulverize by hand. A kitchen rolling pin is good for this. Estimate the soil moisture content on a subsample. Place a 50-g sample of air-dried soil in a 500-mL bottle. Add 100 mL of hexametaphosphate solution and 300 mL of deionized water, stopper bottle, and shake on a mechanical shaker for 24 h. Wash contents of bottle into a 1-L graduated cylinder with deionized water. Add deionized water to within 5 cm of the 1-L mark. Insert

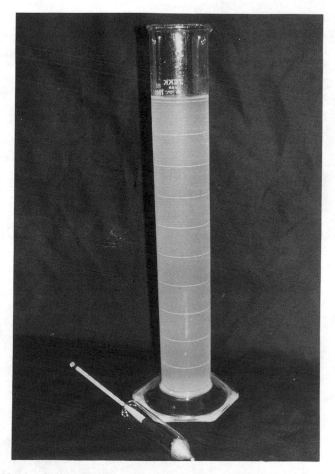

FIGURE 10.14. Soil hydrometer (Bouyoucos style, ASTM Model 152H) and setting cylinder.

the soil hydrometer and fill the cylinder exactly to the 1-L calibration mark. Remove the hydrometer, stopper the cylinder, turn the cylinder end-over-end 10 times, return cylinder to upright position, record time, and set cylinder on laboratory bench. After 20 to 30 sec, insert the hydrometer into the suspension of soil. Read the hydrometer exactly 40 sec after the cylinder was set upright. Repeat the mixing procedure and take another 40-sec hydrometer reading. Immediately record the temperature of the soil suspension. Allow the suspension of soil to settle for 2 h and take hydrometer and temperature readings. Average the two readings taken at 40 sec and correct hydrometer readings for temperature by adding 0.36 g L^{-1} for every 1°C above 20°C or subtracting 0.36 g L^{-1} for every degree below 20°C.

Calculate the percentages of sand, silt, and clay using the temperature-corrected hydrometer readings and the following equations:

FIGURE 10.15. (Upper) Wrist action shaker. (Lower) Rotating table shaker.

$$\text{Oven-dried soil (g)} = \frac{(50 \text{ g}) \text{ (percentage oven-dried weight)}}{100} \quad (10.2)$$

$$\text{Silt} + \text{clay (\%)} = \frac{\text{Corrected 40-sec hydrometer reading}}{\text{Oven-dried soil (g)}} \times 100 \quad (10.3)$$

$$\text{Clay (\%)} = \frac{\text{Corrected 2-h hydrometer reading}}{\text{Oven-dried soil (g)}} \times 100 \quad (10.4)$$

$$\text{Silt (\%)} = \text{(percentage silt} + \text{clay)} - \text{percentage clay} \quad (10.5)$$

$$\text{Sand (\%)} = 100 - \text{(percentage silt} + \text{clay)} \quad (10.6)$$

To obtain the textural class of a soil sample, use the soil texture triangle (Fig. 2.5).

10.5 Bulk Density, Particle Density, and Porosity

The bulk density is the ratio of the dry mass of a soil sample to the bulk volume of the sample. It is the dry weight per unit volume of a soil, and the volume includes the pore space among the soil particles. Data on bulk density are used to calculate the amount of a substance in a volume of soil or to calculate how much of a substance must be incorporated into a given volume of soil to provide a particular concentration. Several techniques for estimating bulk density are given by Blake and Hartge,[13] but the core method is most appropriate for pond aquaculture. Particle density is the density of solid soil particles without pore space included. Sedimentation studies require information on particle density. The amount of pore space in a soil can be estimated from the ratio of bulk density to particle density.

10.5.1 Core Method for Bulk Density

Principle

A core sampler is used to obtain a soil core of specific volume, and the dry weight of the soil core is determined.

Apparatus

Core sampler (5-cm diameter), metal cans or glass weighing bottles, gravity or mechanical-convection drying oven, desiccator with desiccant, and balance.

Procedure

Obtain the soil core from the appropriate depth layer. For general purposes sample the upper 10- or 15-cm layer. Transfer soil cores to tared weighing containers, place them in an oven, and dry to constant weight at 105°C as in the determination of soil moisture. Remove samples from the oven, close lids, and cool in a desiccator. Weigh samples to the nearest milligram. Calculate the volume of the soil core as

$$V = 0.785D^2L \qquad (10.7)$$

where V = volume (cm^3)
D = diameter of soil core (cm)
L = length of soil core segment (cm)

The bulk density is computed with the equation

$$\text{Bulk density (g cm}^{-3}) = \frac{\text{Final weight (g) - tare weight (g)}}{V} \qquad (10.8)$$

10.5.2 Particle Density

Principle

The mass of the soil sample is determined by weighing, and the volume of the particles is estimated from the volume of water that they displace.

Apparatus

Volumetric flasks (25 or 50 mL), semi-microbalance.

Procedure

Pulverize sample to pass a Number 10 (2 mm) sieve. Add about 10 g of air-dried soil to a volumetric flask that has been tared (including stopper). Weigh the flask, contents, and stopper. Determine the moisture content of a duplicate soil sample (see Section 10.3). Fill the flask about half full with distilled water, and wash all soil from the inside of the flask neck. Boil water in the flask for several minutes. Shake the flask gently to prevent soil loss through bubbling or foaming. Cool flask and contents to room temperature and make to volume with freshly boiled and cooled distilled water. Insert the stopper, dry and clean the outside of the flask, and weigh it. Measure the temperature of the water in the flask. Remove the soil and water from the flask and wash the flask. Fill to volume with freshly boiled and cooled distilled water of the same temperature used with the soil. Insert the stopper, dry the outside, and weigh.

Calculate the particle density as follows:

$$\text{Particle density (g cm}^{-3}) = \frac{\rho_w(W_{FS} - W_{FA})}{(W_{FS} - W_{FA}) - (W_{FSW} - W_{FW})} \qquad (10.9)$$

where ρ_w = density of water at observed temperature (g cm^{-3})
W_{FA} = weight of flask filled with air (g)
W_{FW} = weight of flask filled with water at observed temperature (g)
W_{FS} = weight of flask plus oven-dried weight of soil (g)
W_{FSW} = weight of flask containing soil and filled with water (g)

Comment

The densities of water at different temperatures are provided in Table 10.2.

Table 10.2   Densities of Water at Different Temperatures

°C	g cm^{-3}	°C	g cm^{-3}
10	0.9997277	23	0.9975674
11	0.9996328	24	0.9973256
12	0.9995247	25	0.9970739
13	0.9994040	26	0.9968128
14	0.9992712	27	0.9965421
15	0.9991265	28	0.9962623
16	0.9989701	29	0.9959735
17	0.9988022	30	0.9956756
18	0.9986232	31	0.995340
19	0.9984331	32	0.995025
20	0.9982323	33	0.994702
21	0.9980210	34	0.994371
22	0.9977993	35	0.99403

10.5.3 Pore Space

The volume of pore space in a soil (porosity) can be estimated from bulk density and particle density as follows:

$$\text{Pore space } (\%) = 100 - \frac{\text{Bulk density}}{\text{Particle density}} \qquad (10.10)$$

10.6 Organic Matter

The most widely used methods for determining soil organic carbon or organic matter are (1) the Walkley—Black method of sulfuric acid–potassium dichromate oxidation without external heating; (2) the Mebius modification of the Walkley–Black method with external heating; (3) dry combustion in an induction furnace-type carbon analyzer; (4) dry ashing in a muffle furnace.[14] The carbon analyzer technique employs extremely high temperature and converts all soil carbon to carbon dioxide for detection. It does not distinguish inorganic carbon from organic carbon. The Walkley–Black method does not completely oxidize all of the organic matter, and the percentage oxidation varies considerably among samples. The modified Mebius method provides more rigorous oxidation of organic matter than the Walkley–Black technique because external heat is applied, but it is more difficult to perform. Some clays retain considerable water even after drying to constant weight at 105°C. This water is driven off upon ignition and causes high results when soil organic matter is determined by dry ashing. Water loss during ignition can be minimized if ignition temperature does not exceed 350°C.[1] Also, at 350°C, calcium carbonate is not decomposed with release of carbon dioxide.

Ayub and Boyd compared the four methods of soil organic carbon analysis for use on pond soils and found that all methods provided comparable results.[15] The induction furnace carbon analyzer makes rapid analyses, but it is expensive and separate carbonate carbon analyses must be made on some samples. The modified Mebius method is tedious and requires a special reflux apparatus and heater. The Walkley–Black and dry-ashing methods are more suitable for most aquaculture applications.

10.6.1 Walkley–Black Method

Principle

A potassium dichromate–sulfuric acid oxidation technique is used for the determination of the organic carbon content of soil.[14] The general reaction is

$$2Cr_2O_7^{2-} + 3C° + 16H^+ = 4Cr^{3+} + 3CO_2 + 8H_2O \qquad (10.11)$$

The amount of potassium dichromate consumed in this reaction is equivalent to the amount of readily oxidizable organic carbon in the soil sample. A known amount of potassium dichromate is introduced into the digestion mixture. The portion of the dichromate consumed in the oxidation of organic carbon is measured by back titration with ferrous iron:

$$Cr_2O_7^{2-} + 6Fe^{2+} + 14H^+ = 2Cr^{3+} + 6Fe^{3+} + 7H_2O \qquad (10.12)$$

Apparatus

Buret (50 mL), magnetic stirrer and stirring bars, semi-microbalance.

Reagents

Potassium Dichromate, 1.00 N. Dry $K_2Cr_2O_7$ at 105°C and cool in a desiccator. Weigh 49.04 g $K_2Cr_2O_7$, dissolve in distilled water, and dilute to 1,000 mL.

Concentrated Sulfuric Acid, 96% H_2SO_4 or more. If soils are expected to contain appreciable chloride, add 15 g Ag_2SO_4 L^{-1} to the sulfuric acid.

Ferroin Indicator. Dissolve 1.485 g of 1,10-phenanthroline monohydrate and 0.70 g of $FeSO_4 \cdot 7H_2O$ in water and dilute to 100 mL. An alternative is to purchase the 1,10-phenanthroline-ferrous complex already combined as an indicator solution called ferroin (Hach Chemical Company, Loveland, Colorado).

Ferrous Sulfate Solution. 0.5N. Dissolve 140 g of $FeSO_4 \cdot 7H_2O$ in water, add 15 mL of concentrated sulfuric acid, cool, and dilute to 1,000 mL.

Procedure

Grind the dry soil sample to pass a Number 60 sieve (0.25-mm openings). Weigh out 0.50 g or 1.00 g of soil and place in a 500-mL Erlenmeyer flask.

Add 10.00 mL of 1.00 N $K_2Cr_2O_7$, and swirl gently to mix soil with the reagent. Add 20 mL of concentrated H_2SO_4, and swirl vigorously for 1 min. Let flask and contents stand for 30 min. Add 100 mL of distilled water to the flask. Filter the suspension through Whatman No. 1 filter paper. Wash the filter and residue with 100 mL of distilled water. Capture the wash water in the flask with the filtrate. Add 3 or 4 drops of ferroin indicator and titrate with ferrous sulfate solution. The solution will assume a greenish cast and then turn dark green when the titration nears the end point. When this occurs, add ferrous solution drop by drop until the color changes from dark green through blue to red. The ferrous sulfate solution must be standardized daily against 10.00 mL of 1.0 N $K_2Cr_2O_7$ by using this same procedure for organic carbon analysis.

Calculate results with the equation

$$\text{Organic carbon } (\%) = \frac{(\text{meq } K_2Cr_2O_7 - \text{meq FeSO}_4)\,(0.003)\,(100)}{\text{Sample weight (g)}} \quad (10.13)$$

If 10.00 mL of 1.00 N $K_2Cr_2O_7$ are used, the equation may be expressed as

$$\text{Organic carbon } (\%) = \frac{(10 - NV)(0.3)}{W} \quad (10.14)$$

where N = normality of ferrous sulfate
V = volume of ferrous sulfate (mL)
W = sample weight (g)

Comments

The equivalent weight of carbon is 3 in its reaction with $K_2Cr_2O_7$ (Eq. 10.11), because each carbon atom has a valance of 0 in organic matter and is oxidized to carbon dioxide with a valence of $+ 4$. Hence, 12 g carbon per atomic weight divided by four electrons lost per atom in the reaction equals 3 g carbon per equivalent weight. In Equation 10.13, 0.003 g is the milliequivalent weight of carbon. The factor 100 converts results to a percentage.

The amount of organic carbon recovered by the procedure is variable. In terrestrial, surface soils, recovery ranges from 75% to 85%. The organic matter in soil is 48–58% carbon. Some investigators incorporate a recovery factor in the equation for calculating organic carbon. Also, some analysts convert organic carbon to organic matter by assuming a carbon percentage for the organic matter. These factors have been developed for agriculture based on results of studies of percentage recovery of carbon and percentage of carbon in organic matter for specific soils. Such investigations have not been conducted for pond soils. It is probably best to express pond soil results as organic carbon and to not attempt corrections for recovery.

10.6.2 Dry Ash Method

Principle

If a soil sample is ignited at 350°C for several hours, the resulting weight loss will be caused mainly by the oxidation of organic matter to carbon dioxide and water. The temperature must not exceed 350°C to prevent the loss of water from clay minerals and the decomposition of calcium carbonate to calcium oxide and carbon dioxide.

Apparatus

Small porcelain crucibles (10–15 mL), muffle furnace, desiccator and desiccant, and semi-microbalance.

Procedure

Tare a clean crucible and add about 2 g of air-dried soil. Place the sample in an oven at 105°C for 24–48 h. Remove the sample, cool in a desiccator, and weigh. Place the sample in a muffle furnace at 350°C for 8 h, remove from the furnace, cool in a desiccator, and reweigh. Compute the organic matter concentration as follows:

$$OM = 100 - \frac{W_F - W_T}{W_{TS} - W_T} \qquad (10.15)$$

where OM = organic matter concentration (%)
W_T = tare weight of crucible (g)
W_{TS} = tare weight of crucible and oven dry soil (g)
W_F = weight of crucible and soil after ashing (g)

Comment

If it is desired to report the organic carbon concentration instead of organic matter concentration, multiply the organic matter concentration by 0.58. Many authors assume the soil organic matter is 58% carbon, but the concentration varies and is seldom known. It is probably better to report results as ignition loss and assume that the ignition loss is indicative of the organic matter concentration.

10.7 Soil Respiration

Soil respiration is an important consideration because aerobic respiration requires oxygen. When respiration rates become so great that dissolved oxygen is depleted in soil pore water and at the soil–water interface, anaerobic microenvironments develop and products of anaerobic respiration enter the pond water. Methods are given for determining soil respiration *in situ* in full ponds and in empty ponds left fallow between crops. Modification of the biochemical oxygen demand (BOD) procedure may be used to measure soil respiration in the laboratory.[16]

10.7.1 Column Method

Principle

The column method was originally reported by Shapiro and Zur.[17] It is based on measuring the decline in dissolved oxygen concentration in a sample of pond water incubated in contact with pond soil. The decline in dissolved oxygen concentration in a control column without contact with bottom soil must be measured simultaneously to correct for respiration in the water.

Apparatus

Dark PVC columns with tight-fitting caps (columns should be 5 cm in diameter and have lengths 10–15 cm greater than pond depth at the site of measurement), plankton net (50-μm mesh), and polarographic dissolved oxygen meter.

Procedure

Fill pairs of columns with pond water that has been filtered through the plankton net and saturated with dissolved oxygen by thorough agitation for several minutes. Place the top caps on the columns with care to prevent air bubbles. Hold one of the columns vertically in the water with its bottom slightly above the soil. Remove the bottom cap and insert the column into the soil to a depth of 1–2 cm. Remove the top cap and press the column to a depth of 5–10 cm, quickly measure the dissolved oxygen in the column with a polarographic dissolved oxygen meter, and replace the top cap. Repeat the procedure with the control column, but do not remove the bottom cap. After a suitable time, but not more than 24 h, remove the top caps from the pairs of columns and measure dissolved oxygen concentrations at four to five equally spaced depth intervals within the columns. Average the dissolved oxygen concentration for each column.

Estimate the change in dissolved oxygen concentration attributable to soil respiration as follows:

$$\Delta DO_{SR} = (DO_I - DO_F)_{SC} - (DO_I - DO_F)_{CC} \qquad (10.16)$$

where ΔDO_{SR} = change in dissolved oxygen concentration
caused by soil respiration (μg mL^{-1})
DO_I = initial dissolved oxygen concentration in column (μg mL^{-1})
DO_F = final dissolved oxygen concentration in column (μg ml^{-1})
SC = column exposed to soil
CC = control column

The soil respiration rate can be computed with the equation

$$R = \frac{\Delta DO_{SR} V}{T A}$$ (10.17)

where R = soil respiration (μg cm^{-2} h^{-1})
V = volume of water in column (cm^3)
T = time of incubation (h)
A = cross-sectional area of the tube (cm)

Equation 10.17 can be simplified because

$$V = \pi r^2 H$$ (10.18)

where π = 3.1416
r = column radius (cm)
H = column height (cm)

and

$$A = \pi r^2$$ (10.19)

Therefore, we may rewrite Equation 10.17 as

$$R = \frac{\Delta DO_{SR} H}{T}$$ (10.20)

10.7.2 Respiration Chamber Method

Principle

Carbon dioxide released during aerobic respiration in soils can be adsorbed in alkali solution and measured as an index of soil respiration rate. The reaction in which carbon dioxide is adsorbed is

$$CO_2 + 2NaOH = Na_2CO_3 + H_2O$$ (10.21)

The amount of carbon dioxide adsorbed is equivalent to the amount of sodium hydroxide consumed. To determine this, we precipitate the carbonate with barium chloride, and back-titrate the remaining NaOH with standard hydrochloric acid. The reactions are

$$Na_2CO_3 + BaCl_2 = 2NaCl + BaCO_3 \downarrow$$ (10.22)

$$NaOH + HCl = NaCl + H_2O$$ (10.23)

In - Situ Technique

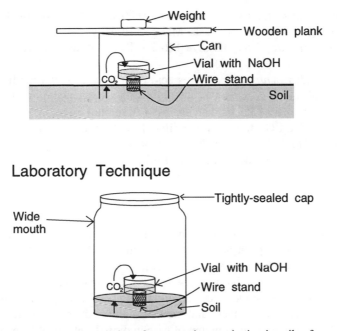

Laboratory Technique

FIGURE 10.16. An *in situ* technique for measuring respiration in soils of empty ponds and a laboratory method for measuring respiration of pond bottom soils. Note: In the *in situ* technique, the wooden plank is placed over the can to provide shade and prevent excessive heating of air and soil under the can.

The amount of NaOH initially present minus the amount remaining at the end of the incubation period is used to compute the amount of carbon dioxide that evolved from the soil, entered the solution, and reacted with sodium hydroxide.

This procedure can be conducted *in situ* in a fallow pond bottom by placing a small vial of alkali solution in a dome above a small area of soil surface to absorb the carbon dioxide released in respiration (Fig. 10.16). The procedure also can be used to measure respiration of soil samples in the laboratory by placing soil samples and vials of alkali solution in sealed jars (Fig. 10.16) that serve as respiration chambers.[18]

Apparatus

Metal cans (25 cm tall × 20 cm diameter), plastic or glass widemouth vials with tightly fitting screw-cap lids (8.5 cm diameter), wire-mesh tubes (5 cm long × 5 cm diameter), light wooden planks (50 cm × 50 cm × 1 cm), stopcock grease, magnetic stirrer, and stirring bars.

Reagents

Sodium Hydroxide, 1.00 *N*. Dissolve 40.00 g NaOH in distilled water and dilute to 1 L in a volumetric flask. Make fresh daily.

Barium Chloride, 3 *N*. Dissolve 126.7 g $BaCl_2$ in distilled water and dilute to 1 L.

Standard Hydrochloric Acid. This acid must have a known normality near 1 *N*. Mix 87.0 mL of concentrated HCl in distilled water and dilute to 1,000 mL in a volumetric flask. Standardize against 1 *N* sodium carbonate solution (5.300 g oven-dried Na_2CO_3 dissolved and diluted to 100 mL in distilled water).

Phenolphthalein Indicator Solution. Dissolve 1 g of phenolphthalein disodium salt in distilled water and dilute to 200 mL.

Procedure

Pipet 10.0 mL of 1.00 *N* sodium hydroxide solution into widemouth vials and seal them. Select a position on pond bottom for the measurement. Carefully stand the wire-mesh tube on the soil, remove the cap from the vial of sodium hydroxide, balance the vial on the wire tube, and place the metal can over the vial. Press the edges of the metal can into the soil. Wooden planks can be placed on tops of cans to shade them and prevent high inside temperatures. Prepare a control for determining the amount of carbon dioxide adsorbed from air initially in the cans. Place a widemouth vial of sodium hydroxide solution on a wooden plank, cover the vial with a can, and seal the area of contact between the can edges and the plank with stopcock grease. Place the plank on the pond bottom and shade the can with another plank. After 24 h, remove cans and immediately seal the widemouth vials of sodium hydroxide. In the laboratory remove the lids from the vials and add an excess of 3 *N* $BaCl_2$ (3 or 4 mL is usually enough), add a few drops of phenolphthalein indicator, and titrate with standard hydrochloric acid until the pink color of the phenolphthalein just disappears. Record the amount of hydrochloric acid used.

Calculate the amount of carbon dioxide evolved with the equation

$$CO_2 \, (mg \, cm^{-2}h^{-1}) = \frac{(B - V) N \times 22}{AT} \qquad (10.24)$$

where B = acid required to titrate control vials (mL)
V = acid required to titrate sample vials (mL)
N = normality of sulfuric acid
A = area of soil surface beneath cans (cm^2)
T = time (h)
22 = equivalent weight of carbon dioxide

Comments

The titration must be done quickly and without appreciable stirring of the precipitate to prevent the acid from reacting with the barium carbonate. The sizes of the metal cans, widemouth vials, and wire tubes are optional. The surface area of the alkali vial should be at least 25% of the cross-sectional area of the can to ensure a large surface for carbon dioxide absorption. It is not necessary to shade the cans, but in hot weather there may be a higher temperature inside the can than outside if shade is not provided. Widemouth, 1-quart Mason jars or 1-L, widemouth plastic jars can be used to contain soil and alkali vials in laboratory studies.

10.8 Soil pH and Lime Requirement

The pH is an index of the hydrogen ion concentration; it is an intensity factor. Soil acidity is a capacity factor that indicates the amount of hydrogen ion held in reserve by the soil. The lime requirement is a measure of the amount of liming material that must be added to a soil to neutralize enough soil acidity to provide a pH near 7.

Three methods for determining the lime requirement of non-acid-sulfate soils are provided. The buffer methods are more accurate and precise than the pH-texture method. One buffer method was developed specifically for soils in Alabama, but it can be used on most red–yellow podzolic soils. The method for acid-sulfate soils is useful only for soils containing iron pyrite.

10.8.1 Soil pH

Principle

Hydrogen ion activity (pH) is measured by direct potentiometry. The electromotive force of a galvanic cell can be related, by means of the Nernst equation, to the activity of hydrogen ion in solution. The voltage produced by the cell is converted by a potentiometer (pH meter) to a current proportional to the pH of the sample. This current is translated into a pH reading on a dial or a digital readout.

The galvanic cell consists of a glass electrode and a saturated calomel reference electrode (Fig. 10.17) that may be immersed in the sample. Sometimes the two electrodes are fused into the same probe body and called a combination pH electrode. The calomel electrode contacts the sample through an asbestos wick. The reaction at the calomel electrode is

$$Hg_2Cl_2 + 2e^- = 2Hg^0 + 2Cl^-$$ (10.25)

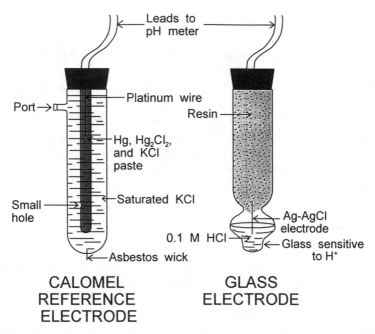

CALOMEL REFERENCE ELECTRODE

GLASS ELECTRODE

FIGURE 10.17. A calomel reference electrode and a glass electrode used in pH measurement.

The potential of the calomel electrode versus a normal hydrogen electrode is +0.242 V. The glass electrode consists of a silver–silver chloride electrode immersed in 0.1 M HCl, which is separated from the sample by a special pH-sensitive glass membrane. The potential of the glass electrode develops because different relative amounts of hydrogen ion are adsorbed by the two sides of the pH-sensitive glass. The adsorption of hydrogen ion on the glass is an ion-exchange process in which hydrogen ion in solution is exchanged with lithium in the glass as follows:

$$Li^+ \text{ glass} + H^+ \text{ solution} = Li^+ \text{ solution} + H^+ \text{ glass} \qquad (10.26)$$

According to Peters et al., the galvanic cell for pH measurement may be written as[19]

$$Ag \left| AgCl_{(s)}, HCl \ (0.1 \ M) \right| \frac{\text{glass}}{\text{membrane}} \left| \text{sample} \right| Hg_2Cl_{2(s)}, KCl_{(sat)} \left| Hg \right.$$

The vertical lines represent phase boundaries across which potentials develop. All of these potentials are constants except for the potential between the outer wall of the glass membrane and the sample and the liquid junction potential that develops between the sample and the potassium chloride solution in the calomel electrode. Because two unknowns are involved, the pH meter must be calibrated against a standard pH solution. The electromotive force produced when the electrodes are immersed in a sample is compared by the meter to the electromotive force produced by the electrodes in the standard pH solution. The difference in electromotive forces between the sample and pH standard is translated by the meter into a pH reading.

Standard pH buffers should be dissolved in distilled water that has been freshly boiled (to expel CO_2) and cooled. Buffer tablets for making solutions of different pH values may be purchased from scientific supply houses. Standard pH buffer solutions that are ready to use also may be purchased.

The manufacturer's instructions should be consulted for use of a pH meter. Before making pH measurements, carefully calibrate the meter with a pH 7 buffer. However, this procedure does not verify that the meter will read other pH values correctly. A second buffer, pH 5 if samples are expected to be acidic or pH 9 if samples are expected to be basic, should be used to determine if the pH meter will read a second pH correctly after it has been calibrated at pH 7.

Soil pH measurements usually are made for a slurry of dry soil and distilled water. The soluble salt content of soils varies, and salts displace exchangeable aluminum ions that hydrolyze to produce hydrogen ions and lower the pH. Some workers use a 0.01 M calcium chloride ($CaCl_2$) solution or a 1 M potassium chloride (KCl) solution instead of distilled water in preparing soil slurries for pH measurement. Ions in these solutions mask differences in soluble salt concentrations among soil samples. The most common soil:liquid ratio used for measuring soil pH is 1:1 (weight soil:volume liquid), but ratios of 1:2 and 1:10 also are employed. The value of soil pH for a given sample will be different, depending on the liquid used (distilled water, 0.01 M $CaCl_2$, or 1 M KCl) and the soil-to-liquid ratio. The procedure given next is for a 1:1 ratio of soil to distilled water.

Apparatus

A pH meter with glass electrode, standard pH buffer solutions.

Procedure

Weigh 20 g air-dried soil into a 50-mL beaker. Add 20 mL of distilled water. Mix thoroughly for 10–15 sec, let stand for 10 min, mix again, and let stand for 10 min more. Calibrate the pH meter at a pH near the expected pH of the sample. Check the validity of the calibration by verifying that the meter will correctly read the pH of buffers with pH values above and below the calibration pH. Insert the pH electrodes into the soil-distilled water slurry. Stir the slurry gently with a glass rod, being careful not to strike the pH probe. Read the pH directly.

10.8.2 Soil pH-Texture Method for Lime Requirement

Principle

Schaeperclaus recognized that the amount of lime needed for ponds increased as the soil pH decreased and as soil texture became finer.[20] He developed a table of lime requirement values based on soil pH and texture.

Apparatus

A pH meter, standard pH buffers, balance.

Procedure

Determine soil texture by feel. Measure the soil pH. Determine lime requirement from Table 10.3.

10.8.3 Buffer Method for Lime Requirement of Red-Yellow Podzolic Soils

Apparatus

A pH meter, standard pH buffers, balance.

Reagents

Buffer. Prepare a *p*-nitrophenol buffer of pH 8.0 ± 0.1 by dissolving 20 g of *p*-nitrophenol, 15 g of H_3BO_3, 74 g of KCl, and 10.5 g of KOH in distilled water and diluting to 1,000 mL in a volumetric flask.

Procedure

Preparation of Sample. Dried soil should be pulverized and the material that passes a Number 10 sieve (2.00-mm openings) used in the test. For most accurate

Table 10.3 *Lime Requirements of Bottom Soils Based on pH and Texture*

Soil pH	Lime Requirement (kg ha^{-1} as $CaCO_3$)		
	Heavy Loams or Clays	Sandy Loam	Sand
<4.0	14,320	7,160	4,475
4.0–4.5	10,740	5,370	4,475
4.6–5.0	8,950	4,475	3,580
5.1–5.5	5,370	3,580	1,790
5.6–6.0	3,580	1,790	895
6.1–6.5	1,790	1,790	0
>6.5	0	0	0

results the proportion of the entire sample that passes the Number 10 sieve should be estimated and used in establishing the liming rate. This step usually is omitted.

Determination of Soil–Water pH. Weigh 20 g of soil into a 100-mL beaker, add 20 mL of distilled water, and stir intermittently for 1 h. Measure the pH of the soil–water mixture with a glass electrode.

Determination of Buffer pH. Add 20 mL of the *p*-nitrophenol buffer to the sample from the determination of soil–water pH. Stir intermittently for 20 min. Make a 1:1 solution of buffer and distilled water. Insert the pH probe and adjust the meter to pH 8.0. Remove the pH probe, wash it with distilled water, and insert it into the buffer–soil mixture. Read the pH while gently stirring the mixture.

Calculation

Use the soil–water pH and the soil–buffer pH to select the appropriate liming rate from Table 10.4. If desired, the liming rate can be adjusted according to the proportion of the original soil sample that passed through the Number 10 sieve.

Comments

If the pH of the soil–water–buffer mixture is below 7.0, repeat the analysis with 10 g of dry soil and double the liming rate from Table 10.4.

10.8.4 General Buffer Method for Lime Requirement

Apparatus

A pH meter, standard pH buffers, balance.

Reagents

Buffer. Prepare a *p*-nitrophenol buffer of pH 8.0 ± 0.1 by dissolving 10 g *p*-nitrophenol, 7.5 g of H_3BO_3, 37 g of KCl, and 5.25 g of KOH in distilled water and diluting to 1,000 mL in a volumetric flask.

Procedure

Prepare sample as described before for the buffer method for red–yellow podzolic soils. Weigh 20 g of dry, pulverized soil into a 100-mL beaker and add 40 mL of buffer. Stir the mixture intermittently for 1 h. After setting the pH meter at pH 8.0 with the buffer, measure the pH of the soil–buffer mixture to the nearest 0.01 pH unit. If the pH is below 6.8, repeat the procedure with 10 g of soil and 40 mL of buffer. For a 20-g sample the equation for the lime requirement is

Table 10.4 Lime Requirements (kg ha^{-1}) of Calcium Carbonate (neutralizing value of 100) to Increase Total Hardness and Total Alkalinity of Pond Water to 20 mg L^{-1} or Above

Soil pH in Water	Calcium Carbonate Required According to Soil pH in Buffer Solution (kg ha^{-1})									
	7.0	7.8	7.7	7.6	7.5	7.4	7.3	7.2	7.1	7.0
5.7	91	182	272	363	454	544	635	726	817	908
5.6	126	252	378	504	630	756	882	1,008	1,134	1,260
5.5	202	404	604	806	1,008	1,210	1,411	1,612	1,814	2,016
5.4	290	580	869	1,160	1,449	1,738	2,029	2,318	2,608	2,898
5.3	340	680	1,021	1,360	1,701	2,042	2,381	2,722	3,062	3,402
5.2	391	782	1,172	1,562	1,953	2,344	2,734	3,124	3,515	3,906
5.1	441	882	1,323	1,765	2,205	2,646	3,087	3,528	3,969	4,410
5.0	504	1,008	1,512	2,016	2,520	3,024	3,528	4,032	4,536	5,040
4.9	656	1,310	1,966	2,620	3,276	3,932	4,586	5,242	5,980	6,552
4.8	672	1,344	2,016	2,688	3,360	4,032	4,704	5,390	6,048	6,720
4.7	706	1,412	2,116	2,822	3,528	4,234	4,940	6,644	6,350	7,056

$$\text{Lime requirement (kg CaCO}_3 \text{ ha}^{-1}) = (8.00 - \text{pH})\,5{,}600 \qquad (10.27)$$

For a 10-g sample double the value obtained with Equation 10.27.

10.8.5 Lime Requirement for Acid-Sulfate Soils

Reagents

Hydrogen Peroxide, 30%. Use a certified analytical reagent grade and store in a refrigerator when not in use.

Phenolphthalein Indicator. Dissolve 1 g of phenolphthalein in 50 mL of 95% alcohol and add 50 mL of distilled water.

Standard Sodium Hydroxide Solution. This solution must be of a known concentration between 0.0100 and 0.0500 *N*. A 0.0100 *N* solution contains exactly 400 mg of NaOH per liter. However, a more concentrated solution of NaOH (0.5 to 1 *N*) should be prepared and an aliquot of it diluted to give the solution for standardization. The solution may be standardized against standard sulfuric acid.

Procedure

Preparation of Sample. Air-dry pond soil and grind with a mortar and pestle to pass a Number 100 sieve (0.15-mm openings).

Oxidation of Sulfide. Place a 5.00-g sample of the pulverized soil in a 500-mL tall form beaker. Add 20 mL of 30% hydrogen peroxide and heat the mixture to about 40°C or until a noticeable reaction similar to boiling occurs. **Caution**: In samples with a high sulfide content the reaction may be vigorous and the mixture may boil over. If this occurs, weigh out a new sample and use less hydrogen peroxide. Once the reaction starts remove the beaker from the heat and let the reaction continue. After the reaction stops add 10 mL of 30% hydrogen peroxide and heat to 40°C again. Continue adding hydrogen peroxide and heating until no reaction occurs at 40°C. Add distilled water to obtain a total volume of about 100 mL and heat the solution for 30 min at 90–95°C to remove any excess hydrogen peroxide.

Titration. After the beaker and contents have cooled, add 5 drops of phenolphthalein and titrate with the standard sodium hydroxide. At the end point, phenolphthalein will produce a faint pink color that persists for at least 30 sec.

Calculation

The following equation is used to calculate the lime requirement resulting from iron pyrite in the soil:

$$\text{Lime requirement (mg CaCO}_3 \text{ 100 g}^{-1}) = \frac{V N (5,000)}{W} \qquad (10.28)$$

where V = volume of NaOH used in titrating to the
phenolphthalein end point (mL)
N = normality of NaOH titrant
W = weight of sample in grams

The potential acidity per 100 g of soil can be used to estimate the liming rate if the weight of the soil to be limed is known.

Comments

Surface soils with a pH below 4.0 (1:1 soil–water mixture) usually are acid-sulfate soils. These soils often have yellowish mottles of the mineral jarosite, and there may be yellow deposits of sulfur and reddish deposits of iron hydroxide on the surface.

Potential acid-sulfate soil may have an initial pH around neutrality when waterlogged. However, if dried, the pH will drop below 3 after a few weeks. For quick identification in the field, a sample can be oxidized with 30% hydrogen peroxide. Place a little of the soil in a few milliliters of hydrogen peroxide and mix well. After 10 min, measure the pH with universal pH paper or with a pH meter. If the soil is a potential acid-sulfate soil, the pH will be below 2.5.

Acid-sulfate or potential acid-sulfate soils also have exchange acidity. The total lime requirement can be estimated by measuring the lime requirement resulting from exchange acidity and from pyrite acidity and summing the two results.

10.9 Nutrients

It is difficult to make a practical assessment of nutrient concentrations in pond soils, because correlation data to relate soil nutrient concentrations to dissolved nutrient concentrations, phytoplankton growth, and aquacultural production are lacking. The most important limiting nutrient in most ponds is phosphorus. To solubilize all of the phosphorus in a soil sample for measurement of total phosphorus, soil samples must be digested in boiling perchloric acid. The fact that such rigorous treatment is required to release all phosphorus bound on soil particles attests to the insolubility and unavailability of a large proportion of the soil phosphorus. Most soil phosphorus analyses rely on chemical extractants or distilled water to extract specific forms of phosphorus or to indicate the amount of soil phosphorus available to plants. Many different extracting solutions have been used, including water, salt solutions, dilute hydrochloric and sulfuric acids, and acidic ammonium fluoride solution. Extraction of phosphorus from pond soils with distilled water or 0.5 N sodium bicarbonate solution allows an indication of phosphorus availability. Soils that yield more than 2 mg L^{-1} of phosphorus to solution when 1 g of soil is extracted with 100 mL of distilled water or 0.5 M sodium bicarbonate have a high phosphorus availability.[21] Methods for determining the concentrations of different chemical forms of phosphorus in soils are provided by Olsen and Sommers.[22]

Nitrogen concentrations in soil samples are usually obtained by the Kjeldahl technique or with nitrogen analyzers. Nitrate and ammonium can be extracted from soils with an electrolyte solution, such as 2 M potassium chloride. Total sulfur in soil samples can be converted to sulfate by ignition, and the sulfate extracted for analysis. Total sulfur also can be measured with sulfur analyzers.

Cations in soils can be extracted with an electrolyte solution, and the concentrations of cations in the extracts determined by atomic absorption spectrophotometry, inductively coupled argon plasma (ICAP) spectrophotometry, or other means.

Soil testing laboratories normally can do soil nutrient analyses quickly, accurately, and inexpensively. Reliance on soil testing laboratories for soil nutrient analyses is highly recommended. Soil test kits are available for pond managers who want approximate concentrations of soil nutrients. The Hach Company, Loveland, Colorado, a well-known manufacturer of water test kits, also manufactures soil test kits.

References

1. Jackson, M. L. *Soil Chemical Analysis*. 1958. Prentice-Hall, Englewood Cliffs, N.J.

2. Page, A. L., R. H. Miller, and D. R. Keeney, eds. *Methods of Soil Analysis: Part 2, Chemical and Microbiological Properties*. 1982. American Society of Agronomy and Soil Science Society of America, Madison, Wisc.

3. Klute, A. *Methods of Soil Analysis: Part 1, Physical and Mineralogical Methods*. 1986. American Society of Agronomy and Soil Science Society of America, Madison, Wisc.

4. Soil Survey Investigations Staff. *Soil Survey Laboratory Methods Manual*. 1991. Soil Survey Investigational Report No. 42, U.S. Department of Agriculture, Washington, D.C.

5. Boyd, C. E. *Water Quality in Ponds for Aquaculture*. 1990. Alabama Agricultural Experiment Station, Auburn University, Ala.

6. Sabbe, W. E., and D. B. Marx. 1987. Soil sampling: Spatial and temporal variability. In *Soil Testing: Sampling, Correlation, Calibration, and Interpretation*, J. R. Brown, ed., pp. 51–69. Soil Science Society of America, Madison, Wisc.

7. Ayub, M., C. E. Boyd, and D. Teichert-Coddington. 1993. Effects of urea application, aeration, and drying on total carbon concentrations in pond bottom soils. *Prog. Fish-Cult.* **55**:210–213.

8. Howe, R. B., and R. H. Meyers. 1970. An alternative to Satterwait's test involving

positive linear combinations of variance components. *J. Amer. Statistical Assoc.* **65**:404–412.

9. Gill, J. L. *Design and Analysis of Experiments*, Volume 1. 1978. Iowa State University Press, Ames, Ia.

10. Gardner, W. H. 1986. Water content. In *Methods of Soil Analysis: Part 1, Physical and Mineralogical Methods*, A. Klute, ed., pp. 493–544. American Society of Agronomy and Soil Science Society of America, Madison, Wisc.

11. Gee, G. W., and J. W. Bauder. 1986. Particle size analysis. In *Methods of Soil Analysis: Part 1, Physical and Mineralogical Methods*, A. Klute, ed., pp. 383–411. American Society of Agronomy and Soil Science Society of America, Madison, Wisc.

12. Weber, J. B. 1977. Soil properties, herbicide sorption and model soil systems. In *Research Methods in Weed Science*, B. Truelove, ed., pp. 59–62. Southern Weed Science Society, Auburn, Ala.

13. Blake, G. R., and K. H. Hartge. 1986. Bulk density. In *Methods of Soil Analysis: Part 1, Physical and Mineralogical Methods*, A. Klute, ed., pp. 363–375. American Society of Agronomy and Soil Science Society of America, Madison, Wisc.

14. Nelson, D. W., and L. E. Sommers. 1982. Total carbon, organic carbon, and organic matter. In *Methods of Soil Analysis: Part 2, Chemical and Microbiological Properties*, A. L. Page, R. H. Miller, and D. R. Keeney, eds., pp. 539–579. American Society of Agronomy and Soil Science Society of America, Madison, Wisc.

15. Ayub, M., and C. E. Boyd. 1994. Comparison of different methods for measuring organic carbon concentrations in pond bottom soils. *J. World Aquaculture Soc.* **25**:322–325.

16. American Public Health Association, American Water Works Association, and Water Pollution Control Federation. 1992. *Standard Methods for the Examination of Water and Wastewater*, 19th ed. American Water Works Association, Washington, D.C.

17. Shapiro, J., and O. Zur. 1981. A simple *in-situ* method for measuring benthic respiration. *Water Res.* **15**:283–285.

18. Boyd, C. E., and S. Pippopinyo. 1994. Factors affecting respiration in dry pond bottom soils. *Aquaculture* **120**:283–294.

19. Peters, D. G., J. M. Hayes, and G. M. Hieftje. *A Brief Introduction to Modern Chemical Analysis.* 1976. W. B. Saunders Co., Philadelphia, Pa.

20. Schaeperclaus, W. *Lehrbuch der Teichwirtschaft.* (Textbook of Pond Management) 1933. Paul Parey, Berlin, Germany.

21. Boyd, C. E., M. E. Tanner, M. Madkour, and K. Masuda. 1994. Chemical characteristics of bottom soils from freshwater and brackishwater aquaculture ponds. *J. World Aquaculture Soc.* **25**:517–534.

22. Olsen, S. R., and L. E. Sommers. 1982. Phosphorus. In *Methods of Soil Analysis: Part 2, Chemical and Microbiological Properties*, A. L. Page, R. H. Miller, and D. R. Keeney, eds., pp. 403–430. American Society of Agronomy and Soil Science Society of America, Madison, Wisc.

INDEX